Rare Earth Materials

Properties and Applications

OTHER BOOKS BY A. R. JHA, PH.D.

MEMS and Nanotechnology-Based Sensors and Devices for
Communications, Medical and Aerospace Applications
ISBN 978-0-8493-8069-3

Next-Generation Batteries and Fuel Cells for Commercial,
Military, and Space Applications
ISBN 978-1-4398-5066-4

Rare Earth Materials: Properties and Applications
ISBN 978-1-4665-6402-2

Solar Cell Technology and Applications
ISBN 978-1-4200-8177-0

Wind Turbine Technology
ISBN 978-1-4398-1506-9

Rare Earth Materials

Properties and Applications

A.R. JHA

CRC Press
Taylor & Francis Group
Boca Raton   London   New York

CRC Press is an imprint of the
Taylor & Francis Group, an **Informa** business

CRC Press
Taylor & Francis Group
6000 Broken Sound Parkway NW, Suite 300
Boca Raton, FL 33487-2742

© 2014 by Taylor & Francis Group, LLC
CRC Press is an imprint of Taylor & Francis Group, an Informa business

No claim to original U.S. Government works

Printed on acid-free paper
Version Date: 20140325

International Standard Book Number-13: 978-1-4665-6402-2 (Hardback)

Library of Congress Cataloging-in-Publication Data

Jha, A. R.
    Rare earth materials : properties and applications / A.R. Jha.
        pages cm
    Summary: "Rare earth materials (REMs) are recognized as a unique group of materials because of their unusual physical and chemical properties. REMs are essential for a diverse array of high-technology applications and key defense systems. This book describes the unique characteristics and applications of 17 REM, as well as their chemical, electrical, thermal, and optical characteristics, with an emphasis on physical and chemical properties. It also covers extraction, refining, and processing"-- Provided by publisher.
    Includes bibliographical references and index.
    ISBN 978-1-4665-6402-2 (hardback)
    1. Rare earths--Evaluation. 2. Rare earths--Industrial applications. 3. Mineral industries.
I. Title.

TA418.9.R37R376 2014
661'.041--dc23   2014007034

Visit the Taylor & Francis Web site at
http://www.taylorandfrancis.com

and the CRC Press Web site at
http://www.crcpress.com

This book is dedicated to my parents, who always encouraged me to pursue advanced research and development activities in the field of science and engineering technology for the benefit of all, regardless of nationality, race, or religion

Contents

Foreword

Rare Earth Materials: Properties and Applications comes at a time when high global demand for rare earth (RE) materials is coupled with an anticipation of a shortage of such materials in the near future. China currently controls the export of close to 94% of the world's rare earth materials and dictates the raw materials cost. In order to reduce dependency on a single procurement source, the author identifies certain countries that are involved in mining, exploring, and refining activities for rare earth materials. It is important to point out that rare earth metals play critical roles in the steel industry, chemical processing technology, nuclear power reactors, all-electric and hybrid-electric vehicles, and a host of other commercial and industrial sectors. Several automobile companies are deeply involved in the procurement of neodymium and samarium rare earth permanent magnets. These magnets are best suited for the electric motors and generators being used in all-electric and hybrid-electric vehicles. These rare earth magnets offer optimum magnetic performance at elevated temperatures, minimum weight and size without compromising reliability, and the lowest aging effects. They also offer optimum magnetic energy products, high reliability, and consistent performance under severe mechanical and thermal environments, none of which is possible with conventional magnetic materials. Test data collected by defense scientists on traveling wave-tube amplifiers (TWTAs), reveal that samarium cobalt permanent magnets yield reliable performance under elevated temperatures close to 300°C. Dr. Jha gives every consideration to the development of commercial, military, and space applications using rare earth materials. He identifies the applications of rare earth materials for commercial and industrial sensors operating under severe thermal, chemical, and humidity environments.

Rare Earth Materials addresses nearly every aspect of rare earth metals and the oxides that are best suited for the components used in automobiles, wind turbines, steam turbines, hydroelectric generators, underwater vehicles, and smoke detection sensors. Studies performed by Dr. Jha indicate that the use of appropriate rare earth materials offers significant improvements in a system's electrical, mechanical, and thermal performance characteristics. His studies further indicate that the use of certain rare earth materials such as cobalt and platinum for cathode electrodes in high-capacity fuel cells offers higher output and efficiency at operating temperatures in excess of 1,000°C. This kind of high-temperature performance is not possible with conventional metals.

Dr. Jha continues his distinguished track record of distilling complex theoretical physical concepts into an understandable technical framework that can be extended to practical applications across a commercial, military, and industrial framework. His "big-picture" approach, which does not compromise the basic underlying science, is particularly remarkable and refreshing. His approach should help undergraduate and graduate students master the difficult scientific concepts with the kind of rare confidence that will be needed for future commercial engineering applications.

This book is well organized and provides mathematical expressions to estimate critical performance parameters. Dr. Jha clearly identifies the cost-effective features, reliability, and safety aspects of system equipment using rare earth materials. In brief, wherever the use of rare earth materials is specified, overall system performance, performance costs, reliability, and safety aspects are clearly mentioned for the user's benefit.

This book provides a treatment of the underlying thermodynamic aspects of systems incorporating rare earth materials. Dr. Jha considers this evaluation of critical importance because it can affect the reliability, safety, and longevity of a system using a rare earth metal or oxide or alloy. He also summarizes the potential advantages for commercial applications. It is interesting to mention that a fuel cell is an electricity generation device that combines an oxidation reaction and a reduction reaction in the fuel cell assembly. In summary, one can state that a fuel cell is an energy-conversion device in which chemical energy is isothermally converted to direct current (dc) energy. Dr. Jha identifies the basic laws of electrochemical kinetics and recommends that a superior rare earth–based electrolyte is essential for generating higher electrical power.

Applications of rare earth materials for underwater vehicles; anti-improvised explosive devices; space sensors for surveillance, reconnaissance, and tracking functions; and battlefield applications are discussed in detail, with particular emphasis on system cost, reliability, safety, aging effects, and longevity. Currently, the dominant end-uses for rare earth elements in the United States are automobile- and petroleum-refining catalysts, in phosphors for color televisions and flat panel displays (cell phones, portable DVDs, and laptops), permanent magnets and rechargeable batteries for hybrid and electric vehicles, and numerous medical devices. Permanent magnets involving neodymium, gadolinium, dysprosium, and terbium materials are used in numerous electrical and electronic components and next-generation generators for wind turbines. Dr. Jha suggests that system designers must consider all aspects whenever they intend to use rare earth materials for specific applications. A wide variety of readers such as undergraduate and graduate students, product design engineers, program managers, and research scientists who are deeply involved in designing portable power plants for future generations will benefit greatly from the book. Technical managers will also find this book most useful for future commercial and industrial applications.

I strongly recommend this book to a broad audience, including students, project engineers, clinical scientists, aerospace engineers, project managers, life science scientists, and project managers involved in the design and development of systems for commercial, industrial, military, and space applications.

Ramesh N. Chaubey, MS

New York University Polytechnic School of Engineering

New York

Engineering Manager and Technical Advisor for Infrared Military Satellites

U.S. Air Force (Retired)

Preface

The publication of this book comes at a critical time when free nations have limited sources of rare earth (RE) materials urgently needed for the development of defense-oriented products. In addition, the current supply of rare earth metals is strictly controlled by China and could potentially be inaccessible during international military conflicts. This book offers alternate sources for the procurement of RE materials and their oxides and compounds at reasonable costs and with no export restrictions. Because rare earth technology is relatively new, its deployment in potential applications is not fully known.

The author investigates the use of RE materials for a variety of applications, such as high-resolution MRI systems and gradiometer sensors, that are most ideal for detecting and treating heart, brain, and lung diseases. The author identifies rare earth samarium cobalt magnets that demonstrate unique capabilities for high-power radar transmitters and wideband, dual-mode traveling wave tubes best suited for electronic warfare applications, where safe and reliable RF performance under harsh mechanical and thermal operating environments is critical. The author identifies RE elements for the development of high-resolution infrared (IR) sensors and planar array detectors that are best suited for reconnaissance, surveillance, and tracking space targets. The author summarizes the performance capabilities of multilayer thin-film cadmium telluride (CdTe) solar cells that offer several benefits over conventional silicon devices.

Research studies performed by the author indicate that CdTe solar cells offer a 50% lower material cost compared to silicon solar cells, high stowage efficiency, higher solar array design flexibility, optimum spacecraft design configuration, and 100 times more space radiation resistance than silicon devices.

The author summarizes the performance of optical detectors using thin films of superconducting material yttrium-barium-copper-oxide (YBCO). These detectors are best suited for measuring IR radiation signals at wavelengths of 13 μm or higher under the cryogenic temperature of 77 K. The YBCO infrared detectors offer noise-equivalent power (NEP) as low as 7×10^{-12} W/Hz$^{0.5}$, a response time better than 50 psec.

Cryogenically cooled focal planar array detectors using thin films of ternary compound alloy Hg:Cd:Te offer improved detectivity, enhanced responsivity, low response time, low dark current, and low NEP. Because of these unique characteristics, these detectors are best suited for detecting very long wavelength infrared (VLWIR) signals or signals emitted by intercontinental ballistic missiles (IBCMs) during boost, cruise, and terminal phases. These IR signals typically vary from 14 to 17 μm. Studies undertaken by the author on these planar array detectors reveal that the focal planar array detectors with optimum detector diameter offer optimum performance at VLWIR signals greater than the 6- to 17- μm range when cooled to a cryogenic temperature of 40 K and foolproof discrimination between decoys and actual targets.

The author identifies potential applications of neodymium- and samarium-based rare earth magnets for various commercial and industrial systems with particular emphasis on all-electric cars and hybrid-electric vehicles. Comprehensive research studies undertaken on these rare earth magnets indicate that such devices are widely used in the design and development of electric motors and generators for all-electric and hybrid-electric vehicles because they offer minimum weight and size as well as higher efficiencies and optimum torque performance under elevated operating temperatures. In summary, rare earth material technology offers countless commercial, industrial, medical, and scientific applications. Note that most applications of rare earth materials require cryogenic temperatures to achieve optimum performance irrespective of commercial or medical applications. In addition to the abovementioned applications involving rare earth metals and their oxides and compounds, the author identifies potential applications of rare earth materials in commercial, industrial, space, and defense system applications. The defense system applications include surveillance, reconnaissance, and tracking of space targets; long-endurance drones for battlefield surveillance; unmanned armed air vehicles equipped with laser-based compact missiles; perimeter security systems; covert high-data-rate transmission equipment; and a host of other sensors that are critical to military and space applications.

This book summarizes important properties, applications, and limitations of RE materials. Potential commercial and industrial applications of RE metals and their oxides are identified with an emphasis on unique device performance, reliability, longevity, and costs of ownership. The role of mischmetal, which is characterized as a complex rare earth compound, is discussed in great detail in relation to the steel and automotive industries; high-power magnets; electric motors and generators; chemical-processing technology; antipollution and smog sensors; bridge construction; earth-digging tractors; offshore drilling towers; jet engine components such as nozzles, blades, and bearings; and a host of other applications where high mechanical strength, resistance to bulking, and resistance to corrosion are required under harsh chemical and climatic environments.

This book contains eight chapters, each of which is dedicated to a specific technology. Chapter 1 is dedicated to the historical aspects and the discovery of rare earth elements and compounds. This chapter describes various scientific research studies that identify the locations of mines for various rare earth materials.

Chapter 2 provides specific details on the unique properties and applications of rare earth oxides, alloys, and compounds. It is important to point out that rare earth oxides are found in raw conditions and require processing based on grain size, growth, shape, and metallurgical characteristics.

Chapter 3 focuses on the properties and applications of superconductive-based rare earth materials. This chapter describes the magnetic properties of rare earth materials as a function of cryogenic temperatures and their applications, which involve forming metallic films, wires, and particles best suited for unique scientific, medical, commercial, and industrial systems and sensors. It should be noted that the thin films of YBCO and thallium-barium-calcium-copper-oxide play critical roles in the design and development of microwave and millimeter-wave components widely used in commercial, scientific, and aerospace and defense-based systems applications, where consistent performance, high reliability, compact packaging, and thermal stability are the principal design requirements. These thin films of superconducting rare earth elements play an important role in the development of low-frequency antennas, yttrium-iron-garnet-tuned filters, and microwave solid-state devices that are widely used in the design of electronic warfare equipment and underwater surveillance sensors. The deployment of rare earth magnets in electric motors and generators has demonstrated significant improvement in weight, size, and reliability. Motors and generators using rare earth magnetic materials are widely used by both all-electric and hybrid-electric vehicles.

Chapter 4 describes potential applications of thin films of rare earth oxides best suited for commercial, scientific, medical, and space and defense system components. These thin films are widely used in the design of solid-state lasers, timing and synchronization clocks, and electro- and magneto-optic instrumentation devices. Rare earth laser crystals are widely used by diode-pumped solid-state (DPSS) lasers operating at most IR wavelengths. Research studies performed by the author indicate that DPSS lasers such as Ho:Tm:VAG and Ho:Tm:YLF lasers can operate at room temperature and offer improved performance in terms of power output, quantum efficiency, spectral bandwidth, and optical beam stability.

Chapter 5 identifies the benefits of rare earth–based alloys when added to basic structural materials such as iron and steel, which are widely used in commercial, industrial, and aerospace system applications. It is important to mention that reliability, safety, and mechanical integrity of the structural materials are of critical importance under harsh mechanical and thermal environments. Metallurgical engineers believe that the overall properties of cast iron are significantly improved by the presence of minute concentrations of elements other than carbon and silicon. Engineers further believe that the slight addition of the rare earth complex alloy mischmetal to steel significantly improves the mechanical properties of the steel in addition to reducing its price per pound.

Chapter 6 is dedicated to the properties and critical applications of rare earth intermetallic compounds such as hydrides and ceramics. Critical electrical and mechanical properties of hydrides and ceramics are summarized with an emphasis on their industrial and commercial applications. In addition, structural aspects of intermetallic rare earth compounds are discussed with particular emphasis on energy conservation, national defense, and medical/dental applications.

Chapter 7 deals strictly with the utilization of rare earth metals and compounds for the development of the glass and polishing industries and crystal technology with an emphasis on cost, process duration, and end-product quality. Rare earth materials responsible for the rapid development of the chemical industry and electro-optic technology with minimum cost and complexity are discussed in great detail.

Chapter 8 describes critical industrial and commercial applications of rare earth metals, alloys, and oxides. This chapter takes a leading role in summarizing the significant benefits of the rare earth compound mischmetal in heavy industrial applications. The addition of a minute portion of mischmetal in high-performance steel offers significant advantages in terms of cost, quality control, and mechanical properties. The benefits of mischmetal, a complex rare earth compound, play a critical role in the development of complex and heavy industrial- and defense-related products such as tanks, jet fighters, gas turbines, and certain rocket components. It will be of great interest to system managers to know that Adolf Hitler deployed cutting-edge innovation technology that incorporated the rare earth mischmetal compound in the design and development of the first jet fighter in April 1945. In the chapter summary, the author highlights significant improvements in steel quality and strength, chemical processing technology, glass polishing, smog detection, and other industrial processes (with minimum cost and complexity)—all as a result of the deployment of rare earth materials.

I want to express my sincere gratitude to Ed Curtis at Taylor & Francis and Marc Johnston at Cenveo Publisher Services for their meaningful suggestions in incorporating changes to the text, completing the book on time, and seeing everything through to fruition—all of which they did with remarkable efficiency and coherency.

I also want to thank to my wife Urmila Jha, my daughters Sarita Jha and Vineeta Mangalani, and my son Captain Sanjay Jha of the U.S. Army for their genuine support, which inspired me to complete the book on time despite a tightly prescribed production schedule.

Chapter 1

History and Discovery of Rare Earth Materials

1.1 Introduction

This book comes at a critical time when the deployment of rare earth materials is beneficial in various commercial, industrial, medical, military, and space applications. The use of rare earth elements or oxides has resulted in significant improvements in the performance of these applications, including efficiency, longevity, and reliability, in addition to a considerable reduction in the weight and size of components and/or overall systems. These improvements are strictly due to the unique chemical, electromagnetic, and magnetic properties of rare earth materials.

Rare earth materials are often characterized by various names, such as rare earth elements (REEs), rare earth metals (REMs), rare earth oxides, or yttrium-based rare earth material. Essentially, rare earth materials are classified into two distinct categories: light rare earth elements (LREEs), which is known as the cerium group, involving lanthanum-europium, and heavy rare earth elements (HREEs), also known as the yttrium group, consisting of gadolinium-lutetium-yttrium elements. Based on pertinent scientific papers, published reports, and chemical elements, REMs (or REEs) are composed of a set of 17 chemical elements as listed in the periodic table [1]. Furthermore, the 17 rare earth materials include 15 lanthanides plus scandium and yttrium. Scandium and yttrium are considered to be rare earth elements because they tend to occur in the same core deposits as lanthanides and exhibit identical chemical properties.

Some rare earth materials are named for the material scientists who discovered their chemical properties, and some are named based on their geographical discovery locations. Despite their names, rare earth elements (with the exception of the radioactive promethium) are plentiful in the earth's crust. For example, cerium is the 25th most abundant element at 68 parts per million, similar to copper. However, because of their unique geochemical properties, rare earth elements are typically dispersed randomly in all directions; they are not often found in concentrated and economically exploitable forms. Therefore, the few economically exploitable deposits in the earth's crust are known as rare earth materials—so named because of the scarcity of these minerals.

The 17 original rare earth elements and their chemical properties are summarized in Table 1.1. The REE with the lowest atomic weight and atomic number is listed first, while the element with highest atomic weight and atomic number is listed last [1]. Note that the abbreviations are often used in various reports on rare

Table 1.1 Seventeen Rare Earth Elements and Their Properties

Element (Symbol)	Atomic Weight	Atomic Number	Valance
Scandium (Sc)	45	21	3
Yttrium (Y)	88.9	39	3
Lanthanum (La)	139	57	3
Cerium (Ce)	140	58	3, 4
Praseodymium (Pr)	141	59	3
Neodymium (Nd)	144	60	3
Promethium (Pm)	145	61	3
Samarium (Sm)	150	62	2, 3
Europium (Eu)	152	63	2, 3
Gadolinium (Gd)	157	64	3
Terbium (Tb)	159	65	3
Dysprosium (Dy)	162	66	3
Holmium (Ho)	164.9	67	3
Erbium (Er)	167	68	3
Thulium (Tm)	169	69	3
Ytterbium (Yb)	173	70	2, 3
Lutetium (Lu)	175	71	3

earth materials in scientific journals or in this particular book. These abbreviations can be very useful to both students and design engineers who are deeply involved in the development of high-technology components or systems deploying rare earth materials.

1.2 Early History of Rare Earth Materials

Scientists from Sweden, Norway, and other European nations contributed significantly to the early history of rare earth element mining operations, their oxides, and their important properties, such as chemical, metallurgical, magnetic, electro-optical, and nuclear properties [1]. An overview presented by the Austrian scientist Ekkehard Greinacher in 1981 [2] provided an early history of rare earth materials and their potential commercial, industrial, medical, and military applications. European scientists, such as Gadolin (at the Swedish University in Abo), Mosander, and von Welsbach, provided useful information on rare earth elements and their oxides. In fact, von Welsbach already had a patent for a lanthanum-zirconium element that was well suited for consumer applications.

The first rare earth mineral discovered was gadolinite, which is a compound consisting of cerium, yttrium, iron, silicon, and other elements. The black mineral was first discovered at a quarry in the village of Ytterby in Sweden and was thus called *ytterbite*, but it was subsequently renamed gadolinite by Carl Axel Arrhenius in 1794 [1]. Interestingly, many of the rare earth elements bear names derived from this particular village in Sweden. The rare earth element research studies performed by Johan Gadolin at the Royal Academy discovered an ore that was later called *ytteria*. Later, in 1803, Swedish scientists J.J. Berzelius and Wilhelm Hisinger discovered a white oxide and called it *ceria*.

It took another 30 years for scientists to discover the other elements that were contained in ytteria and ceria with similar properties. In 1839, Carl Gustav Mosander separated ceria by heating the nitrate and dissolving the product in nitric acid. He called the oxide of the soluble salt *lanthana*. It took three more years to further separate the lanthana into didymium and pure lanthana. In 1842, Mosander also separated ytteria into three oxides: pure ytteria, terbia, and erbia. These names are derived from the village name of Ytterby. The rare earth material yielding pink salts was called *terbium*, whereas the one yielding yellow peroxide was called *erbium*.

By 1842, the number of the rare earth elements reached six: yttrium, terbia, lanthanum, didymium, erbium, and terbium. Johan Berlin and Marc Delafontaine tried to separate out the crude ytteria and found the same substances that Mosander obtained earlier. There were no further discoveries for the next 30 years, and didymium was listed in the periodic table of elements with a molecular mass of 138. In 1879, the new element samarium was isolated by Paul-Émile Lecoq de Boisbaudran from the raw mineral samarskite.

The samaria rare earth material was further separated by Boisbaudran in 1886. Similar results were obtained by Jean Charles Galissard de Marignac through direction isolation from samarskite. This element was named *gadolinium* after the Swedish scientist Johan Gadolin; the oxide was named *gadolina*. Further spectroscopic analysis by various scientists between 1886 and 1901 discovered several new spectroscopic lines that demonstrated the presence of an unknown element. The fractional crystallization of the oxide then yielded the europium element in 1901.

Rare earth scientists believe that the third source for rare earth elements was established in 1889. This is a mineral similar to gadolinite, now called *samarskite*. This particular mineral from the southern Ural Mountains of Russia was discovered by Swedish inventor Gustave Rose. However, the Russian chemist R. Harmann proposed that the new element should be called *ilmenium*.

The exact number of rare earth elements that existed at that time was not clear. Therefore, a maximum number of 25 was estimated, which was approved by other mineral scientists and research scientists. The use of x-ray spectra obtained by x-ray crystallography by Henry Moseley made it possible to assign atomic mass numbers to the elements discovered. Moseley found that the exact number of lanthanides had to be 15 and that element 61 had yet to be discovered. Using these facts about atomic mass numbers based on crystallography, Moseley also demonstrated that hafnium (element number 72) would not be a rare earth element. Unfortunately, Moseley was killed in 1915 during the First World War, years before hafnium was discovered. Therefore, the claim of Gorges Urbain that he discovered element number 72 was untrue. Hafnium lies immediately below zirconium in the periodic table, with both elements having similar chemical and physical characteristics.

During the 1940s, Frank Spedding and others developed a chemical ion exchange for separating and purifying rare earth elements during the research activities of the Manhattan Project. The chemical ion exchange technique was first applied to the actinides (i.e., any element in a series of elements with increasing atomic numbers, beginning with actinium of atomic number 89) in order to separate plutonium-239 and neptunium from uranium, thorium, actinium, and the other actinide rare earth materials produced in nuclear reactors. Plutonium-239 was extremely desirable because it is a fissionable material [2].

The principal sources of rare earth elements are the minerals bastnäsite, monazite, and lateritic ion-absorbing clays. Despite their high relative abundance, rare earth minerals are more difficult to mine and extract than equivalent sources of transition metals, making rare earth elements very expensive. Their industrial applications were very limited until efficient separation techniques were developed during the late 1950s and early 1960s to minimize cost and complexity, such as ion exchange, fractional crystallization, and liquid extraction.

1.2.1 Early History of Rare Earth Oxides

The early history of applications for rare earth elements and their oxides can be divided into the following three distinct periods of applications, starting in 1788:

- 1788 to 1891: In the most preliminary period, rare earth elements were strictly scientifically examined but were not yet technically used for specific applications.
- 1891 to 1930: In the second period, rare earth elements were first identified for industrial applications.
- 1930 to 1960: In the third period, the systematic properties of rare earth elements were established, particularly for wide applications.

In particular, the period from 1940 to 1960 is distinguished by the systematic discovery of chemical and physical properties for methods of separation and use of rare earth elements as byproducts of various atomic research programs in the Western industrial countries, including the United States, England, and Germany [2]. The global production of rare earth oxides over the period from 1950 to 2000 is illustrated in Figure 1.1.

The following are the most popular rare earth oxides, which have a variety of commercial, industrial, and scientific applications:

- Thulium oxide (Tm_2O_3)
- Ytterbium oxide (Yb_2O_3)
- Cerium oxide (Ce_2O_3)
- Dysprosium oxide (Dy_2O_3)
- Europium oxide (Eu_2O_3)
- Gadolinium oxide (Gd_2O_3)
- Holmium (H_2O_3)

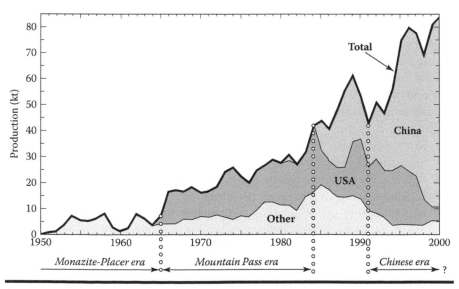

Figure 1.1 Global production of rare earth oxides from 1950 to 2000. (Based on http://pubs.usgs.gov/fs/2002/fs087-02.)

- Lanthanum (La_2O_3)
- Lutetium (Lu_2O_3)
- Neodymium (Nd_2O_3)
- Praseodymium (Pr_2O_3)
- Samarium (Sa_2O_3)
- Scandium (Sc_2O_3)

Material scientists have integrated these rare earth oxides into cutting-edge applications such as wind turbines, gas turbines, jet engines, scramjet engines, superconducting microwave filters, traveling wave tube amplifiers (TWTAs) in offensive and defensive weapon systems, high-capacity rechargeable electrodes, catalysts for self-cleaning ovens, fluid catalysts for oil refineries, chemical oxidizing agents, masers, infrared lasers, welding glass screens, nuclear batteries, nuclear magnetic resonance relaxing agents, magnetic resonance imaging contrast agents, scan detection devices, x-ray machines, and various kinds of steel products for industrial and commercial applications.

1.2.2 Key Processing Requirements for Rare Earth Elements and Their Oxides

Raw rare earth elements and their oxides have very limited use in any commercial or industrial application. Therefore, the processing of these elements and their oxides is essential. The processing of oxides may involve several steps, which can be very time-consuming and expensive.

The prices of rare earth metals and their oxides are shown in Figure 1.2. As shown, the prices of the HREEs are higher than their oxides. In addition, the prices of LREEs are much lower for lanthanum, cerium, praseodymium, and neodymium because of their lower processing costs. The costs of thulium (Tm) and europium (Eu) are higher because of their wide commercial and industrial applications.

Heavy rare earth metals require a number of complex processing steps before they can be used in any meaningful application, including the following:

- Mining rare earth ore from mineral deposits
- Separating the rare earth ore into individual rare earth oxides
- Refining the rare earth oxides into metals with different purity levels
- Forming the metals into rare earth alloys
- Manufacturing the alloys into components, such as permanent magnets (which are widely used in defense and commercial applications)
- Quality control, which includes such steps as visual inspection and surface parameter verification

Note that the unprocessed ore from the mines is part and parcel of the mining operation. Once the ore is separated from the waste, it is handed over for material

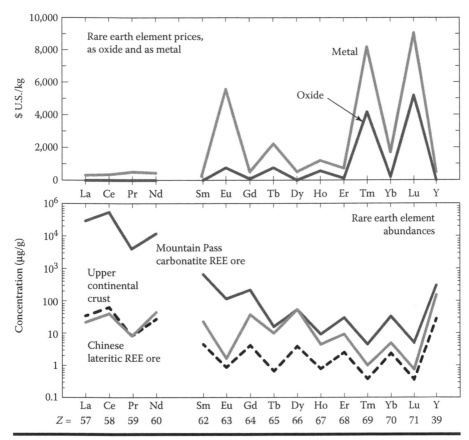

Figure 1.2   Prices in U.S. dollars per kilogram and abundance of rare earth elements.

processing, which is the most costly and cumbersome operation. Approximately 75% of the total of the rare earth element cost is involved in mining and processing efforts. The correct location of geological sites, procurement of new supplies, use of latest extraction techniques, improvements in refining procedures, and advancements in processing technology are all necessary to bring the costs for rare earth elements down to affordable levels.

1.2.3 Worldwide Rare Earth Material Deposits

This section will address the REE abundance in various layers of the earth. (The use of the term *abundance* reflects unfamiliarity rather than true rarity.) The more abundant rare earth elements have crustal concentrations that are similar to common industrial metals, such as nickel, chromium, copper, zinc, tin, tungsten, and molybdenum, as shown in Figure 1.3. The relative abundance of the rare earth

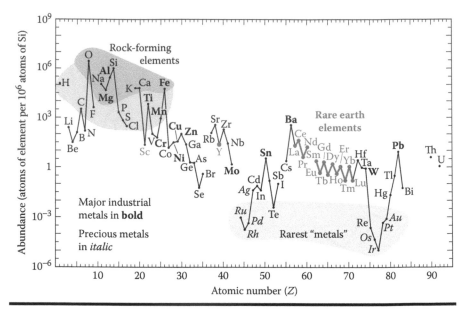

Figure 1.3 Abundance of the chemical elements in earth's upper continental crust.

elements are normally found in the earth's upper continental crust as a function of atomic number (Z). The rare earth elements with atomic numbers ranging from 45 to 90 are the rarest metals. Many of the elements are classified into overlapping categories, such as rock-forming elements; rare earth elements, such as lanthanides (La, Lu, and Y); major industrial metals; precious metals, such as gold and platinum; and the nine rarest metals, consisting of six platinum group elements plus gold (Au), rhenium (Re), and tellurium (Te). Even though gold has the highest atomic number (79) compared to Re and Te, which are considered to be rare earth materials, it is called a noble metal.

Extraction of rare earth materials and their oxide deposits requires comprehensive mining surveys of prospective sites. Mining efforts to extract rare earth materials have been undertaken in several countries, including China, the United States, Brazil, South Africa, Australia, Sweden, Russia, and other European countries. Particularly, China and the United States have made significant investments in the mining and refinement of rare earth materials. Between 1975 and 1978, large mining efforts were undertaken by China in the southern Tibetan region to extract rare earth materials. Deposits of rare earth materials may also be present in the Indian state of Arunachal Pradesh, which has similar geological features and terrain aspects. Therefore, it may be fruitful for the Indian government to undertake mining efforts in this region to locate the presence of rare earth materials.

Recent studies by the author seem to indicate that the geographical distribution and rare earth material production are mostly confined to China, which essentially allows China to exclusively control both the supply and cost of rare earth materials. Based on a mining survey, the Lynas Corporation won approval for mining operations in Malaysia for a rare earth mineral that is capable of supplying two-thirds of China's current demand within next few years. The same survey further revealed that China controls the export of approximately 94% of the rare earth material market worldwide. China has also signed a mining and exploration contract for rare earth materials with the Afghan government that is worth more than $2 billion. Of note, in 2009, China blocked the export of rare earth materials to Japan, thereby forcing Japanese companies to extract rare earth elements from used electronic components to compensate for the shortage. Because of such unpredictable Chinese behavior, Japan and Vietnam signed an agreement to supply rare earth elements to each other in case of emergency.

Rare earth scientists and mining experts believe that Mountain Pass in California is the major U.S. source for bastnäsite ore, whereas China is the major source for lateritic ore (Figure 1.4). Bastnäsite ore is dominated by the elements lanthanum, cerium, and neodymium, with europium through lutetium plus yttrium totaling only 0.4% of the world supply. The lateritic ion-absorption ore is dominated by yttrium.

Immediate and appropriate efforts are needed to break the monopoly control on rare earth materials. Geological site surveys and mining efforts should be given immediate attention to locate the presence of rare earth materials. These materials must be extracted and refined at various locations to establish secondary sources for rare earth materials. As mentioned, the actual processing of rare earth materials involves complex procedures, such as the extraction and screening of raw materials, recycling, refining, visual inspection, quality control techniques, identification of

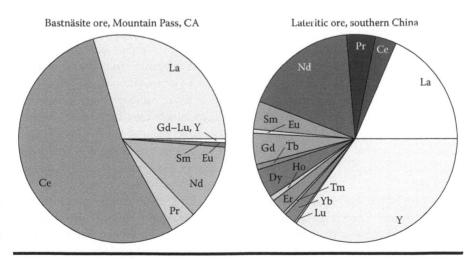

Figure 1.4 Proportions of individual rare earth elements in bastnäsite ore and lateritic ion-absorption ore.

critical spectroscopic parameters, and separation of rare earth elements based on demand and specific applications.

1.2.4 Initial Worldwide Mining Operations for Rare Earth Materials

This section discusses the initial mining efforts to explore for deposits of rare earth materials and their oxides worldwide. Before World War II, only one or two countries showed any interest in rare earth mining operations. However, in the 1950s and 1960s, the deployment of rare earth elements in jet engines, wind turbines, and a host of other commercial and industrial applications compelled other countries to start rare earth material mining operations.

The United States was the first country to start rare earth material mining operations at Mountain Pass in 1965. These operations were discontinued in 2002 and then resumed in 2007. China started mining operations at least by the year 2000, possibly earlier. China has initiated mining operations in various regions of the country to extract rare earth materials. The main Chinese source of rare earth materials is the southern part of Tibet. It is believed that no other rare earth material mine outside China could compete with China on cost advantage.

Countries such as Afghanistan, Brazil, India, South Africa, Australia, Japan, and the United States are actively engaged in establishing mining operations. Several projects are in the advanced stages of development, such as Mountain Pass in the United States; the Dubbo Zirconia Project in Australia; Zeus-Kipawa, Nechalacho, and Strange Lake in Canada; and Steenkampskraal in South Africa. As of December 2010, 14 such projects had been fully recognized as rare earth mineral sources.

Quasi-mountain regions are considered to be prime locations for rare earth mining operations. The Mount Weld mining facility in Western Australia, which was under development for 10 years, was scheduled to become fully operational in 2013. Engineers at the Technology Metals Research Organization at the Mountain Pass mining facility in California are expected to make rapid progress in extracting rare earth materials in the coming years. The Chinese have focused on mining exploration in the southern part of Tibet, starting as early as 1992. The Chinese mines located in that region not only produce the highest percentage of rare earth materials, but they also control the highest percentage of exported rare earth metals and oxides.

Because of the wide applications for rare earth materials in defense products, both the United States and Japan have bought sufficient rare earth materials to last for 30 to 40 years. Their facilities will not only explore for rare earth materials but also will be equipped for refining and quality control procedures to obtain rare earth elements. Japanese mining experts are aggressively pursuing massive underwater mining efforts to discover rare earth elements and new rare earth metal deposits that can break the Chinese monopoly on rare earth metals. Japanese geologists have recently found huge concentrated deposits of rare earth materials in the Pacific seabed that could exceed 100 billion tons. In just a single square mile of seafloor, the quantity would be enough to cover nearly half of the world's annual demand for rare earth metals. The undersea rare earth material deposits were located near Tahiti and Hawaii in international waters at a depth ranging from 11,500 to 20,000 feet. Deep-sea mining of copper and nickel is already underway in the Pacific Ocean. Japanese scientists are planning to undertake feasibility studies to explore the mining of rare earth materials. These deep-seabed mining efforts must use mini-robot technology. Japan and Vietnam are planning to explore gas, oil, and rare earth minerals in the seabed of the South China Sea, where six countries have declared vested interests.

The current production status of these mining facilities is not fully known. Mining experts and geological scientists believe it takes approximately 8 to 10 years to obtain feasibility studies, environmental statements, and financial analyses before any mining source is confirmed as a real supply source for rare earth materials. It is predicted that China's monopoly on rare earth materials will not last much longer because of these worldwide mining efforts. Furthermore, due to rising prices, there is more incentive to develop mines outside China to weaken the Chinese control on rare earth materials. Incidentally, China has already indicated they will not hesitate to use rare earth metals as political leverage against Japan or any other country. They have threatened to cut off the export of rare earth metals to the United States, Japan, and other European countries if political situations worsen.

As can be seen, rare earth metals are not only economic assets but also strategic assets. A report by Mark Humphries [3] revealed that the rare earth metal reserves of China, the United States, Russia, Australia, and other countries amount to approximately 36%, 13%, 19%, 3.5%, and 5.5% of the world's total reserves, respectively. However, China's REM output is approximately 97% of the world's total [3].

Once reserves are proven, without any pre-existing infrastructure, it takes approximately 3 to 5 years until first production for one or more of the HREE-rich deposits. Sometimes, rare earth–rich deposits can provide an appreciable amount of heavy rare earth metals before the light rare earth metals are available. The procurement of new supplies of heavy rare earth metals occurs less frequently than the light rare earth metals. Note that heavy rare earth metals are used in far less quantities and are currently critical to a large number of technologies.

Preliminary information on worldwide mining operations was summarized by the U.S. Geological Survey in 2002 [4]. This report made several critical observations. First, although rare earth ore deposits are geographically diverse, current capabilities to process the ores into finished rare earth materials are limited exclusively to Chinese sources. It is important to mention that the United States previously performed all stages of the rare earth material supply chain, but now the Chinese have a dominant position that could affect worldwide REM sources and their prices. This dominant position of China is politically very dangerous. If any country has any conflict or disagreement with China, the Chinese would not hesitate to either cut off the supply of REMs to that country or charge extremely high prices.

Industrial estimates by mining experts predict that building a U.S. rare earth supply chain could take 10 to 15 years and is dependent on several factors, such as securing capital investments in pricing structure, developing new technologies that take additional time, and acquiring patents. These patents are currently held by international companies. Acquiring material patents is not only unpredictable but also very costly, extremely complex, and time consuming [4].

The U.S. Department of Defense (DOD) is in the early stages of assessing its dependency on rare earth ore materials. Both government and industry experts have identified a wide variety of defense systems and components that are dependent on rare earth elements that can be provided by lower-tier subcontractors in the REM supply chain. Defense-based systems and components will be the most likely to depend on rare earth materials based on their life cycles and lack of substitute materials. It is not appropriate to identify defense systems or components that deploy rare earth elements available from Chinese sources that are currently provided by lower-tier subcontractors [4].

Rare earth ore reserves are believed to exist in China, the United States, Australia, Greenland, Canada, South Africa, India, Brazil, Russia, and other regions in Europe. Mining and industrial experts believe that rare earth deposits in the United States, Canada, Australia, and South Africa could be mined and available by 2014 [4]. This U.S. Geological Survey report [4] identified the worldwide rare earth material reserves and the mining production output in tons, which are briefly summarized in Table 1.2. From the tabulated data, one can see that southern China has most of the heavy rare earth ore reserves as well as refining facilities capacity close to 97%.

1.2.5 U.S. History of Rare Earth Materials Development

According to the U.S. Government Accountability Office report on rare earth materials in the defense supply chain from April 2010 [5], U.S. rare earth manufacturing capabilities began to decline over the period from 1998 to 2005. Currently, the United States has no rare earth material mining and processing facilities. The Mountain Pass facility currently does not have the capability to refine oxides into pure rare earth metals [5].

The development history of rare earth materials in the United States can be summarized as follows:

- 1949 to 1951: Large rare earth material deposits were discovered in the United States. Furthermore, large amounts of LREEs were discovered at the Mountain Pass facility.

Table 1.2   Worldwide Rare Earth Material Reserves and Their Mining Production Output in 2009

Country	Rare Earth Oxide Reserves (tons)	Mine Output of Rare Earth Oxides (tons)
China	36,000,000	120,000
Other countries	22,000,000	NA
Commonwealth countries	19,000,000	NA
USA	13,000,000	0
Australia	5,400,000	0
India	3,100,000	2,700
Brazil	48,000	650
Malaysia	30,000	380
World	99,000,000	124,000

Source: http://pubs.usgs.gov/fs/2002/fs087-02.

- 1951 to 1964: There is no record of any technological progress on rare earth materials over this time period.
- 1965 to 1985: Mountain Pass is the only reliable source of rare earth elements. The United States performs all stages of rare earth material processing, including screening, refining, and quality control.
- 1998: The Mountain Pass separation plant closes after regulatory problems with the main wastewater pipeline.
- 2002: The Mountain Pass rare earth material facility suspends operations.
- 2003: Magnequench, who produces a neodymium-iron-boron (Nd-Fe-B) permanent magnet, closes their U.S.-based plant and moves the equipment to China. Furthermore, the German company that was producing Nd-Fe-B permanent magnets closes its operation at Elizabethtown, Kentucky.
- 2005: Hitachi Magnet Corporation, who produced Nd-Fe-B permanent magnets, closes its production operation at Edmore, Michigan.
- 2007: Some U.S. production of rare earth elements at the Mountain Pass rare earth separation plant resumes. The Mountain Pass facility separates bastnäsite concentrate from the stockpiles produced before the Mountain Pass facility shutdown.

In summary, large amounts of rare earth deposits were discovered in the United States around 1950. U.S. rare earth manufacturing began to decline in the period from 1998 to 2005. Some U.S. production of rare earth materials resumed in 2007, including the operation of the Mountain Pass rare earth separation plant.

Currently, most of the world's rare earth material processing occurs in China, including approximately 97% of rare earth oxide production. Full operation of the Mountain Pass facility resumed in 2012, with particular emphasis on the mining and separation of cerium, lanthanum, neodymium, and praseodymium oxides. However, the Mountain Pass facility currently does not have the capability to refine oxides into pure rare earth metals [5].

1.3 Potential Commercial and Industrial Applications of Rare Earth Materials

In this section, potential applications of rare earth metals and their oxides are identified with significant benefits in terms of efficiency, reliability, and longevity. Note that military and space applications may require more stringent processing requirements than commercial applications. These stringent quality control requirements for defense applications will make the procurement costs significantly higher compared to commercial applications.

High-technology and energy applications of REEs have grown dramatically in diversity and importance since the 1960s. Furthermore, because many of these applications are highly specific and substitutes for the REEs are inferior or unknown, REEs have acquired a level of technological significance that is much greater than what was expected from their relative obscurity.

Although they are actually more abundant than any similar industrial metals, rare earth elements are less likely to become concentrated in exploitable ore deposits. As a result, most of the world's supply of REEs comes from only a few sources, which could be consumed over a short period. For example, the United States had a generally sufficient supply of REEs for several decades but has since become dependent upon imports from China, which has changed the strategic situation.

1.3.1 Commercial Applications

The commercial applications of rare earth materials are summarized in Table 1.3. Some rare earth materials are currently used in nuclear power reactors to generate electricity.

1.3.2 Industrial Applications

Rare earth materials are widely used in aerospace systems and components. Rare earth elements are used in jet engines, scramjets, battery electrodes for high-power batteries, and a host of other aviation products. Samarium-cobalt magnets are best suited for high-power radiofrequency sources, TWTAs, and microwave filters. Samarium-cobalt magnets are used in single-mode and dual-mode TWTAs in

Table 1.3   Examples of Commercial Products That Use Rare Earth Materials

Rare Earth Element	Commercial Product
Neodymium, praseodymium, terbium dysprosium	Mobile phones, computer hard drives, cameras
Neodymium, praseodymium, dysprosium lanthanum, cerium	Hybrid electric vehicles, high-capacity batteries, infrared lasers
Europium, yttrium, terbium, lanthanum	Energy-efficient light bulbs
Cerium, lanthanum, neodymium, europium	Glass additives
Europium, yttrium, erbium	Fiber-optic lines, fiber-optic amplifiers
Promethium	Portable x-ray equipment
Scandium	High-intensity flood lights for stadiums
Samarium-cobalt	Permanent magnets for electric motors, widely deployed in hybrid electric vehicles
Uranium, plutonium	Fuel rods for nuclear power plants

which improved electron beam focusing, higher efficiency, and minimum weight and size are critical design requirements. Yttrium-iron-garnet filters are widely used in satellite and airborne systems in which sharp cutoff frequency, minimum pass-band loss, and high attenuation in stop-band regions are the principal design requirements. High-performance diode-pumped solid-state infrared lasers employ rare earth elements such as holmium, thulium, yttrium, and erbium to achieve superior laser performance at infrared frequencies. Critical performance capabilities of these lasers are summarized in Section 4.3.5.1.

1.4 Defense System Applications

Approximately 4 to 30 rare earth materials are currently used in a variety of defense and military systems [4], including the following:

- Lasers for illumination of targets
- Precision-guided munitions
- Covert communication systems
- Electronic warfare equipment
- Radar systems

- Avionics (mission computer)
- Night-vision goggles
- Surveillance and reconnaissance satellites
- Computer hard drives
- Samarium and neodymium permanent magnets for high-efficiency and high-torque electric motors
- Focusing magnets for TWTAs

Samarium-cobalt magnets are best suited for high-temperature environments, whereas neodymium magnets are preferred for applications where a high magnetic field is the principal requirement. Samarium-cobalt permanent magnets have been deployed in navigation systems of advanced battlefield tanks because they maintain reliable magnet performance even at temperatures as high as 300°C [4]. These magnets are best suited for battlefield weapons operating under severe thermal and vibration environments. Furthermore, these permanent magnets provide the highest energy product $B_m H$, where B_m stands for the maximum flux density and H stands for magnetic field intensity.

In the case of mission computers [5], rare earth materials have been used to provide reliable performance even under atmospheric radiation and nuclear radiation environments. Essentially, the electronic circuits and discrete semiconductor components used in mission computers are radiation-hardened. Therefore, the reliability of these components will not be reduced or compromised.

Even commercial-off-the-shelf (COTS) products containing rare earth elements have been included in the defense system supply chain, such as mission computers and hard drives, because these COTS products meet the stated reliability, cost-effective performance, and demands of harsh operating conditions.

1.4.1 Requirements for Building a U.S. Supply Chain

New processing technologies are likely needed to rebuild a U.S. supply chain of rare earth materials to compete with China or any other rare earth supplier. According to the estimates of scientists and planners, it will take at least 5 years to rebuild the defense supply chain, which will also involve large startup costs. Furthermore, some foreign countries, such as Japan, hold key technology patents for manufacturing neodymium-iron-boron (Nd-Fe-B) permanent magnets that expire after 2014. If U.S. companies wish to enter the Nd-Fe-B permanent magnet market, they must wait for the patents to expire. If any other country in the meantime files a patent for such magnet materials, then U.S. companies will experience further delays. The development of alternatives to rare earth materials could reduce the dependence on existing rare earth materials, but the new materials may not meet the magnet performance or current application requirements.

The Mountain Pass mine in the United States has the largest rare earth deposits in the free world, but it lacks manufacturing capabilities and facilities to process the rare earth ores into finished components, such as permanent magnets with specific performance specifications. The Mountain Pass mining facility does not have substantial amounts of heavy rare earth elements, such as dysprosium, which are best suited for heat-resistance capabilities and are essential for permanent magnets deployed in various industrial and defense system applications. Other U.S. rare earth deposits exist in Montana, Colorado, Idaho, Utah, and Wyoming; however, these deposits are still in the early exploratory stages and need intense development efforts.

Once a company has the necessary capital to start mining operations, it can take at least 10 years for the mining facility to become fully operational. Most of that time is required to comply with state and federal mining regulations. Mining experts believe that it would require at least 3 years to develop a pilot plant capable of refining the oxides into rare earth metals using new technologies. Mining companies with existing infrastructures say they will not be able to restart rare earth metal production without a steady source of oxides within North America. In addition, the environmental concerns associated with some rare earth materials, such as thorium and radium, make both the extraction and processing very time consuming, difficult, and costly. U.S. rare earth mines and processing facilities must comply with state and federal environmental regulations.

Many defense systems and components will continue to deploy rare earth metals in the future based on their life cycles and the shortage of effective substitute metals. For example, the Aegis Spy-1 radar system, which has a life cycle of 35 years, has used samarium-cobalt-based magnetic components that will need to be replaced during the radar's life cycle. The components that deploy the rare earth materials will experience performance degradation and need to be replaced to meet the radar's performance specifications. It is possible that future generations of defense system components, such as the transmit/receive modules used in the radar systems, will continue to depend on rare earth metals. In addition, both current and future radar and electronic warfare systems might strictly depend on rare earth materials.

1.4.2 Corrective Measures to Overcome the Shortage of Rare Earth Metals for Defense Products

Appropriate corrective steps must be taken to mitigate the effects of a shortage of critical rare earth elements in current and future defense systems and components. Self-sufficiency in the production and the availability of rare earth materials is vital for defense products and must be given serious consideration. For airborne systems and components, the U.S. Air Force Research Laboratory should focus on the availability of widely used rare earth materials and their potential suppliers. Airborne radar transmitters and electronic warfare systems require samarium-cobalt and Nd-Fe-B permanent magnets due to their unique thermal and magnetic characteristics. U.S. dependency on Chinese rare earth materials poses a security risk to national defense products. The U.S. Army Armament Research, Development and Engineering Center should identify their current and future requirements for rare earth materials in their defense systems and should plan to establish reliable supply sources within North America.

For naval warfare systems, the U.S. Navy should take appropriate steps to look for alternate sources for the procurement of Nd-Fe-B permanent magnets, which are currently only available from Chinese sources. The Electronic Naval Systems Command and Naval Surface Warfare Center should actively try to secure domestic sources for rare earth materials and award research and development contracts for the supply of rare earth materials for use in traveling wave tubes in the Aegis radar.

Studies performed by the author on various defense products indicate the DOD must initiate programs for the domestic production of rare earth materials as soon as possible to be widely deployed by offensive and defensive weapon systems. The DOD should seriously consider funding the mining and processing efforts at the Mountain Pass mining facility because this facility can be put into operation in the shortest time and with a minimum investment. Time is of critical essence and no further delay is acceptable if the United States wants to reduce dependency on Chinese rare earth materials.

Other U.S. government agencies should start addressing the shortage of rare earth materials and the security risks involved due to their dependency on Chinese supplies of these materials. The Department of Energy, the Department of Commerce, and the Office of Science and Technology should address the domestic production of rare earth materials to eliminate dependency on Chinese sources for rare earth materials. Interagency meetings should be held to discuss concerns about the Chinese monopoly on the rare earth materials and the possibilities of various domestic sources for rare earth materials.

1.4.3 Impact on Defense Product Performance due to Lack of Appropriate Substitutes

The author has investigated the performance degradation of some defense components that are strictly dependent on the use of rare earth materials due to the lack of appropriate substitutes. For example, TWTAs, which are widely used by electronic warfare and radar systems, must use permanent magnets with unique magnetic performance in high-temperature environments. Samarium-cobalt magnets offer the highest magnetic flux density under operating temperatures as high as 300°C. High magnetic density is essential to maintaining improved electronic beam performance regardless of harsh thermal and vibration environments. No other substitute is currently available to replace this particular magnetic material.

Take the case of another component—the fin actuator that is widely used in precision-guided munitions that are specially designed around the performance capabilities of neodymium-iron-boron rare earth magnets. This particular rare earth material offers the highest energy product of flux density and magnetic field intensity. No other rare earth material or any conventional metal offers such magnetic performance. Therefore, a substitute will not be able to perform similarly to this rare earth magnet.

A DOD inquiry [5] raised concerns by the industry due to the shortage of domestically produced rare earth materials. A DOD study from September 2010 addressed the vulnerabilities in the defense supply chain and identified recommendations to mitigate any potential risks involved in the supply chain. This study was also intended to transform the National Defense Stockpile so that the materials not produced domestically would be made available to support defense needs for specific defense programs. The 2009 National Defense Stockpile configuration report identified cerium, lanthanum, europium, and gadolinium elements, which have already caused weapon system production delays. The report recommended further studies to determine the severity of the delays. Risks and delays due to the shortage of rare earth materials must be avoided at all costs, particularly in the case of defense weapon systems.

1.4.4 Identification of Security Risks due to Nonavailability of Rare Earth Elements and Alloys

According to the criteria specified in the *Defense Acquisition Guidebook* [5], the program office should elevate supplier base concerns when an item is produced by a single source supplier and meets one or more of the following criteria:

1. The item is used by three or more programs.
2. The item represents an obsolete, emerging, or enabling technology.
3. The item requires 12 months or more to manufacture.
4. The item has limited surge production capability.

In general, industrial policy can help DOD planners address a supplier gap or vulnerability when requested. Although not related to rare earth materials, industrial policy allowed the U.S. Army's Hellfire Missile Program to procure a chemical from China that was no longer produced in the United States. This allowed the U.S. Army to explore a longer-term solution to develop a domestic source for the chemical. This type of action will eliminate the delay in completion of a critical weapon system for national defense.

It is important to mention that a sole source supplier can pose serious problems for the defense procurement agency in terms of price, reliability, on-time delivery, and quality control. When a product is required by a specific time and date, the supplier can demand higher prices. If a rare earth product, such as a samarium-cobalt permanent magnet, is not delivered on time, it can delay the delivery of TWTAs, which will affect the testing and delivery of electronic warfare equipment. This can be a serious matter, particularly if the country is involved in a military conflict. Furthermore, if the rare earth material supplier is in alliance with the enemy, single-source procurement is not advisable, especially when the product is an integral part of a current weapon system. To avoid these problems, procurement of defense-based products must be negotiated from at least two or three reliable sources.

A GAO report on the national defense of rare earth materials in the defense supply chain [5] revealed that the United States is not currently producing enough neodymium-iron-boron permanent magnets and is depending on China for delivery. The report further revealed that the United States has only one local supplier of samarium-cobalt permanent magnets used in TWTAs, which are critical components for electronic warfare equipment, microwave filters with sharp skirt performance levels, and other defense-related systems. Industrial experts and defense scientists indicate that only one Chinese company is producing and selling commercial quantities of rare earth materials. Some Japanese companies are producing rare earth metals in limited quantities; they do not offer these materials as a finished product but use them to produce alloys and magnets. The Japanese strictly rely on rare earth oxides and metals from China. China has threatened to cut off the supply of rare earth materials to Japan and reduce the export of rare earth metals in order to demand higher prices. China can completely cut off the supply of rare earth metals supplies if there are regional political disagreements or military conflicts. Lack of manufacturing assets and refining facilities in the country pose serious problems for the conversion of rare earth oxides and appropriate rare earth metals for use in defense-oriented products. Therefore, the availability of finished rare earth metals is of critical importance, and dependency on other countries is not a reliable option.

1.4.5 Processing Efforts to Convert Rare Earth Ores and Oxides into Metals for Defense System Applications

Very complex, time-consuming, and costly procedures are involved in the extraction of rare earth metals from mined ores and oxides. Transportation, labor, and quality control inspection efforts are the major cost factors. Unskilled workers remove the rare earth ores from the mines, then the ores are separated into rare earth oxides. Separation of oxides from ores may require special screening processes. Complex and time-consuming processes are used to refine the ores to rare earth metals or elements. Complicated methods are used to convert rare earth metals into rare earth alloys such as samarium-cobalt, which is used in the manufacture of samarium-cobalt permanent magnets. Similarly, neodymium rare earth metal is converted into a neodymium-iron-boron alloy, which is used to manufacture neodymium-iron-boron permanent magnets with high-energy products.

China produces more than 75% of neodymium-iron-boron permanent magnets and controls the export and pricing of these magnets, which are widely deployed in defense products. The United States produces small amounts of samarium-cobalt permanent magnets from samarium-cobalt alloy supplied by England. Only one English company produces a limited quantity of samarium-cobalt alloy; however, it relies on rare earth metal and oxide supplies from China. Any Chinese export restrictions on rare earth materials will create a real shortage for U.S. and North Atlantic Treaty Organization defense products.

1.4.6 Market Dominance by China Could Affect the Supply of Rare Earth Materials for Defense Use

According to government and industrial scientific data, the future availability of some rare earth materials, such neodymium, dysprosium, and terbium, is strictly controlled by Chinese suppliers. Furthermore, China has adopted domestic production quotas on rare earth materials and reduced its export quotas, which will rapidly increase the prices of rare earth materials. Recently, China has raised prices to a range of 15 to 25%, which will increase rare earth material prices for materials from non-Chinese sources. China may plan on greater vertical integration in the rare earth materials market in the immediate future, which will unquestionably enhance China's market power and dominance. Because China is currently exporting rare earth oxides and metals, China may exclusively export finished rare earth material products in the future, thus fetching much higher prices.

China's dominance in the rare earth material supplies market presents serious problems and restrictions on the availability of these materials for many commercial and industrial applications, including emerging energy technologies, electronic devices, hybrid electric cars, defense products, and a host of national security applications. In addition, an unrestricted supply of rare earth materials is highly essential to U.S. national security and economic well-being.

1.4.7 U.S. Demand for Rare Earth Materials for Various Defense System Applications

U.S. demand for rare earth elements and materials was projected to rise in 2011 and 2012, according to a published report by the U.S. Geological Survey [6]. For example, permanent-magnet demand was expected to grow by 12 to 18% per year. Demand for automobile catalysts, petroleum catalysts, electric motors, and generators for hybrid electric vehicles and jet engines was expected to increase between 8 to 10% each year. Demand for rare earth materials will increase for use in flat-panel television displays, rechargeable batteries, defense products, and medical devices. In addition, prices for rare earth elements and alloys will increase when the demand for these materials is high.

Prices of selected rare earth oxides are shown in Table 1.4. Examination of the data in Table 1.4 indicates that the prices of half of the materials have doubled, even within 1 year. There could be further price increases if China puts restrictions on the export of rare earth elements and their alloys.

Table 1.4   Prices of Selected Rare Earth Oxides

Rare Earth Material	Price ($ per kg)		
	2007	2008	2010
Cerium	2.50	4.35	50
Dysprosium	85	110	295
Europium	300	475	625
Lanthanum	3.10	7.75	53
Neodymium	29	27	80
Terbium	555	650	605
Yttrium	7	15	56

1.5 Industrial Applications Using Mischmetal

The first industrial application of rare earth materials was in 1891 when Auer von Welsbach obtained patents for the Auer incandescent mantle, which is composed of 99% thorium oxide and 1% cerium oxide. Since then, more than 10 billion incandescent mantles have been produced and consumed all around the world. The light produced from such a source is cheaper than conventional electric bulbs. In addition, even today, this method of light production remains superior to electric lighting systems in remote areas. Gadolinite and bastnäsite from the Swedish mines were recognized as the first raw materials for the rare earth elements, along with thorium. Later, new raw materials were found. "Carolina sand" was found in the United States and an inexhaustible reserve of monazite was discovered in Brazil, which guaranteed raw material supplies far into the future.

Welsbach accumulated so much thorium in the processing of rare earth materials for the production of lanthanum-zirconium incandescent mantles that he had to look for other applications. Welsbach had already established that thorium oxide provided light radiation at elevated temperatures, and that this light radiation or intensity became poorer as the thorium oxide became poorer. Welsbach further established that cerium was the main impurity, and it was not difficult to determine from this the optimum dosage of thorium oxide with cerium oxide. However, many problems remained to be solved by research scientists, application technicians, device/product engineers, inventors, and developers in the rare earth industry.

Welsbach was granted patents for the discovery of a pyrophoric metallic alloy called "flintstone," which was composed of 70% mischmetal and 30% iron. In 1908, Welsbach was successful in producing pore-free mischmetal using fused salt electrolysis. He proposed to use the mischmetal-iron flints for ignition in gasoline engines—an idea that is perhaps still relevant today in view of the efforts to make engines and cars more compact and lighter.

The third major invention using rare earth elements was the addition of rare earth fluoride as a wick in arc light carbons, which was used for a variety of lighting schemes, such as cinema light sources and searchlights. This use of rare earth compounds is based on the intensive arc light source developed by a German scientist in 1910. German inventors produced more than 1,250 tons of flints between 1908 and 1930 [2], which was considered an important rare earth product. This required the consumption of about 1,300 to 1,800 tons of rare earth chloride, meaning that roughly 1.3 tons of rare earth materials were needed to produce 1 ton of oxide.

Between 1930 and 1940 extensive development activity was focused on various applications involving rare earth materials. Particularly successful endeavors included the production of neophan for sunglasses, polishing agents using rare earth oxides to replace iron oxide, and decolorization using cerium oxides. Later, cerium oxides were widely used for various commercial and industrial applications.

1.5.1 Properties of Rare Earth Materials Best Suited for Industrial Applications

The property requirements of rare earth elements or alloys are strictly dependent on specific applications. One set of properties of rare earth materials may be essential for commercial applications, whereas another set of properties may be best suited for industrial applications. The most important characteristics of rare earth materials are as follows, regardless of application:

■ Chemical properties
■ Metallurgical properties
■ Electro-optical properties
■ Magnetic properties
■ Nuclear properties

1.5.1.1 Chemical Properties

The chemical properties of rare earth materials are particularly important for polishing applications. The rare earth element or its oxide is deeply involved in the polishing process, which is essential for the extensive chemical reactions on the surface of glass. In other words, these chemical reactions are needed to remove the material by a mechanical abrasion process. A specific polishing agent may be required for a particular material surface (e.g., metal or glass) and even for individual kinds of glass. The rare earth oxides, especially cerium oxide, are considered to be best suited for polishing glass surfaces.

Rare earth elements are considered to be physiologically inert and therefore present no danger to the environment. This is the most outstanding benefit of the rare earth oxide for polishing applications. Rare earth oxides offer two unique pharmaceutical applications that are based on anions or corresponding salts rather than the effect of rare earth metals—cerium oxalate as a treatment for seasickness and neodymium (Nd)-isonicotinate as a possible treatment for thromboses. These treatments have important pharmaceutical byproducts.

1.5.1.2 Metallurgical Properties

Metallurgical properties play a key role in the development of rare earth alloy-based compounds as they are best suited for the development of a special product with unique performance capabilities. The rare earth elements operate as scavengers for oxygen, sulfur, and other deleterious elements as well as being boundary surface-active substances. This is particularly true in two of the most economically important fields of applications: production of nodular graphite castings through spheroidization of graphitic components and treatment of steel for sulfide inclusion shape control. In both cases, most beneficial effects are achieved with minimum cost and complexity.

It is important to mention that rare earth metals are rarely used for the fixing of oxygen and sulfur in light metals for production of conductive copper and conductive aluminum. However, the use of rare earth elements as magnesium hardeners remains important. In this particular case, the rare earth metals serve by precipitation of intermetallic compounds of high thermal stability.

1.5.1.3 Electro-Optical Properties

The electro-optical properties of rare earth elements are of critical importance in their applications. Because of the atomic structure of the shell, there are narrow and sharp absorption and/or emission lines in the visible range of the infrared spectrum that may be used in various ways or applications. In addition, the oxides and/or oxide-based components in glass provide a high index of refraction with low dispersion characteristics.

The first applications of rare earth elements were in the optical field—namely, Auer incandescent mantles and arc light carbons. As a result of comprehensive laboratory investigations in 1964 by Levine and Patilla, the use of truly rare and expensive europium together with yttrium as red phosphors for color television screens was a major leap forward for the rare earth industry. Due to the strong and sharp emission line of europium element at 610 A° without a yellow component, which is perceived by the eye as a wonderfully saturated red color tone, it was possible to achieve an evenly colored picture in color television. One can say that 1964 was the birth of color television.

Gadolinium oxide (Gd_2O_3) plays an important role in the field of x-ray intensifiers. When a rare earth oxide is placed on the x-ray film, it is excited by the x-rays and emits in the visible range of the spectrum. Due to presence of the rare earth oxide, the x-ray film can be exposed with a minimum of radiation. This indicates the extreme sensitiveness of the rare earth oxide.

The coloring of glass using neodymium or praseodymium is strictly based on their selective absorption capability in the visible range of the spectrum. In the decolorizing of glass surfaces, the oxidation effect of four-valent cerium is combined with the absorption of small quantities of neodymium or praseodymium. Similarly, cerium concentrates, which always contain some praseodymium and neodymium, are added to glass melts. One can achieve decolorization by the selective absorption of didymium through optical compensation. In doing so, a combination of chemical and optical properties is used.

The most important application of praseodymium is the production of very beautiful high-temperature-resistant lemon-yellow pigments that are widely used in industrial ceramic applications. Praseodymium is built into the zirconium silicate lattice, thereby yielding full optical splendor. Lanthanum oxide is used in glass applications because of its high index of refraction and lack of this color. Even today, optical glasses use up to 40% lanthanum oxide to provide corrosion-resistant capabilities. In addition, rare earth elements serve as activators in glass-based lasers. Among such lasers, the neodymium-based laser is well known and widely used when efficiency, optical stability, and coherency are principal requirements.

1.5.1.4 Magnetic Properties

The atomic structures of rare earth elements exhibit interesting magnetic properties that lead to various applications for which minimum weight and size are principal design requirements. Rare earth elements such as samarium, neodymium, and praseodymium are most ideal for the development of permanent magnets for electric motors and generators.

Rare earth elements play an important role in the production of hard magnetic materials. Studies performed by the author revealed that hard magnets offer consistent performance, reliability, and longevity under severe thermal and mechanical operating environments. The studies further indicate that most prominent are the alloys of samarium with cobalt in the atomic ratio of 1:5 or 2.17. Therefore, samarium-cobalt magnets are widely deployed in TWTAs because of their superior magnet performance and uniform optical beam over the entire length of the tube. These TWTAs are best suited for radar and electronic warfare applications.

Future development of rare earth magnets will likely be pursued for other rare earth elements, such as neodymium, praseodymium, lanthanum, and other heavy rare earth elements capable of achieving particular effects. The development of magnetic bubble memories is strictly based on the use of gadolinium-gallium garnets. These magnetic bubble memories are used for the storage of information because of their large storage density. The use of gadolinium as a heat pump may be another magnetic application in which the Curie point, which happens to be at room temperature for this material, can be very well utilized. Performance capabilities of rare earth magnetic materials will be summarized later in Section 3.6.

1.5.1.5 Nuclear Properties

Major research programs involving rare earth materials were given serious consideration for the generation of nuclear energy around 1960. These programs resulted in interesting aspects for the use of rare earth elements, particularly oxalate to combat seasickness and nausea during pregnancy and neodymium to combat thromboses. All of these applications had only relatively small use during this transition period as compared to the large quantities of rare earth materials that continually became available. Note that thorium oxide (ToO_2) is best suited as a catalyst production of gasoline by the Fischer-Tropsch process. After a few years, thorium oxide was replaced by magnesium oxide (MgO) in this application for economic and scientific reasons.

The discovery of nuclear fission in 1939 was a major event in the scientific world because it opened the prospect of an entirely new source of power generation using the internal energy of the atomic nucleus—a key aspect of nuclear technology. In the 1950s (first in the United States and England, then later in other European countries), major programs related to atomic energy were carried out. The basic materials in the 1950s were uranium and thorium, which can be used to release nuclear energy by fission. Thorium was selected as the first atomic fuel for atomic and nuclear power plants. After the large stockpile purchases of thorium by the atomic countries as a feed material for atomic breeder reactors, huge quantities of rare earth byproducts were left behind in the late 1950s and early 1960s. This disequilibrium was eliminated by the middle of 1960s after the termination of the large stockpile programs. Since then, thorium has been accumulating in large inventory stocks at all monazite processors—and will continue to accumulate until a new use for this material is found.

The large atomic programs initiated in the 1950s provided a great advantage for rare earth materials as well as for industry. The rare earth elements that occurred in large quantities during the fission products of atomic reactors were extensively examined using sophisticated scientific methods at great expense, and their separation from each other was carefully pursued. Therefore, by the end of the 1950s, a large volume of scientific data and the properties of rare earth metals were known to physicists and atomic scientists.

Professor Frank H. Spedding was responsible for establishing the rare earth center at the Iowa State University, which developed several nuclear research programs with a principal objective to generate huge electric power using nuclear technology. Building on this stable scientific foundation, a wide range of applications for rare earth elements developed between 1960 and the present, with exponentially increasing consumption. If one rare earth element is selected for a specific application, the remaining elements can be investigated with minimum time and complexity for other applications.

At the beginning of the 1960s, there was a market for lanthanum in the optical glass industry, for cerium in polishing technology, and for praseodymium/neodymium in the glass industry for coloring and decolorization. However, there was no interest in samarium and europium at that time. An industrial research center had accumulated large stocks of these materials in the form of concentrates and high purity with a book value of zero [2]. This changed suddenly in 1965 when europium was used in the United States as a red phosphor in color television sets; however, even then samarium oxide continued to remain behind. These stocks were first reduced by the development of rare earth–based magnet materials in recent years.

A balance between the various applications must be achieved so that the costs for each application do not rise excessively. With the great broadening of the use of rare earth materials, however, this balance will become easier to achieve in the near future. There are enough applications today for the mixtures of rare earth elements, and it makes little difference in catalytic cracking that samarium and europium were previously extracted. Therefore, flexibility exists in the applications of rare earth elements, such as those using samarium and europium.

To bring about systemization among the variety of applications, it is necessary to break down the uses of rare earth elements into five groups of properties, as mentioned earlier: chemical, metallurgical, optical, magnetic, and nuclear. In the applications of chemical properties, the high affinity of rare earth metals for oxygen is primarily involved. This leads to their applications as flints, wherein their highly exothermic reaction with oxygen present in air is used. On the same properties rests their application as getter metals, wherein residual oxygen (e.g., in amplifier tubes) is bound up.

The chemical-ceramic properties of rare earth elements result from their high affinity for oxygen, which yields highly stable oxides that can be used in high-temperature materials. Likewise, the high melting points of sulfides are of great interest. However, the use of rare earth elements in this particular field is limited because of their sensitivity to carbon dioxide and water vapor.

In another example, yttrium oxide has become increasingly important in the stabilization of zirconium oxide (ZrO_2) in the cubic phase. Zirconium oxide is used as a high-temperature material. At high temperatures, this oxide becomes conductive. The yttrium oxide–stabilized zirconium oxide serves also as an electrode, which can be used in the high-temperature electrolysis of water; in addition, yttrium-lanthanum oxides serves as a solid electrolyte that is best suited for high-power storage batteries and fuel cells. It is interesting to mention that zirconium oxide/yttrium oxide has a similar function as an electrode in the so-called lambda sensors. These lambda sensors can be used to determine the oxygen content in the exhaust gases of automobiles, jet engines, and gas turbines.

As mentioned earlier, the chemical properties are also used in a variety of applications for rare earth elements as catalysts. Most important are cracking catalysts for the petroleum industry. Rare earth elements are combined into molecular sieves (Y-zeolite) and serve in fluid-bed or fixed-bed reactors to increase the yield of gasoline. In addition, they are used as combustion catalysts for automobiles and for air pollution control.

1.5.1.6 Effect of Impurities on the Quality of Rare Earth Materials

In the design and development of samarium-cobalt permanent magnets, the impurities have a dilution effect. One can therefore use a 99.9% pure samarium rare earth element to avoid any deterioration in the magnetic performance of the device under consideration. A rare earth metal is significantly cheaper at about 90% purity, perhaps even at 80%, with the balance composed of other rare earth materials. In any case, the composition of the other rare earth elements should be held constant, which is not always easy or even possible. The impact due to impurities in other magnetic materials, such as neodymium-based magnets, is unclear. The author plans to investigate the impact on the magnet performance, if any, due to impurities in other rare earth magnet materials.

In polishing applications, ceric oxide is the active element in the polishing processing media. Most polishing plants are satisfied with 50% pure ceric oxide, which is readily available in the natural mixture of the light rare earth elements as they are extracted either from bastnäsite or monazite. Sometimes suppliers intentionally add impurities in order to keep the cost of the product low.

1.5.1.7 Impact due to Substitution on Rare Earth Materials

The substitution of rare earth elements by other cheaper processes or other substances can play an important role in future applications. It should be perfectly clear that rare earth elements were never cheaper materials, although they occur abundantly in nature with rare earth materials. The cost of rare earth elements rises rapidly if a specific rare earth material of high purity is required. Under these conditions, a design engineer must look for a cheaper material for an established field of application—a danger that the rare earth industry has learned to fear. This situation has led to the characteristic rise and fall of economic success for rare earth enterprises.

Typical defined procedures include the following:

- Replacement of Auer incandescent gas mantles by conventional incandescent lamps
- Replacement of rare earth elements for sulfide inclusion shape control by extreme steel desulfurization with the aid of calcium (in the correct amount)
- Replacement of rare earth–bearing arc light carbons by high-pressure argon arc lights, particularly in cinema film projectors and searchlights

Procedures due to the introduction of other substitutions include the following:

- Use of magnesium instead of mischmetal for production of modular graphite castings
- Use of hafnium, which is associated with the zirconium compound (its presence is roughly 5% of the compound material), instead of europium for use in atomic submarines due to its high absorption of thermal neutrons (about 500 times that of zirconium) and lower density (about half that of zirconium)
- Replacement of ceric oxide by tin oxide or zirconium oxide as pacifiers for enamels
- Use of titanium-iron alloy for hydrogen storage instead of lanthanum nickel ($LaNi_5$)

1.5.1.8 Role of Rare Earth Materials in Nuclear Reactor Performance

Studies performed by the author on nuclear power reactors seem to indicate that some rare earth elements play key roles in capturing cross-sections of thermal neutrons in nuclear reactors. In this regard, rare earth elements such as europium, samarium, gadolinium, and dysprosium are best suited for capturing cross-sections of thermal neutrons associated with these materials. Studies further indicate that the elements—namely, gadolinium and dysprosium—can be used as so-called burnable poisons to efficiently achieve a uniform neutron flux during the lifetime of a nuclear element. In fact, europium shows remarkable properties as a high-capture cross-section of the natural isotope combined with an uninterrupted series of five isotopes. These isotopes arise upon capture of a neutron and all have high-capture cross-sections. Therefore, the capacity of such as a neutron absorber is significant.

In addition, there was an attempt to equip American atomic submarines with europium-based fuel control rods. In the early days, this was not possible because of an insufficient supply of europium. However, ultimately U.S. Navy nuclear submarines were equipped with europium-based control rods. Later, yttrium was deployed as a tubing material for molten salt nuclear reactors because it has a low-capture cross-section for thermal neutrons. In other words, yttrium can be used when designing a thermal shield in a nuclear reactor. In addition, cerium and yttrium hydrides were successfully tried as neutron moderators due to their remarkable temperature stability. These rare earth elements play key roles in the design of control rods and thermal shielding for nuclear reactors.

1.5.1.9 Effects on Specific Applications due to Variations in Properties

For specific applications, rare earth elements can be used in various purities according to whether the general properties of the rare earth elements or specific properties of individual elements are required. The rare earth industry generally distinguishes three grades of purity as follows:

- The group of unseparated rare earth elements in a composition that occurs naturally in ores obtained directly from a mine
- Concentrates producible by simple chemical precipitation reactions, which in general contain 60 to 90% of the individual element required
- Pure rare earth elements that contain between 98 and 99.999% of a rare earth oxide

The price of rare earth materials increases over several orders of magnitude in the corresponding series. The price increase is strictly dependent on demand and supply, the material's purity, and its critical use for specific applications. Table 1.5 presents rare earth metal prices and their typical applications.

A comprehensive survey by the author revealed that the steel industry has large-scale applications for rare earth materials. These materials are needed as catalysts, naturally occurring mixtures, and concentrates and for polishing glass. For other commercial and industrial applications involving electronic products in great demand, higher prices are quoted. Therefore, the application of rare earth elements requires careful consideration and close cooperation between the producers and users so that an optimum balance between desired effects, purity, and product costs can be found for each specific application. The examples in this section explain the balance between the properties of rare earth elements for specific applications.

1.5.1.9.1 Examples of Substitution

In the production of red phosphors for color television, each screen requires approximately 5 to 10 g of yttrium oxide and 0.5 to 1.0 g of europium oxide. In most cases, high requirements are placed on the purity and use of a specific rare earth element. As a result, the prices for rare earth elements are quoted typically in dollars per gram.

In general, substitution should not be ignored or feared in the case of chemical, optical, or magnetic properties. However, for metallurgical and nuclear properties, there is always a danger that a more economical solution for the problem can be squeezed out of rare earth elements. In certain applications, such as polishing media, flints, catalysts, phosphors, magnets, optical glass components, coloring and decoloring of glass, pigment formers, and x-ray intensifiers, the work or the application is not threatened by the substitution procedure.

1.5.1.9.2 Impact of Market Surveys on Substitution

The world demand for rare earth elements is in the range of approximately 25,000 tons per year. This estimate is based on calculations for rare earth oxides of various rare earth elements. These materials are available through the production facilities at the Mountain Pass Mine in California; from the monazite mines in Australia, Brazil, and India; from Molycorp mines in the Mojave Desert in California; and from monazite sites as a byproduct from the production facilities of tin ore, rutile, and various

Table 1.5   Typical Prices and Applications of Selected Rare Earth Elements

Rare Earth Element	Price ($/lb.)	Typical Applications
Cerium	75	Self-cleaning ovens
Dysprosium	315	Fuel control rods in nuclear reactors
Europium	4,000	Lasers, fuel control rods in nuclear reactors
Erbium	500	Fiber-optic amplifiers, metallurgical process
Gadolinium	250	Microwave filters, thermal neutron blankets
Hafnium	175	Masers, nuclear fuel control rods
Lanthanum	7	Carbon lights, special optical glasses
Lithium	8	Anode material for batteries
Lutetium	8,000 (highest)	Catalyst, hydrogenation
Neodymium	120	Magnets, lasers, colored glasses
Palladium	650	Catalyst, hydrogenation
Praseodymium	225	Lasers, colored glasses for welding operations
Samarium	250	Permanent magnets for motors and generators
Scandium	7,500	Alloys for aerospace components, space missiles
Thallium	12	Photo cells, lasers
Thulium	1,200	Infrared lasers
Ytterbium	300	X-ray sources
Yttrium	200	Lasers, catalysts

Source: Weast, R.C. 1971. *Handbook of Chemistry and Physics,* 51st ed. Boca Raton, FL: CRC Press.

Note that this table does not reflect the percentage purity of the rare earth material.

heavy minerals. Huge reserves of rare earth materials are also found in the People's Republic of China. Material scientists believe that a tenfold increase in the demand for rare earth materials could be satisfied over a period of several years. As a matter of fact, mine experts believe that the byproducts of existing hematite and magnetite mines are operational in the autonomous regions of Outer Mongolia and southern Tibet, which is adjacent to the Indian state of Arunachal Pradesh. These mining operations are fully accessible through infrastructures and railroad lines. The classic problem in the processing of the ores can be solved where cheap labor is available and high-technology techniques using sophisticated equipment are not required.

Furthermore, it is believed that the future supply of ore will be a reasonable quantity that presents no problems to the rare earth industry. However, a matter of serious concern is the price for processing and refining the ore. In addition, there is a single dominant supplier that does not believe in fair trade practices. As far as the production rate of rare earth elements is concerned, it is strictly dependent on exploration, mining, refining and quality control operations by rare earth material companies. It takes 8 to 10 years to complete all feasibility studies, environmental statements, and financial/economical analyses before any mining source is confirmed as a reliable source for rare earth materials and their oxides. China has already indicated it will not hesitate to limit exports of rare earth metals as political leverage against Japan. To neutralize the Chinese threats, other countries, including the United States, Australia, and Japan, are looking for mining and refining operations within their own borders.

1.5.1.9.3 Impact of Supply and Demand on the Cost of Rare Earth Elements

The supply of rare earth materials is strictly dependent on the demand and consumption of the materials needed for various applications. The consumption of rare earth elements can be divided into four groups of applications. Approximately 98% of rare earth elements, with respect to quantity, are used in the following fields of application:

- Metallurgy
- Chemical/catalysts
- Glass-based products
- Polishing media

However, if one looks at the value of the products, the picture is completely different. Phosphors may be the most important because they are widely used in many applications. However, the importance of metallurgy in the steel and glass-ceramic industries is increasing, whereas the use of rare earth elements as catalysts and chemicals is declining. It is possible this trend in applications will change in the future because a partial substitution for rare earth metals, known as mischmetal, is being used in metallurgy. In addition, an increased demand for rare earth materials is expected in European countries.

The demand for rare earth elements is also affected by general economic conditions, particularly for steel, glass, phosphors, and electronics; however, the demand for rare earth catalysts seems to be independent of the economy. Furthermore, because of an increased demand for oil refinery operations in Europe for the production of gasoline, an increased demand for rare earth materials is expected. In addition, the production of pyrophoric alloys is determined by the large number of smokers and the results of their attempt to give up smoking habits. Therefore, flints are quite independent of the state of economy.

The demand for rare earth elements and their alloys will significantly increase as the world population increases. Both the prices of rare earth materials and rare earth material demand by applications change with time, regardless of the country of origin. Variations in rare earth demand for applications in the United States and the world are shown in Table 1.6 as a function of time.

The applications and percentage of rare earth materials used in various applications are based on the following statements:

■ Worldwide demand for rare earth materials is estimated to be approximately 136,000 tons per year, with global production of approximately 133,600 tons in 2010, which is expected to rise to 185,000 per year by 2015. The difference is expected to be covered by previously mined ground stocks for rare earth

Table 1.6 Rare Earth Material Demand by Application in the United States and the World

Application	Demand (%)			
	2010 (U.S.)	*2010 (World)*	*2015 (U.S.)*	*2015 (World)*
Catalysts	60	20	43	15.5
Glass	7	9	4	6
Polishing	7	15	17	15
Metal alloy conversion	7	18	8.5	19
Permanent magnets	3.3	21	13	26
Phosphors	3.3	7	4	6
Ceramics	10	5	8.5	5.5
Others	3.3	5	2	5.5

Source: Congressional Research Service, 101 Independence Ave., SE, Washington, DC.

materials. Mineral scientists predict that an additional mine at Mount Weld in Australia will close the short-term gap for raw materials.

- New mining projects could take anywhere from 10 to 12 years to reach the desired production level. However, the U.S. Geological Survey predicts that global rare earth material reserves and undiscovered resources are large enough to meet the worldwide materials demand.
- The major applications of rare earth elements include automobile catalytic converters; crack catalysts in petroleum refining; phosphors in color television screens and flat-panel displays widely used for cell phones, portable DVD players, and laptops; permanent magnets and rechargeable batteries for hybrid electric vehicles and electric vehicles; generators for wind turbines; and a host of medical portable equipment and implantable medical devices.
- Major defense applications include jet engines, missile guidance systems, antimissile systems, space-based surveillance and reconnaissance satellites, and communication systems.
- The United States was once self-reliant on domestically produced rare earth elements; nevertheless, since 1995 or so, the country has become 100% dependent on rare earth materials imported solely from China because of low-cost mining and processing operations.
- Even though China currently has a monopoly on rare earth elements due to their enormous mining and processing experience and cheap labor, early contributions to the knowledge on rare earth elements and oxides was made by European scientists. In addition, remarkable contributions were made by Europeans in the 1970s, 1980s, and 1990s, and credit must be given to them [1]. Rare earth elements became known to the world with the discovery of the black mineral "ytterbite" (which was renamed gadolinite in 1800 by Carl Axel Arrhenius) at a quarry in the village of Ytterby in Sweden. However, complex processing techniques and applications of rare earth materials were defined by the physicists and engineers born in the early nineteenth century.

1.5.1.10 Applications of Lanthanides

There are 17 REEs within the chemical group known as lanthanides, plus yttrium and scandium. The selected end uses of lanthanides are summarized in Table 1.7.

1.6 Impact of International and World Trade Organization Guidelines on Supply of REEs

Rare earth materials have demonstrated superior device performance for permanent magnets, rechargeable batteries, catalysts, hybrid electric vehicles, smartphones, medical devices, lasers, self-cleaning ovens, oil refineries, jet engines, cryogenic devices, antimissile systems, permanent magnets that are capable of providing reliable performance at temperatures as high as 300°C, satellite communication

Table 1.7 Selected End Uses of Lanthanides

Element	Major Uses
Light rare earth elements	
Lanthanum	Metal alloys, hybrid engines
Cerium	Catalysts, petroleum refining metal alloys
Praseodymium	Magnets
Neodymium	Catalysts, petroleum refining, hard drives, headphones, engines
Samarium	Permanent magnets
Europium	Computer screens, red color for televisions
Heavy rare earth elements	
Terbium	Phosphors, magnets
Dysprosium	Permanent magnets, hybrid engines
Erbium	Phosphors, lasers, fiber-optic amplifiers
Yttrium	Fluorescent lamps, red color, ceramics, metal alloy agents
Holmium	Lasers, glass coloring
Thulium	Medical x-rays, lasers
Lutetium	Catalysts for petroleum refining, infrared lasers
Ytterbium	Lasers, steel alloys
Gadolinium	Permanent magnets

Source: http://pubs.usgs.gov/circ/1993/0930n/report.pdf.

equipment, space-based surveillance and reconnaissance systems, and a host of other commercial, industrial, and defense systems. Therefore, any Chinese restrictions or reductions of export quotas for rare earth elements to the United States will not only be a violation of World Trade Organization (WTO) rules but also create a shortage of REEs for the deployment and manufacture of defense-based weapon systems. This can pose a serious threat to national defense.

The Chinese monopoly on REEs forced the United States, Japan, and the European Union to bring the sensitive issue before the WTO. However, defense analysts and weapon manufacturers believe the Chinese will not respond to this matter. China has had at least 16 border discussions with India with no resolution of the border issue. The Chinese also had at least three discussions with Tibetan leaders with no useful resolution on the critical political issues between them. Even U.S. President Barack Obama is in talks with China to defend America's manufacturing capability and energy security industry. The president will not allow the country to break the WTO's rules and regulations.

Furthermore, China controls the global mining and processing of rare earth metals that are widely used in wind turbines, powerful permanent magnets for high-power radiofrequency sources, and traveling wave tube amplifiers that are vital for electronic warfare equipment and other commercial, industrial, and defense products. The Chinese have argued that these rare earth material export quotas are aimed to protect the resources and environment of the country and are in accordance with WTO rules and regulations. However, U.S. officials allege that Chinese quotas, export duties, and other practices to control the production and supply of REEs have artificially increased prices to put pressure on the United States and other countries to move operations, jobs, and technologies to China.

The United States won a WTO complaint against Chinese export restraints on nine other industrial components. However, experts at the Peterson Institute for International Economics viewed this WTO request on rare earth materials as being too little and too late. According to the Peterson Institute, the prices of rare earth materials have come down recently. American and other foreign companies have increased their production of rare earth materials, and some manufacturing companies have moved their operations to China, where the materials are readily available.

The United States believes the first step in court proceedings will lead to a full legal case and the establishment of a WTO dispute settlement panel, if the matter is not resolved through talks within 60 days. White House officials believe they have demonstrated they are serious in pursuing alleged unfair trade practices, which will put pressure on China for a resolution. According to a *Los Angeles Times* report, President Obama brought trade cases against China at nearly twice the rate of the George W. Bush administration [7]. Obama was successful in halting a surge in imports of cheaper Chinese tires, which protected the jobs of more than 1,000 American tire workers, although this is in dispute with some U.S. business groups. Obama set up a Trade Enforcement Unit (TEU) and charged it with aggressively rooting out unfair trade practices around the word. The president stated that he will not hesitate to take action against any country if U.S. workers and businesses are being subjected to unfair trade practices. In the near future, the TEU will look into unfair trade practices by China with regard to rare earth materials.

1.7 Critical Role of China in the Development of the Rare Earth Element Industry

Chinese scientists and engineers have been involved in the research and development of REEs since 1952 [8]. Their long-term outlook and investment has yielded significant results for China's rare earth industry. China has two state-run laboratories: one for rare earth material chemistry and applications, which has focused strictly on rare earth separation techniques and is affiliated with Peking University, and another for rare earth resource utilization, which is associated with the Changchun Institute of Applied Chemistry. Additional research laboratories that concentrate on rare earth elements include the Baotou Rare Earth Research Institute, which was established in 1963 and is the largest rare earth research institution in the world, and the Beijing General Research Institute for Nonferrous Metals, which was established in 1952. These research and development laboratories have contributed significantly to the Chinese rare earth industry and China's monopoly on rare earth materials.

The Chinese are trying to extract rare earth materials from the iron deposits at Bayan Obo in Inner Mongolia, which contain significant amounts of rare earth elements in iron ore mines. China has made a national policy that will use Bayon Obo as the center of production, research, and development for REEs. REEs are currently produced in several other provinces, such as Baotao (Inner Mongolia), Shangdong, Jiangxi, Guangdong, Hunan, Guangxi, Fujian, and Sichuan.

Published reports indicate that China's annual production of REEs increased by more than 40% between 1978 and 1989. The same reports further indicate that the Chinese export of REEs increased in the 1990s, which drove down the prices of the elements. China had 130 neodymium-iron-boron magnet producers in 2007 with a total capacity of 80,000 tons. Chinese output grew from 2,600 tons in 1996 to 39,000 tons in 2006 [8].

Spurred by sudden economic growth and increased consumer demand for rare earth materials, China is ramping up for higher production of wind turbines, consumer electronic devices, and other commercial and industrial sectors that require more rare earth materials for domestic use. However, safety and environmental conditions related to the production of REEs will increase the cost of operations in China, which will increase the cost for the export of REEs. The purity, actual material content, output inspections, and quality control procedures for REEs selected for export may be of concern. China has set up REE manufacturing to meet its surging demand for consumer electronic devices, such as cell phones, laptops, toys, and devices that are widely used for green energy technology products. Green technology experts predict that China is anticipating an increase from 12 gigawatts (GW) of wind energy production in 2009 to 100 GW in the year 2020.

Around 2005, several solar-powered systems and wind turbines were put into pipelines by the Chinese government to meet the growing electricity requirements of its citizens. To meet the domestic requirements for rare earth elements, China introduced policy initiatives to restrict the export of rare earth materials, particularly terbium, dysprosium, thulium, yttrium, lutetium, and other heavy rare earth elements. Industry experts believe that China's principal goal is to build out and serve its domestic manufacturing industry as well as attract foreign investors to locate foreign-owned facilities in China in exchange for excess rare earth materials, metals, alloys, oxides, and access to the emerging Chinese market. In this way, the Chinese will have complete control of foreign money as well as advanced design and development technology without using their own money. In addition, the Chinese do not have to worry about WTO rules and regulations.

1.8   Summary

This chapter presented the early history and discovery of rare earth materials in great detail. Seventeen rare earth elements and their properties were summarized, with emphasis on their atomic numbers. Early historical facts on rare earth ores and oxides were presented. Identification of rare earth oxides for various commercial and industrial applications were identified. Key processing requirements for REEs and their oxides were discussed, with emphasis on quality control and processing costs.

Mining efforts, processing requirements, and separation costs to obtain finished rare earth elements were addressed. The separation steps to recover rare earth oxides from rare earth ores are the most complicated and costly processes. Refining the ores into rare earth alloys with different purity levels is required to convert metals into rare earth alloys. Neodymium-iron-boron is a classic example of a rare earth alloy that is widely used in the manufacture of permanent magnets, which are best suited for applications where higher operating temperatures and high energy product are the principal design requirements.

REE mining operations and REE metal deposits were described in great detail, identifying their mining production outputs worldwide. Countries with rare earth ore reserves, active mining operations, and production capabilities (including processing facilities) were identified. U.S. facilities for the development of rare earth materials, such as Mountain Pass, were briefly mentioned, with emphasis on mining outputs of rare earth materials.

Potential commercial, industrial, and defense applications of rare earth elements and their alloys were described. For example, a samarium-cobalt magnet, which uses an REE alloy, is best suited for TWTAs, which are critical elements of electronic warfare equipment and radar systems. This rare earth permanent magnet offers unique magnetic properties under elevated temperatures as high as 300°C, with no compromise in magnetic performance and electronic beam focusing. Neodymium-iron-boron is another rare earth alloy that is used for magnets when higher operating temperatures and high energy product are the basic requirements. These two examples provide evidence of the importance of rare earth alloys in permanent magnets.

Steps are needed to build up the U.S. supply chain of rare earth materials for various applications. Corrective measures were discussed to avoid a shortage of REEs for defense product applications. Security risks for the United States due to nonavailability of critical rare earth elements for defense applications were summarized.

The properties of rare earth materials that are best suited for industrial and commercial applications were highlighted. Metallurgical, chemical, electro-optical, magnetic, and nuclear properties of rare earth materials were summarized. Effects due to the presence of impurities in REEs were identified with regard to device performance level.

The typical prices of rare earth elements and their critical applications were briefly summarized. The effects of supply and demand of rare earth materials on commercial, industrial, and defense applications were identified, with emphasis on element cost and its availability. Worldwide and U.S. demand for rare earth elements were projected. Effects due to environmental issues and World Trade Organization guidelines were discussed, with emphasis on the cost and timely availability of rare earth materials. The critical role of China in the industrial development of rare earth elements and their alloys for various applications were discussed.

References

1. Wikipedia. *Rare earth element.* http://en.wikipedia.org/wiki/Rare_earth_element#Discovery_and_early_history.
2. Greinacher, E. 1981. History of rare earth applications. *American Chemical Society Symposium Series* 164:3–17.
3. Humphries, M. 2011. *Rare earth elements: The global supply chain.* September 6.
4. U.S. Geological Survey. 2002. *Rare earth elements—Critical resources for high technology: Fact Sheet 087-02.* http://pubs.usgs.gov/fs/2002/fs087-02.
5. U.S. Government Accounting Office. 2010. *Rare earth materials in the defense supply chain* (GAO-10-617R). http://www.gao.gov/assets/100/96654.pdf.
6. U.S. Geological Survey. 2007. *Minerals yearbook: Volume I—Metals and minerals.* http://minerals.usgs.gov/minerals/pubs/commodity/myb.
7. Parson, C. 2012. *Los Angeles Times,* March 2012.
8. Hurst, C. 2010. *China's Rare Earth Industry: What Can the West Learn?* Washington, DC: Institute for the Analysis of Global Security, 8–10.

Chapter 2

Properties and Applications of Rare Earth Oxides, Alloys, and Compounds

2.1 Introduction

This chapter focuses on the properties and applications of rare earth oxides, alloys, and compounds for the most prominent materials. Potential commercial, industrial, and defense applications are identified. The important characteristics of the materials are given serious consideration.

It is important to mention that oxides of rare earth elements (REEs) are found in raw condition and require processing based on grain size, growth, and characteristics. Prioritization is based on different refining techniques and potential applications of the oxides. Rare earth materials or elements have limited applications and therefore require processing that involves several steps, which can be time consuming and costly. Preliminary studies performed by the author revealed that the prices per pound for the heavy rare earth elements (HREEs) are higher than those of their oxides. The studies further revealed that the prices of the light rare earth elements (LREEs) are much lower for lanthanum (La), cerium (Ce), praseodymium (Pr), and neodymium (Nd) because of low processing costs. The studies indicate that costs for REEs such as thulium (Tu) and europium (Eu) are relatively high due to their wide commercial and industrial applications. All rare earth elements, including the LREEs, are shown in Figure 2.1.

KEY TO CHART

Atomic number →	50 +2 +4
Symbol →	Sn
Atomic weight →	118.69
	−18−18−4

← Oxidation states

← Electron configuration

Transition elements

Group 8

Periodic table of the elements

1a	2a	3b	4b	5b	6b	7b	8			1b	2b	3a	4a	5a	6a	7a	0	Orbit
1 +1 −1 H 1.00797 1																	2 0 He 4.0026 2	K
3 +1 Li 6.939 2−1	4 +2 Be 9.0122 2−2											5 +3 B 10.811 2−3	6 +2 +4 −4 C 12.01115 2−4	7 +1 +2 +3 +4 +5 −1 N 14.9994 2−3	8 −2 O 15.9994 2−6	9 −1 F 18.9984 2−7	10 0 Ne 20.183 2−8	K−L
11 +1 Na 22.9898 2−8−1	12 +2 Mg 24.312 2−8−2											13 +3 Al 26.9815 2−8−3	14 +2 +4 Si 28.086 2−8−4	15 +3 +5 −3 P 30.9738 2−8−5	16 +4 +6 −2 S 32.064 2−8−6	17 +1 +5 +7 −1 Cl 35.453 2−8−7	18 0 Ar 39.948 2−8−8	K−L−M
19 +1 K 39.102 −8−8−1	20 +2 Ca 40.08 −8−8−2	21* +3 Sc 44.956 −8−9−2	22 +2 +3 +4 Ti 47.90 −8−10−2	23 +2 +3 +4 +5 V 50.942 −8−11−2	24 +2 +3 +6 Cr 51.996 −8−13−1	25 +2 +3 +4 +6 +7 Mn 54.9380 −8−13−2	26 +2 +3 Fe 55.847 −8−14−2	27 +2 +3 Co 58.9332 −8−15−2	28 +2 +3 Ni 58.71 −8−16−2	29 +1 +2 Cu 63.546 −8−18−1	30 +2 Zn 65.37 −8−18−2	31 +3 Ga 69.72 −8−18−3	32 +2 +4 Ge 72.59 −8−18−4	33 +3 +5 −3 As 74.9216 −8−18−5	34 +4 +6 −2 Se 78.96 −8−18−6	35 +5 −1 Br 79.904 −8−18−7	36 0 Kr 83.80 −8−18−8	−L−M−N
37 +1 Rb 85.47 −18−8−1	38 +2 Sr 87.62 −18−8−2	39* +3 Y 88.905 −18−9−2	40 +4 Zr 91.22 −18−10−2	41 +3 +5 Nb 92.906 −18−12−1	42 +6 Mo 95.94 −18−13−1	43 +4 +6 +7 Tc (97) −18−13−2	44 +3 Ru 101.07 −18−15−1	45 +3 Rh 102.905 −18−16−1	46 +2 +4 Pd 106.4 −18−18−0	47 +1 Ag 107.868 −18−18−1	48 +2 Cd 112.40 −18−18−2	49 +3 In 114.82 −18−18−3	50 +2 +4 Sn 118.69 −18−18−4	51 +3 +5 −3 Sb 121.75 −18−18−5	52 +4 +6 −2 Te 127.60 −18−18−6	53 +1 +5 +7 I 126.9044 −18−18−7	54 0 Xe 131.30 −18−18−8	−M−N−O
55 +1 Cs 132.905 −18−8−1	56 +2 Ba 137.34 −18−8−2	57* +3 La 138.91 −18−9−2	72 +4 Hf 178.49 −32−10−2	73 +5 Ta 180.948 −32−11−2	74 +6 W 183.85 −32−12−2	75 +4 +6 +7 Re 186.2 −32−13−2	76 +3 +4 Os 190.2 −32−14−2	77 +3 +4 Ir 192.2 −32−15−2	78 +2 +4 Pt 195.09 −32−16−2	79 +1 +3 Au 196.967 −32−18−1	80 +1 +2 Hg 200.59 −32−18−2	81 +1 +3 Tl 204.37 −32−18−3	82 +2 +4 Pb 207.19 −32−18−4	83 +3 +5 Bi 208.980 −32−18−5	84 +2 +4 Po (209) −32−18−6	85 At (210) −32−18−7	86 0 Rn (222) −32−18−8	−N−O−P
87 +1 Fr (223) −18−8−1	88 +2 Ra (226) −18−8−2	89 +3 Ac (227) −18−9−2																−O−P−Q

Lanthanides	58* +3 +4 Ce 140.12 −20−8−2	59* +3 Pr 140.907 −21−8−2	60 +3 Nd 144.24 −22−8−2	61* +3 Pm (145) −23−8−2	62* +2 +3 Sm 150.35 −24−8−2	63* +2 +3 Eu 151.96 −25−8−2	64* +3 Gd 157.25 −25−9−2	65* +3 Tb 158.924 −27−8−2	66* +3 Dy 162.50 −28−8−2	67* +3 Ho 164.930 −29−8−2	68* +3 Er 167.26 −30−8−2	69* +3 Tm 168.934 −31−8−2	70* +2 +3 Yb 173.04 −32−8−2	71* +3 Lu 174.97 −32−9−2	−N−O−P
Actinides	90 +4 Th (232) −18−10−2	91 +5 +4 Pa (231) −20−9−2	92 +3 +4 +5 +6 U (238) −21−9−2	93 +3 +4 +5 +6 Np (237) −22−9−2	94 +3 +4 +5 +6 Pu (244) −24−8−2	95 +3 +4 +5 +6 Am (243) −25−8−2	96 +3 Cm (247) −25−9−2	97 +3 +4 Bk (247) −27−8−2	98 +3 Cf (251) −28−8−2	99 Es (254) −29−8−2	100 Fm (257) −30−8−2	101 Md (256) −31−8−2	102 No (254) −32−8−2	103 Lw −32−9−2	−O−P−Q

104

Numbers in parentheses are mass numbers of most stable isotope of that element.

Figure 2.1   Periodic table of the elements. Light rare earth elements are indicated by asterisks.

The availability of rare earth elements or materials is of critical importance. Because the Chinese government is attempting to restructure the rare earth mining industry and establish export quotas, countries such as Japan, South Korea, and the United States are building strategic stockpiles of rare earth materials. A *Technology Metals Research* report from June 2011 indicated that the level of stockpiling could have a dramatic impact on availability, particularly for the HREEs [1].

Chapter 1 discussed the mining sources from which rare earth materials are available or can be procured. This chapter describes a unique and complex way to obtain the isotopes of various rare earth elements. When a nuclear power reactor is shut down for repair or maintenance, the reactor requires several days to cool down the reactor core. The cooling period is dependent on the amount of nuclear fuel used, effectiveness of thermal shielding, thermal neutron flux density, and radiation shield efficiency. Preliminary research studies performed by various authors on the time needed to cool down a nuclear power reactor core with a power generation capability of 1,000 mW is at least 100 days. After the cool-down period, isotopes of various rare earth elements are seen in the reactor core [1]. Isotopes of these rare earth elements, their percentage weight, and their distribution in the reactor can be summarized as follows [2]:

- Group 2: Strontium (3.8%) and barium (4%)
- Group 3: Yttrium (2.3%), lanthanum (4%), and cerium (10.1%)
- Group 4: Zirconium (9.9%)
- Group 5: Niobium (4.7%)
- Group 6: Tellurium (1.2%)
- Group 7: Iodine (0.6%)
- Group 8: Ruthenium (5%)

Under exposure and cooling conditions, these elements constitute approximately 45% (by weight) of the fission products. Other rare earth elements may be present in significant amounts, including neodymium (10%), xenon (10%), caesium (10%), molybdenum (8%), praseodymium (4.5%), samarium (3%), and technetium (2.7%). However, they either make a relatively small contribution to the radioactivity or are gaseous; therefore, they are not considered here.

Some of the rare earth isotopes are nontoxic but radioactive; therefore, the half-lives of these rare earth materials are important. The half-lives of the most common rare earth–based isotopes are summarized in Table 2.1.

2.1.1 Processing Requirements for Rare Earth Materials

The cost and complexity of processing rare earth materials are strictly dependent on the type of rare earth material. For example, HREE materials require a large number of complex processing procedures before they can be used in any meaningful

**Table 2.1 Half-Lives of the Most
Common Rare Earth Isotopes**

Isotope	Half-Life
Ce^{144}	290 days
Pr^{144}	17.5 minutes
Eu^{156}	15.4 days
La^{140}	40 hours
To^{232}	1.39×10^{10} years
Pu^{239}	2.41×10^{4} years
U^{233}	1.62×10^{5} years

Source: From Glasstone, S. 1955. *Principles
of Nuclear Reactor Engineering.*
New York: Van Nostrand.

application. In general, the steps involved in mining and processing can be summarized as follows:

1. Mining of rare earth ores is possible from the predicted mineral deposits of a specific location.
2. Extraction of rare earth materials and their oxides from the deposits requires a comprehensive mining survey of the proposed sites [2].
3. Separation of rare earth ore into individual rare earth oxides is necessary.
4. Refinement of rare earth oxides into rare earth materials with different purity levels is not only complex but also costly. Higher purity levels will further increase the cost.
5. Converting rare earth oxides into alloys requires specific procedures for different applications.
6. Conversion of alloys into rare earth compounds that can be used for various commercial, industrial, and defense applications requires specific procedures. Deployment of rare earth alloys such as neodymium-iron-boron has potential commercial and industrial applications. These magnets are widely used in the design and development of electric motors and generators deployed in electric and hybrid electric vehicles because of their light weight and compact size.
7. Quality control procedures that involve visual inspections, verification of parametric values, and reliability data for the rare earth material are necessary for some defense applications.

In this chapter, many of the popular rare earth oxides with significant commercial, industrial, space, and defense applications are identified, including the following [4]:

- Ytterbium oxide (Yb_2O_3)
- Cerium oxide (Ce_2O_3)
- Dysprosium oxide (Dy_2O_3)
- Gadolinium oxide (Gd_2O_3)
- Lanthanum oxide (La_2O_3)
- Europium oxide (Eu_2O_3)
- Neodymium oxide (Nd_2O_3)
- Praseodymium oxide (Pr_2O_3)
- Samarium oxide (Sa_2O_3)

Among these oxides, ytterbium oxide, lanthanum oxide, and neodymium oxide have critical commercial applications. Defense experts believe that dysprosium (Dy) is the most critical rare earth material in terms of import reliance [6]. Therefore, a continuous and unrestricted supply of this particular rare earth material is essential as far as defense applications are concerned. A country should not depend on the export quotas from hostile countries or suppliers. One classic example involves the Chinese export ban on shipments of rare earth materials and oxides to Japan on September 22, 2010 [1].

2.1.2 Mining and Surveying Requirements

The following mining and surveying efforts are essential to locate rare earth elements and their oxides:

1. Reliable mining and surveying efforts are necessary to secure mineral rights agreements with the host government. The approval of contracts involves long and complex negotiations and international obligations. Furthermore, geopolitical considerations must be carefully examined and evaluated when securing rare earth material mining rights.
2. Close coordination between rare earth scientists and mining experts is essential to locate the rare earth oxide deposits.
3. Mining officials must make sure that the processing of minerals and their oxides will be done at the mining location to keep the overall mining expenses and oxide processing costs to a minimum.
4. If processing of rare earth materials at the mining site is not possible, the cost of transportation from the mining facility to the processing location must be carefully evaluated.

2.2 Production and Availability of Rare Earth Oxides

Leading mining experts and material scientists believe that significant quantities of rare earth oxides are generally found in tailings accumulated from 50 years of accumulating uranium ore and operating mines in the vicinity of Estonia. The author believes that, due to the rising prices of rare earth oxides, extraction of oxides has become economically viable. Therefore, extraction of rare earth oxides offers a cost-effective technique to obtain rare earth oxides at the lowest prices. Estonia currently exports about 30,000 tons of rare earth oxides per year, which amounts to approximately 2% of the worldwide production capacity.

Preliminary studies performed by the author indicate that nuclear rod reprocessing offers another source of rare earth oxides and other rare earth elements. The nuclear fission of uranium or plutonium produces a full range of rare earth oxides and elements, including isotopes that are best suited for medical treatments and diagnosis. However, due to the radioactivity characteristics of many of these isotopes, it is unlikely that the extraction of isotopes from the mixture can be done safely and economically. Recycling, environmental considerations, rare earth material prices, and geopolitical considerations play important roles in obtaining rare earth oxides and their associated elements, as described in the following sections.

2.2.1 Recycling Issues

Recycling is a critical issue that is deeply involved in the extraction of rare earth oxides and associated elements from mines. The most efficient recycling technology is the one that separates rare earth oxides from electronic waste and other byproducts that have potential applications in commercial and industrial products.

Recent advancements in recycling technology have made the extraction of rare earth oxides and elements from rare earth ores more feasible in terms of cost and speed. Several recycling facilities are currently operating to full capacity in Japan; they have supplied approximately 300,000 tons of rare earth oxides stored in unused electronic components. In France, the Rhodia Group has installed two recycling plants in La Rochelle and Saint-Fons that are currently producing more than 200 tons per year of rare earth oxides that are best suited for fluorescent light sources.

2.2.2 Environmental Considerations

The mining, refining, and recycling of rare earth materials can have serious environmental consequences if the facility is not properly managed and not operating reliably. A particular hazard is mildly radioactive slurry resulting from the common presence of thorium and uranium in the rare earth element ores. In addition, toxic acids are generally used during the refining process. Improper handling of these substances could result in extensive environmental damage. According to a May 2010 public report, China announced a major five-month crackdown on illegal mining operations to protect the environment and its resources. The campaign was particularly focused in southern China, where illegal mining operations are carried out in rural regions due to a lack of policing by environmental officials. These illegal mining operations tend to release the toxic wastes into the general water supply, thereby creating environmental and health-related problems. Even the major mining operations in Baotou in Inner Mongolia, where much of the world's rare earth materials are refined, have caused not only major environmental damage but also serious health problems for the local population.

The United States provided $100 million in 2011 for the cleanup of the Bukit Merah mine in Malaysia, which presented a health hazard to the local population. Residents of that town blamed the rare earth refinery for birth defects and several cases of leukemia within 5 years in a community of 11,000 people. Seven of the individuals with leukemia died. The local residents claimed that there was no leukemia in the town before the rare earth mining facility was in operation. The facility management had to remove about 11,000 truckloads of radioactively contaminated materials in the summer of 2011. In addition, the management was asked to move more than 80,000 steel barrels of radioactive waste material to a hilltop repository. A Mitsubishi contractor was assigned for the cleanup job. A director of Asian Rare Earth said that the company might have sold a few bags of calcium phosphate on a trial basis as it sought to market byproducts. A former resident of Bukit Merah said that cows who ate the grass grown with the fertilizer all died, which indicates the presence of radioactivity.

Another environmental issue was highlighted after the Fukushima Daiichi nuclear disaster in Japan. Widespread protests took place in Kuantan over the Lynas refinery incident and the radioactive waste resulting from it. Originally, the ore was to be processed with very low levels of thorium. Lynas founder and chief executive Nicholas Curtis said there was absolutely no environmental or public health risk. However, Dr. T. Jayabalan, who had been monitoring and treating patients affected by the Mitsubishi plant, had great doubt about this and was wary of the assurances given by Lynas. According to the doctor and other medical experts, the argument that low levels of thorium are present in the ore does not make the environment safe for the public because radiation exposure from the radioactive source is cumulative. Because of environmental concerns, the construction of the facility was halted until an independent United Nations International Association for Environmental Act (IAEA) panel investigation was completed. The IAEA report did not find any instance of any noncompliance with international, radiation safety standards in this project. Because the IAEA investigation was satisfactorily completed, no further construction was halted. However, after this incident, new restrictions were announced by the Malaysian government in late June 2011.

2.2.3 Geopolitical Considerations

Geopolitical considerations must be taken into account before selecting a particular mining operating site. The mine operator will have rough time dealing with a totalitarian regime. Such a regime can later impose operational restrictions or conditions that may not be economically feasible for mine operators.

China has cited resource depletion and environmental concerns as the reasons for a nationwide crackdown on its rare earth mineral production sector. Cutting down their exports of rare earth metals and oxides will push Chinese manufacturers up the supply chain, so they can sell the valuable finished goods to the world (e.g., samarium-cobalt or neodymium-iron-boron permanent magnets) at much higher prices rather than cheap raw materials. In one classic example, the General Motors division that was dealing with miniaturized magnet research was forced to shut down its U.S. office and move its entire staff to China in 2006. China's export quotas only apply to rare earth elements or metals, not to the products made from these metals, such as permanent magnets. In this way, the Chinese profit from both the raw materials and the finished products (in this case, the permanent magnet).

On September 2, 2010, *The Economist* reported on a July 2010 announcement by China on the latest in a series of export reductions, amounting to 40% or approximately 30,000 tons. Soon after, on September 22, 2010, China instituted an export ban on the shipments of rare earth oxides to Japan in response to the detainment of a Chinese fishing boat captain by Japanese coast guard officials [1]. In the same year, the U.S. Department of Energy *Critical Materials Strategy* report identified dysprosium as the most critical rare earth element in terms of import reliance.

In 2011, a report by the U.S. Geological Survey (USGS) and the U.S. Department of the Interior revealed the Chinese rare earth industry, outlined industry trends with China, and examined national policies that may guide the future of rare earth material production [6]. The report further revealed that the Chinese production of rare earth minerals has accelerated over the past two decades. In 1990, China accounted for only 27% of such minerals. According to rare earth minerals experts, the percentage accuracy of rare earth minerals as quoted by the Chinese is questionable. The worldwide production of minerals was 132,000 metric tons, with China producing 129,000 of these tons.

According to a recent Chinese government report, China will reduce the export of rare earth minerals to the world. This policy must not be underestimated because it poses a serious problem regarding a reliable supply of rare earth minerals for the rest of the world. Because of international politics and an increase in domestic Chinese demand, a serious reduction in the export of rare earth minerals can be expected.

In 2006, China allowed 47 domestic rare earth producers, including 12 foreign rare earth producers and traders, to export such minerals. Rare earth scientists believe that export controls have tightened because only 22 domestic rare earth producers and 9 Sino-foreign rare earth producers and traders were authorized to handle the export of rare earth materials by 2011. It appears that the Chinese government will keep strict controls on the export of rare earth materials in place. According to a Chinese rare earth development plan, annual rare earth production may be limited to between 130,000 and 140,000 metric tons during the period from 2009 to 2015. The export quota for rare earth products may be roughly 35,000 metric tons. The Chinese government may allow about 20 domestic rare earth producers and traders to handle the export of rare earth materials and finished products.

Because of unpredictable and unreliable Chinese export polices, the United States must look for alternate sources of rare earth materials to protect the national security of the country. China's principal objective is to protect its rare earth industry, with emphasis on the national policies and security of the nation. This policy will retain its control and dominance over the strategic materials of the country. China will not hesitate to cut off rare earth materials to any country that goes against its national policy. Because of China's unpredictable behavior and hostile attitude, the USGS is actively surveying the southern Afghanistan region for possible exploration of rare earth deposits under the protection of U.S. military forces. Since 2009, the USGS has conducted remote sensing surveys as well as field work to verify Russian claims that volcanic rocks containing rare earth metals exist in the Helmand province near the village of Khanashin. The USGS study team has located a sizeable area of rocks in the center of an extinct volcano containing LREEs, including cerium and neodymium, which are well suited for fiber-optic amplifiers and permanent magnets, respectively. The U.S. team has mapped about 1.3 million metric tons of desirable rock, which is sufficient for a 10-year supply of these two rare earth materials at the current demand levels. Defense officials have estimated its monetary value at about $7.4 billion.

2.2.4 Rare Earth Pricing Considerations

Rare earth elements are not exchange-traded in the same way as precious elements (gold or silver) or nonferrous metals (nickel, copper, tin, and aluminum). Rare earth elements are generally sold on the private market, which makes their procurement prices very difficult to monitor or track. However, their prices are published periodically on some websites, such as http://www.mineralprices.com. The 17 rare earth elements described in Chapter 1 are not usually sold in their natural form; instead, they are distributed in mixtures of various purity levels, such as neodymium metal that is greater than or equal to a 99.5% purity level. Therefore, the pricing of a rare earth element can vary based on the purity level and the quantity required by the end user's application. Table 2.2 summarizes the prices of rare earth elements. The prices in Table 2.2 were quoted more than four decades ago. More current prices were not available from suppliers.

2.3 Description, Properties, and Applications of Rare Earth Elements, Oxides, Alloys, and Compounds

This section summarizes the important properties and potential applications of rare earth metals, oxides, alloys, and compounds that are widely used in commercial, industrial, and defense sectors. To limit the material in this section, the author has

Table 2.2 Estimated Prices of Rare Earth Metals or Elements

Rare Earth Element	Symbol	Cost ($/lb.)
Cerium	Ce	75
Dysprosium	Dy	19
Erbium	Er	456
Europium	Eu	4,000
Gadolinium	Gd	250
Holmium	Ho	500
Lanthanum	La	70
Lutetium	Lu	8,000
Neodymium	Nd	120
Praseodymium	Pr	225
Samarian	Sm	250
Scandium	Sc	7,500
Scandium	Sc	750
Yttrium	Y	200
Ytterbium	Yb	300

Source: Weast, R.C. 1971. *Handbook of Chemistry and Physics,*
51st ed. Boca Raton, FL: CRC Press.

selected the best-suited and most widely deployed rare earth materials. Preliminary studies undertaken by the author indicate that important properties of lutetium, dysprosium, samarium, cerium, terbium, ytterbium, neodymium, and a few other elements and their oxides, alloys, and compounds deserve serious consideration because of their multiple applications.

2.3.1 Rare Earth Oxides

The rare earth oxides that are most widely used in commercial, industrial, and defense applications can be summarized as follows:

- Thulium oxide (Tm_2O_3)
- Cerium oxide (Ce_2O_3)
- Ytterbium oxide (Yb_2O_3)

- Dysprosium oxide (Dy_2O_3)
- Europium oxide (Eu_2O_3)
- Lanthanum oxide (La_2O_3)
- Lutetium oxide (Lu_2O_3)
- Neodymium oxide (Nd_2O_3)
- Praseodymium oxide (Pr_2O_3)
- Samarium oxide (Sm_2O_3)

2.3.2 Properties of Rare Earth Elements, Oxides, and Compounds

2.3.2.1 Lutetium

This particular rare earth element was first discovered by George Urbain in 1907. Lutetium is the last member of the rare earth series. It has two isolates with atomic mass numbers of 174.94 and 175.94. This material is available in the form of pure metal, oxide, or a compound such as lutetium tantalate ($LuTaO_4$), which is most ideal for x-ray phosphors. This compound is used as a dopant in matching the lattice parameters of certain garnet crystals, such as indium-gallium-garnet crystals. Elemental or metallic forms include pellets, rods, wires, and granules for evaporation sources. Its oxide is used to provide nanoparticles and nanopowders, which yield ultrahigh surface areas that are best suited for nanotechnology research and development activities.

Lutetium oxides are available as powders and dense pellets that are most ideal for optical coatings and thin-film applications. Its oxides can be found in insoluble and soluble forms, including chlorides, nitrates, and acetates. This element is not toxic. The metal is available in purities ranging from 99.00 to 99.99%. It is available as foil, a sputtering target, or a rod. Unlike most rare earth materials, lutetium lacks a magnetic moment and has the smallest metallic radius. It is abundant in earth. Its electrical conductivity at room temperature is 79 $\mu\Omega$-cm and its thermal conductive is 16.4 W/(m·K).

Medical research and development are investigating lutetium for use in brain tumor treatment, longitudinal imaging techniques, laparoscopic management and treatment of hepatic tumors, molecular imaging, imaging and treatment for prostate tumor cells, and treatment for long-term stabilization of malignant gas trimmers.

Laser scientists have able to design and develop an efficient diode-pumped laser operation for a $TmLu_2O_3$ laser operating at 27 μm. This particular laser is best suited for receiver applications where detection of long-wave signals are of critical importance. Its chemical characteristics are as follows:

- Atomic number: 71
- Molecular weight: 174.97 g/mol

- Density: 9.7 g/cm³ at room temperature
- Energy of first ionization: 522.7 kJ/mol

2.3.2.2 Dysprosium

Dysprosium was discovered by Paul-Émile Lecoq de Boisbaudran in 1886. Its atomic mass is 155.929. Its seven isotopes are found in abundance on Earth; their atomic masses are summarized in Table 2.3. This material is widely used in mobile phones, smartphones, and computer tablets.

Dysprosium is commonly used in the production of neodymium-iron-boron permanent magnets to enhance the mechanical strength of the magnets under harsh mechanical and thermal environments. It is widely used in special ceramic compounds in which high performance over a wide temperature range is the principal requirement. This element can also be used in dysprosium-iron-garnet optical crystals, which are best suited for electro-optics systems.

This element is added to various advanced optical formulations because of its ability to emit in the 470 to 500 nm and 570 to 600 nm infrared spectral ranges. The high electron beam characterization of the dysprosium-based laser is most suited for clinical therapy, tumor targeting, and other clinical applications. Furthermore, due to its high susceptibility, it is widely deployed in dysprosium-based compounds, which are best suited for data storage applications, such as high-density compact discs.

Dysprosium comes in various shapes and forms. Its oxides are available as powders and dense pellets. This rare earth element is best suited for high-performance optical coatings and thin-film applications. Dysprosium is available in soluble forms such as chlorides, nitrates, and acetates. This rare earth metal is moderately toxic; therefore, care should be taken during road, sea, and air transportation. Its electrical conductivity at room temperature is 57 $\mu\Omega$-cm and its thermal conductivity is roughly 10.7 W/(m·K).

Table 2.3 Atomic Mass and Abundance of Dysprosium (Dy) Isotopes

Isotope	Atomic Mass	Abundance on Earth (%)
Dy-156	155.82	0.16
Dy-158	157.82	0.10
Dy-160	158.83	2.34
Dy-161	160.93	18.90
Dy-162	161.93	25.50
Dy-163	162.93	24.90
Dy-164	163.93	28.12

Table 2.4 Atomic Mass and Abundance of Neodymium (Nd) Isotopes

Isotope	Atomic Mass	Abundance on Earth (%)
Nd-142	141.91	27.13
Nd-143	142.91	12.19
Nd-144	143.91	23.90
Nd-146	145.91	17.19
Nd-148	147.92	5.76
Nd-150	149.92	5.64

2.3.2.3 Neodymium

Neodymium was first discovered by Carl Auer von Welsbach in 1885. This rare earth element is the most abundant among rare earth materials after cerium and lanthanum. It is available as metal or oxide, with purity ranging from 99 to 99.99%, or as a compound. The metal is available in the form of a foil, rod, or sputtering target. Its compounds are available as submicrons and nanopowders. Primary applications of this metal include infrared lasers, glass coloring and tinting, neodymium-iron-boron permanent magnets (best suited for electric motors and generators), protective goggles for welding, cathode ray tube displays (to enhance contrast between reds and greens), dielectric coatings, and multilayer capacitors (widely used in radiofrequency components). Its metallic forms include wires, rods, pellets, and granules. Its oxides are available in powder and dense pellet forms. The fluoride oxides of this material are most ideal for metallurgy, chemical and physical vapor deposits, and optical coatings. Each of neodymium's naturally occurring isotopes are summarized in Table 2.4.

The electrical conductivity of the neodymium element is 64 $\mu\Omega$-cm at room temperature, whereas the thermal conductivity is 16.5 W/(m·K). The latest research and development activities undertaken on neodymium indicate its usefulness as a spectroscopic analytical tool, a neodymium-yttrium-aluminum-garnet (Nd-YAG) laser for glaucoma diagnosis, environmental remediation, an upgraded multipulse laser, and a long-pulse Nd-YAG laser for curing cysts.

2.3.2.4 Samarium

Samarium was first discovered by Paul-Émile Lecoq de Boisbaudran in 1879. Samarium can be found in oxide form, elemental or metal form, or compound form. The metallic version is available in various forms, such as rods, films, and sputtering targets. Its oxides come in nanopowder and dense pellet forms. These oxides are best suited for optical coatings and thin-film applications. Its insoluble fluorides are widely deployed in metallurgy, chemical and vapor deposition, and specialized optical coatings. Its soluble oxides include chlorides, nitrates, and acetates.

Samarium compounds are available in submicron and nanopowder forms. Samarium-titanate has demonstrated remarkable dielectric properties; it is most ideal for precision optical coatings and high-performance capacitors at microwave frequencies. Its metallic forms include rods, wires, pellets, and granules. Its nanopowders and nanoparticles have high surface areas with unique characteristics. The room-temperature electrical conductivity of this rare earth element is about 88 $\mu\Omega$-cm and its thermal conductivity is 13.3 W/(m·K).

Oxides and compounds of this rare earth material have several commercial, industrial, and defense applications. Samarium is the principal rare earth material used in the production of samarium-cobalt (Sm_2Co_7) permanent magnets. These magnets are widely deployed in the design and development of electrical motors and generators for electric and hybrid electric vehicles. This particular rare earth permanent magnet retains all the magnetic properties, even at temperatures as high as 300°C. No other permanent magnet has such unique magnetic performance under harsh mechanical and thermal environments. Samarium-cobalt permanent magnets are widely used in the design and development of traveling wave-tube amplifiers (TWTAs), which are the backbones of radar and electronic warfare systems. This particular rare earth permanent magnet offers a significant reduction in the weight and size of the magnetic structure while maintaining a uniform radiofrequency (RF) beam over the entire length of the TWTA at elevated temperatures.

Samarium-cobalt permanent magnets are best suited for applications in which weight, size, reliability, and high operating temperatures are the limiting factors. Its potential applications include electric motors and generators, aerospace components, servo motors, and high-power microwave sources. Samarium-cobalt permanent magnets have replaced the more expensive neodymium-iron-boron magnets. Because of their weak spectral absorption band, samarium-cobalt magnets are used in filter glass on Nd-YAG solid-state lasers to surround the laser rod in order to achieve improved laser efficiency by absorbing stray emissions.

This rare earth material is widely used in research activities dealing with selective separation of samples based on synergistic extraction techniques, interactions between metal ions and carbohydrates, syntheses and spectroscopic studies, and chemical synthesis evaluation and characterization of some aluminum-based rare earth metallic compounds. Its medical research applications include use in molecular imaging, as a contrast agent, and in radioisotope medical research studies, with an emphasis on diagnosis and clinical treatments for various diseases. Both neodymium and samarium are considered light rare earth elements.

2.3.2.5 Cerium

Cerium was first discovered by Wilhelm von Hisinger in 1803. Cerium is the most abundant of the rare earth materials. This material is chemically characterized by two valence states: the +3 ceros state and the +4 ceric state. Cerium is available as a metal, oxide, or compound. This rare earth material is strongly acidic, moderately toxic, and a strong oxidizer. Its most common applications are in metallurgy, glass, polishing, ceramics, catalysts, and phosphors. Its most common industrial application is in the steel manufacturing process. It is also considered the most efficient glass polishing agent.

Cerium is widely used in manufacturing medical glassware and aerospace windows due to its excellent mechanical strength properties. It is particularly suited for manufacturing high-quality ceramics, including dental compositions and phase stabilizers in zirconium-based products. This rare earth material is also well suited for high-performance optical components. Cerium metal has been used in alloys to design and develop permanent magnets, but these magnets were not found to be cost-effective. Furthermore, the magnets were suitable for harsh mechanical and thermal environments. The alloys of this rare earth element are best suited for auto-motive power-train components.

Cerium is available in metallic forms, such as rods, pellets, wires, and granules. The nanoparticles and nanopowders provide ultrahigh quality surface areas, which can be used to develop components with unique performance capabilities. As men-tioned earlier, its oxides can be used in optical coatings and thin-film applications. Its insoluble oxides are fluorides and its soluble oxides are chlorides, nitrates, and acetates. Cerium's compounds are best suited for components that have potential applications in the commercial, industrial, and defense fields. Its stable and non-radioactive naturally occurring isotopes have various medical and clinical applica-tions. The details of its isotopes are summarized in Table 2.5.

Table 2.5 Atomic Mass and Abundance of Cerium (Ce) Isotopes

Isotope	Atomic Mass	Abundance on Earth (%)
Ce-136	135.907	0.19
Ce-137	136.908	Negligible
Ce-138	137.906	0.25
Ce-139	138.907	Negligible
Ce-140	139.905	88.48
Ce-141	140.908	Negligible
Ce-142	141.909	11.08

Cerium's electrical conductivity at room temperature is 75 $\mu\Omega$-cm, whereas its thermal conductivity is 11.4 W/(m·K). Recent research and development investigations by various scientists have demonstrated excellent catalytic properties, evidence of an induction step during cobalt oxidation, presence of fluorescent carbon dots, and formation of cerium oxide nanocrystals.

2.3.2.6 Ytterbium

Ytterbium was first discovered by Jean de Marignac in 1878. Its atomic mass number is 168, and it has only one isotope (designated as Yb-168) with atomic mass of 167.934. The electrical conductivity of this rare earth material at room temperature is 29 $\mu\Omega$-cm, whereas its thermal conductivity is 34.9 W/(m·K). The room-temperature thermal conductivity can be expressed using the CGS or MKS system as shown in Table 2.6. This isotope has unique properties for several commercial, industrial, and medical applications.

Ytterbium oxides are available in several forms, such as nanopowders and dense pellets. Its oxides are best suited for optical coatings and thin-film applications. Its insoluble oxides are chlorides that are widely used in metallurgy and chemical, and physical vapor deposition, which is most ideal for high-performance optical coatings. Ytterbium oxides are also available in soluble forms, including chlorides, nitrates, and acetates. These are also used in producing chemical compounds for industrial applications. Its nanoparticles and nanopowders provide high-quality surface areas that are widely used in nanotechnology research and development components, with unique properties and benefits. These components have many commercial, industrial, and scientific applications. Its metallic forms come in wires, pellets, and granules for evaporation source material processes, which are best suited for commercial and industrial applications.

As shown in the periodic table (Figure 2.1), ytterbium has two valence states—+2 and +3. All elemental metals, compounds, and solutions may be synthesized at ultrahigh purity levels (99.999%) for scientific laboratory standards,

Table 2.6 Thermal Conductivity of Various Metals in CGS and MKS Systems

Metal	Thermal Conductivity	
	W/(cm·°C)	W/(m·K)
Aluminum	2.18	2.37
Copper	2.94	3.98
Cobalt	0.69	0.73

Note: Values shown are strictly dependent on the purity and the surface conditions of the metals.

advanced electronics components, thin-film deposition techniques using sputtering targets and evaporation materials, metallurgy, optical materials, and other high-technology applications. This rare earth material has a stable and nonradioactive isotope that is best suited for medical research in conjunction with laser sources. Metallic ytterbium compounds are soluble in organic or aqueous solvents. The isotope can be used to acquire vital physical and chemical analytical techniques and properties, such as x-ray diffraction, surface area analysis, and other critical parameters of certain industrial materials.

Ytterbium is considered fairly toxic and therefore must be carefully handled during sea, air, and road transportation. Ytterbium safety data and biological aspects are strictly dependent on the ytterbium metal, nanoparticles, and its compounds. Hazard formation, toxicity, and mode must be given serious consideration during laboratory testing and transportation, regardless of mode of transportation.

Ytterium compounds have no biological role. The first and second ionization energy levels are 603.44 KJ/mol and 1,174.82 KJ/mol, respectively. Ionization energy is defined as the least required energy to release a single electron from the atom in its ground state during the gas phase.

Ytterium-based lasers offers excellent beam quality and low noise levels. Noise performance of a Yb-doped single frequency is outstanding. Fiber-optic amplifiers yield excellent overall performance over wide spectral regions when operated in conjunction with a Yb-based fiber laser emitting at 976 nm. The fabrication and characterization of Yb-doped-zirconium-germano-alumino silicate phase-separated nanoparticle-based fibers yield outstanding optical performance. Comprehensive research and development activities have been focused on fingermark detection capability, synthesis, photo acoustic microscopes, and energy transfer.

Current research and development activities deal with fingerprint detection using YVO(4):Er-Yb luminescent upconverting particles; synthesis, structure, and reactivity of supramolecular ytterbium-aqua; energy transfer and enhanced 1.54-μm emission in erbium-ytterbium thin films; photo darkening of rare earth–doped silica; mode-locked 0.5 μJ fiber lasers emitting at 976 nm; electrochemical spectroscopic investigations of the interactions of a ytterbium complex with DNA; generation of 578-nm yellow light over 10 mW using second harmonic generation of an 1,156-nm external-cavity diode laser; thermal effects in a kilowatt all-fiber optical source; and white emission of lithium ytterbium nanocrystals.

2.3.2.7 Thorium

Thorium was first discovered by Jöns Berzelius in 1818. This element and its isotopes occur in abundance on Earth. All elemental metals, compounds, and solutions may be synthesized in ultrahigh purity for laboratory standards, advanced electronic components, metallurgy, thin-film deposition using sputtering targets, and optical materials best suited for high-technology applications. Thorium is a block F, group 3, period 7 element. The number of electrons in each of thorium's shells is 2.

Thorium is available as a metal, compound, or oxide. Thorium is a lanthanide rare earth materials and is primarily deployed in nuclear power plant applications. Thorium metal comes in the form of foil, rod, or sputtering target; its compounds comes in the form of submicrons and nanopowder. This rare earth material is widely used as a tungsten coating in electronic components because of its high emission characteristics. Its fluoride and oxide are used in advanced electro-optical device applications because of their high refractive indexes. Thorium can be used in other high-temperature glass applications, such as in the mantles of lamps and to produce crystal growth crucibles.

The name thorium originates from the Scandinavian god Thor, the Norse god of war and thunder. No wonder it is widely used in nuclear reactors [3]. The molecular formula, safety data, research information and properties are available for many specific states, forms, and shapes. Its metallic forms include rods, pellets, wires, and granules as evaporation source material. Its nanoparticles and nanopowders yield ultrahigh surface areas with unique properties best suited for research and development activities.

The oxides of thorium are available in forms of powders and dense pellets, which are widely used for optical coatings and thin-film applications for which performance, reliability, and longevity are the principal design requirements. Its insoluble oxides are fluorides, which have potential applications where oxygen is undesirable, such as metallurgy, chemical and physical vapor deposition, and some optical coatings. Soluble oxides of thorium include chloride, nitrates, and acetates.

Thorium is highly radioactive and can collect in bones, which can cause bone cancer even several years after exposure. Furthermore, breathing this material in substantial amounts could be lethal. All elemental metals, compounds, and solutions can be synthesized in ultrahigh purity for laboratory evaluations, advanced electronic components, thin-film deposition using sputtering targets and evaporation materials, metallurgy and optical materials, and other high-technology applications. Thorium isotopes are stable (nonradioactive) and are used for medical research studies. Organo-metallic thorium compounds are soluble in organic solvents. For chemical and physical analysis techniques, such as x-ray diffraction and surface area analysis, contact American Elements sources.

Thorium and its two isotopes (Th-229) and (Th-230), with atomic masses of 229.0317 and 232.038, respectively, are naturally available in abundance on Earth. Thorium's room temperature electrical conductivity is 13 $\mu\Omega$-cm, whereas the thermal conductivity is 54 W/m·K. Because of its unique properties, some nuclear physicists suggest the use of liquid fluoride in thorium-based reactors for nuclear power generation.

The safety data for thorium metal, nanoparticles, and its compounds are strictly contingent on the form. The first ionization energy level for thorium (the least required energy to release a single electron from the atom in its ground state in the gas phase) is approximately 608.51 kJ/mol. This material is highly radioactive, toxic, and hazardous; therefore, it must be carefully handled during land, sea, and air modes of transportation.

This material has a molecular weight of 232.04 g/mol. Because this particular rare earth material is often used under extreme thermal environments, its thermal properties are of critical importance. Important thermal properties of this material are summarized as follows:

- Heat of fusion: 19 kJ/mol
- Heat of vaporization: 514 kJ/mol
- Heat of atomization: 599 kJ/mol
- Melting point: 1,750°C
- Boiling point: 4,790°C

Research and development studies undertaken for thorium can be briefly summarized as follows:

- Synthesis and characterization of thorium sulfates
- Synthesis, structure, reactivity, and computational studies on thorium oxo and sulfido metallocenes
- The discoveries of uranium-235 and symmetric fission
- Matrix infrared spectroscopic and density functional theoretical investigations on thorium and uranium atom reactions with dimethyl ether
- Interaction of thorium with nitrate in aqueous solution
- Use of liquid thorium for nuclear power reactors
- A cryogenic beam of refractory, chemically reactive molecules with expansion cooling
- Gamma-ray laser capable of emitting in the optical spectral range
- Gamma-spectrometric analysis of high-salinity fluids
- Screening of plant species for phytoremediation of uranium, thorium, barium, nickel, strontium, and lead-contaminated soils from a uranium mining tailings repository in South Africa
- Background radiation and individual dosimetry in the coastal region of Tamil Nadu state of India
- Dust concentration analysis in non-coal-mining areas and its impact on human health
- Exposure evaluation based on measurements undertaken for occupational hygiene at research laboratories in Poland from 2001 to 2005

2.3.2.8 Gadolinium

Gadolinium was first discovered by Jean de Marignac in 1880. Gadolinium and its isotopes (Table 2.7) are available in abundance on Earth. Its atomic structure, ionization energy, electrical conductivity, and thermal properties are best suited for magnetic resonance imaging (MRI) and other commercial research applications. This element is particularly useful as an injectable contrast agent for patients undergoing MRI. It is important to mention that its high magnetic moment can significantly reduce the relaxation times, leading to improved signal intensity. Gadolinium can be used as a host for x-ray cassettes and in scintillator materials for computer tomography and other commercial applications. It is also widely used as a phosphor.

Gadolinium is available as an oxide or as a compound. The metallic forms of this rare earth element include rods, wires, pellets, and granules. Its nanoparticles and powders offer excellent surface areas with unique properties and outstanding benefits that are best suited for commercial and industrial research applications. Its oxides are available in powder and dense pellet forms and are most ideal for optical coatings and thin-film applications. Its insoluble oxides are fluorides and are best suited for metallurgy research, chemical and physical vapor deposition, and fine optical coatings. Soluble oxides of gadolinium include chlorides, nitrates, and acetates. Compounds of this material can be manufactured as per a specific application. Gadolinium compounds have no biological role.

Gadolinium itself is very toxic. However, the isotopes shown in Table 2.7 are stable or nonradioactive; they are generally used for medical research studies and analytical investigations such as x-ray diffraction and surface area analysis. Gadolinium compounds are soluble in organic solvents. The safety data for gadolinium metal, particles, and compounds can vary widely depending on the form of the material involved. Gadolinium's electrical conductivity is 140.5 $\mu\Omega$-cm at room temperature, whereas its thermal conductivity is 10.5 W/(m·K). The heat of fusion is 15.5 kJ/mol and the heat of vaporization is 301 kJ/mol.

Research and development activities on this rare earth material are focused on the following disciplines, with primary emphasis on medical diagnosis and treatment [2]:

■ Feasibility studies using magnetic resonance enterography for the assessment of terminal inflammatory activity in children
■ MRI findings of the parotid gland
■ MRI and computed tomography (CT) evaluation of congenital pulmonary vein abnormalities

Table 2.7 Atomic Mass and Abundance of Gadolinium (Gd) Isotopes

Isotope	Atomic Mass	Abundance on Earth (%)
Gd-152	151.920	0.20
Gd-154	153.921	2.18
Gd-155	154.922	14.80
Gd-156	155.923	21.47

- MRI characterization of progressive cardiac dysfunction
- Ultrasonography, CT, and MRI techniques for specific medical treatments
- Effectiveness of combined MRI and contrast-enhanced CT techniques for specific medical treatments
- Potentiality of the gastric motility drug lorglumide in prostate cancer imaging
- Assessment of distribution and evolution of mechanical dyssynchrony of a myocardial case
- Hyperpolarized spectroscopy to detect early-stage tumors

Gadolinium oxide sputtering targets, gadolinium-selenide sputtering targets, magnesium-gadolinium sputtering targets, gadolinium-telluride sputtering targets, and zirconium-gadolinium sputtering targets play key roles in some commercial and industrial applications. Ultrathin gadolinium thin foils are best suited for scientific applications in which reliability and precision measurements are the principal requirements.

2.3.2.9 Terbium

Terbium was first discovered by Carl Mosander in 1843. This particular material is widely used in phosphors, fluorescent lamps, and as a high-intensity green emitter that is used in projection televisions, such as the terbium yttrium-aluminum-garnet. This material is available as a metal, oxide, or compound with various purity levels. The metal can be in the form of foils, rods, and sputtering targets. Its metallic form includes wires, rods, pellets, and granules as evaporation source material. Its compounds are found as submicrons and nanoparticles. Nanoparticles and nanopowders offer high surface areas with remarkable properties. Its insoluble oxides are fluorides, which are best suited for metallurgy, chemical and physical deposition, and optical coatings. Its isotope Tb-159 has an atomic mass of 158.925 and is 100% abundant on Earth. Its electrical conductivity at room temperature is 118 $\mu\Omega$-cm and its thermal conductivity is approximately 11.1 W/(m·K).

Terbium responds efficiently in x-ray excitation and is widely used as an x-ray phosphor. Terbium alloys are used in magneto-optic recording films as terbium-iron-cobalt. Recent research and development efforts have concentrated on energy transfer between the terbium-binding peptide and the red fluorescent proteins, interactions between metal ions and carbohydrates, detection of bacterial endospores in soil, use of the terbium complex as a luminescent sensor, and development of a terbium complex–based luminescent probe for imaging of endogenous hydrogen-peroxide generation in plant tissues.

2.3.2.10 Yttrium

Yttrium was first discovered by Johan Gadolin in 1794. It is found in abundance on Earth and is contained in other minerals. Yttrium occurs in nearly all rare earth minerals. This rare earth material was first used by the National Aeronautics and Space Administration (NASA) for lunar rock sample collection during the Apollo II mission. Yttrium has a silvery metallic luster and it is fairly stable in air. However, finely divided yttrium is very unstable in air.

Yttrium oxide is considered to be one of the most important oxides due to its vast commercial applications. Yttrium compounds are widely used for critical commercial and industrial applications, such as for making yttrium vanadium oxide (YVO_4), which is a very useful rare earth–based oxide. Europium and yttrium oxide (Y_2O_3):europium phosphors both yield red color in television tubes, and many hundreds of thousands of pounds of this compound have been used for this application. It is evident that both the YVO_4 and Y_2O_3 rare earth oxides have potential commercial applications. Note that the yttrium material has a boiling point of 1,495°C, whereas the vanadium rare earth material has a boiling point of 1,800°C.

Yttrium oxide is used to manufacture yttrium-iron-garnet (YIG), which is widely used in microwave filters where sharp cutoff performance in stop-band regions is essential [8]. YIG is highly efficient as both a transmitter and transducer of acoustic energy. Small amounts of yttrium (0.1–0.2%) can be used to reduce grain size in chromium, molybdenum, zirconium, titanium, manganese, and aluminum alloys to improve the mechanical strength of the alloys. This rare earth material can be used as an oxidizer for vanadium and other nonferrous metals. Yttrium is also finding applications in the design and development of lasers operating in the infrared spectral regions. This element is used as a catalyst for ethylene polymerization. It has wide applications in ceramic and glass formulas because its oxide offers a very high melting point, improved shock resistance, and low expansion characteristics on glass.

Natural yttrium has only one isotope—Y-89, with an atomic mass number of 89. Preliminary studies by early scientists indicate that 20 other unstable nuclides and isomers have been characterized. This rare earth material can be produced commercially by reduction of its fluoride oxide with calcium metal. It is commercially available with a purity level of 99% or more at an estimated cost of $200 per pound. Yttrium is widely used in the design and development of microwave filters with sharp cutoff characteristics, infrared lasers at various emitting wavelengths, and TWTAs (critical elements of electronic warfare systems).

The thermal conductivity of yttrium is 11.5 W/(m·K) and its melting point is 1,523°C. A high melting point is of critical importance for lasers, magnets, and other components operating under elevated temperatures over long durations. Yttrium plays a critical role in the design and development of diode-pumped solid-state (DPSS) lasers operating at various infrared wavelengths. In the case of a DPSS laser system, the light from the laser-diode pump excites the atoms in the rare earth–doped materials (e.g., thulium, erbium, holmium) when added to the solid-state host laser crystals (e.g., yttrium-aluminum-garnet [YAG], yttrium-lithium-fluoride).

Studies undertaken by the author on various solid-state lasers indicate that trivalent rare-earth dopants are essential in designing infrared (IR) lasers emitting at specific wavelengths. Trivalent rare-earth dopants (e.g., erbium, thulium, holmium) emit IR radiation at 2.94 μm, 2.10 μm, and 2.04 μm, respectively. Simulated energy then extracts optical energy from the excited atom as a laser beam. Strong absorption lines at the desired wavelengths can be obtained by adding appropriate amounts of rare earth dopants to the solid-state laser host crystal to absorb the diode-pumped laser light and to transfer its energy to the dopant atom.

2.3.2.11 Thulium

Thulium was first discovered by Per Teodor Cleve in 1879. Thulium occurs in all quantities along with other rare earth materials in a number of minerals. Its melting point is 1,545°C, and it has valences of 2 and 3. Thulium is the least abundant of the rare earth elements, but as new sources of rare earth materials are discovered, it is being considered like gold, silver, or cadmium. This rare earth element costs approximately $1,200 per pound. It has 16 isotopes with atomic masses ranging from 161 to 176. Natural thulium (Tm-169) is a very stable isotope and has potential applications in commercial products. Tm-171 is potentially useful as an energy-producing source. Natural thulium is widely used as a ferrite (ceramic magnet materials). These ferrite materials are best suited for designing acousto-optic devices (e.g., tunable optical oscillators) and microwave components (e.g., ferrite isolators and circulators) that are widely deployed to protect radar receivers and transmitters against microwave spikes. Thulium is low to moderately toxic and therefore should be handled with care.

According to studies performed on diode-pumped lasers using solid-state doped crystals, erbium is generally selected to absorb the pump light and thulium is used to transfer optical energy from erbium to holmium. The addition of chromium to the host crystal permits strong absorption of the visible wavelengths emitting below 800 nm. The optical power output from the laser will be less than 100 mW under the best operating conditions. A tripled-doped, lamp-pumped Er:Tm:Ho:YAG solid-state laser has demonstrated a continuous wave power output of more than 15 W at a 2-μm wavelength, which is otherwise not possible with such high power output.

A solid-state laser that is doped with rare earth material requires cryogenic cooling to keep the population of the lower power level from blocking the laser operation at the desired wavelength. The lifetimes and energy levels of potential trivalent rare earth ions must be investigated to determine the optical energy storage level in the solid-state laser crystal, such as YAG crystal or yttrium-lithium-fluoride crystal. When doped with yttrium, these two last crystals yield optimum laser performance at infrared wavelengths with improved efficiency, high optical stability, and uniform laser beams.

Both the ferrites and garnets characterized as spinel structures play critical roles in the design and development of sophisticated microwave components, particularly for defense applications. The general formula for a spinel structure is MFe_2O_4, where M is the rare earth material selected. The general formula for a garnet can be written as $Y_3Fe_5O_{12}$, where Y is yttrium. The formula for the yttrium-iron-garnet can be written as $3Y_2O_3 \cdot 5Fe_2O_3$.

2.3.2.12 Thallium

Thallium was first discovered spectroscopically by Sir William Crookes in 1861. This metal was isolated both by Crookes and Claude-Auguste Lamy in 1862. Thallium can be obtained from the smelting of lead and zinc ores. Extraction of this rare earth material is very complex and depends strictly on the source of the thallium. The metal is very soft and malleable. This metal can be cut with a knife. It has 20 isotopes, with atomic masses ranging from 191 to 210. Natural thallium is a mixture of two isotopes. This element and its compounds are toxic and therefore must be handled carefully. Special precautions must be taken during its transportation by air, land, or sea. Skin contact with this material must be avoided.

During the melting of this material, adequate ventilation should be provided. Its compound thallium sulfate is widely used as a rodenticide and ant killer. It is odorless and tasteless and gives no warning of its presence. Thallium sulfate is also widely used in photocells. Thallium bromide crystals have been used as infrared detectors. Thallium oxide has been used to produce glasses with a high index of refraction in scientific investigations and for various commercial and industrial applications. Thallium and its two isotopes are found in abundance on Earth, as shown in Table 2.8.

Thallium's most important properties include its atomic structure, first ionization energy, electrical conductivity, and thermal conductivity. The electrical conductivity of thallium sulfide changes with exposure to infrared light. This particular characteristic thallium sulfide makes the compound extremely useful for manufacturing photocells. Low-melting glasses are formed when thallium is used with sulfur or selenium and arsenic. These glasses have similar room-temperature properties to ordinary glass, but they are durable due to their superior mechanical strength. Thallium is available as a rod, film, foil, pellet, or sputtering target. Its compounds are available as submicrons and nanopowders. The nanoparticles and nanopowders yield high-quality surface areas with unique characteristics.

Thallium oxides are available as powders or dense pellets; they are best suited for precision optical coatings and thin-film applications. The fluorides are insoluble

Table 2.8 Atomic Mass and Abundance of Thallium (Tl) Isotopes

Isotope	Atomic Mass	Abundance on Earth (%)
Tl-203	202.972	29.52
Tl-205	204.974	70.48

oxides; they are widely used in metallurgy, chemical and physical vapor deposition, and some special optical coatings. Soluble oxides of thallium include chloride, nitrates, and acetates. Thallium alloy is reported to freeze at −60°C, which is approximately 20°C below the freezing point of mercury.

The most important physical properties of thallium can be summarized as follows:

- Melting point: 1,800°C
- Specific density: 11.5 g/cm^3
- Electrical conductivity: 18 μΩ-cm
- Thermal conductivity: 46.1 W/(m·K)
- Freezing point of thallium alloy: −60°C
- Energy of first ionization: 589.36 kJ/mol

Research and development studies undertaken by various medical experts on this rare earth material have observed significant benefits, including brain spectral analysis using thallium-cadmium-zinc-telluride, thalium-201 in scintigraphy to distinguish malignant from benign soft-tissue tumors, impact of exercise started within 2 weeks after acute myocardial infarction on myocardial perfusion and left ventricular function, biological safety of nasal thallium-201 administration, head-to-head comparison of contrast-enhanced cardiovascular magnetic resonance and thallium-201 single-photon-emission computed tomography for prediction of reversible left ventricular dysfunction in heart disease, dose-enhancement studies in microbeam radiation therapy, and the effects of chemical and radioactive properties of thallium-201 isotope on a new cadmium-zinc-telluride cardiac camera and a spectroscopic camera.

2.4 Availability of Rare Earth Materials in Various Forms and Their Principal Applications

This section addresses the various forms in which rare earth materials can be procured. The principal applications of each rare earth material are identified. The various categories of each rare earth material as a metal, oxide, or compound are described in detail.

2.4.1 Rare Earth Metals

Rare earth metals can be purchased in various forms to convert into alloys, which can be used in coatings and thin-film chemical vapor deposition (CVD) techniques, such as thermal and electron beam (e-beam) evaporation, low-temperature organic evaporation, atomic layer deposition, and metal-organic chemical vapor deposition (MOCVD) for specific applications such as fuel cells and solar cells. Fuel cells are

Table 2.9 Various Forms of Rare Earth Metals

Rare Earth Metal	Forms					
	Pellet	*Powder*	*Sputtering Target*	*Rod*	*Foil*	*Granule*
Cerium	X	X	X	X	X	X
Dysprosium	X	X	X	X	X	X
Erbium	X	X		X	X	X
Europium	X	X		X	X	X
Gadolinium	X	X		X	X	X
Holmium	X	X		X	X	X
Lutetium	X	X		X	X	X
Praseodymium	X	X		X	X	X
Samarium	X	X		X	X	X
Terbium	X	X		X	X	X
Thulium	X	X		X	X	X
Ytterbium	X	X			X	X

widely used as portable alternate energy sources in buses or cars, whereas solar cells are deployed to convert solar energy into electrical energy.

Rare earth materials are available in various forms such as metals, pellets, discs, powders, granules, nanoparticles, ingots, discs, sputtering targets, wires, foils, and rods (Table 2.9).

2.4.2 Rare Earth Oxides

Rare earth oxides are available in various forms, including pellets, powder, tablets, sputtering targets, and nanopowder (Table 2.10). Pellets and sputtering targets are best suited for coatings and thin-film applications, CVD, and physical vapor deposition processes, including thermal and electron beam evaporation techniques for commercial and industrial ceramic product applications.

2.4.3 Rare Earth Compounds

Solutions for rare earth compounds are available in bulk quantities for research and commercial applications. The various solutions used for producing rare earth compounds are shown in Table 2.11.

Table 2.10   Various Forms of Rare Earth Oxides

Rare Earth Oxide	Forms				
	Pellet	*Powder*	*Tablets*	*Sputtering Target*	*Nanopowder*
Cerium oxide	X	X	X		X
Dysprosium	X	X	X	X	X
Erbium	X	X	X	X	X
Europium	X	X	X	X	X
Gadolinium		X	X	X	X
Holmium	X		X	X	X
Lutetium	X	X	X	X	X
Neodymium	X		X	X	X
Samarium	X				X
Terbium	X		X	X	X
Thulium	X		X	X	X
Ytterbium	X	X		X	

2.4.4 Rare Earth Organometallics

Rare earth organometallics provide rare earth sources that are soluble in organic solvents. Examples of commercially produced organometallic rare earth materials are shown in Table 2.12.

Recent research and development activities have been directed towards structural and spectroscopic properties of LaOF:Eu(3+) nanocrystals, structural investigation of the negative thermal expansion in yttrium and rare earth molybdates, high-pressure phase transitions in the rare earth metal erbium to 15 GPa (pressure), molecular nitrides with titanium and rare earth metals, risk to patients from the administration of iodine and gadolinium preparations in diagnostic clinical studies, assessment of radionuclide and metal decontamination in a thorium-rich area in Norway, and mechanical decoupling between myofibroblasts and cardiomyocytes to slow electric conduction in fibrotic cells.

2.5 Rare Earth Materials Used in the Development of Fuel Cells

A fuel cell is an energy conversion device in which the chemical energy is isothermally converted into direct current electricity. It can convert chemical energy

Table 2.11   Various Solutions Used to Produce Rare Earth Compounds

Rare Earth Element	Solution Type				
	Acetate	Chloride	Nitrate	Sulfate	Bromide
Cerium	X	X	X	X	X
Dysprosium	X	X	X	X	X
Erbium	X	X	X	X	X
Europium	X	X	X	X	X
Gadolinium	X	X	X	X	X
Holmium	X		X	X	X
Neodymium	X	X	X		X
Samarium	X	X	X	X	X
Terbium	X	X	X	X	X
Thulium	X	X	X	X	X
Ytterbium	X		X		X

directly into electrical energy without involving the thermodynamic relationship, as demonstrated by the Carnot cycle to limit the efficiency of heat engines. Certain rare earth compounds play a key role in the design and development of fuel cells in which high electrical energy and enhanced electrochemical efficiency are the principal design requirements.

Table 2.12   Examples of Commercially Produced Organometallic Rare Earth Materials

Rare Earth Material	Acetylacetonate	Diethylhexanoate	Trifluoromethanesulfonate
Cerium	X	X	X
Dysprosium	X	X	X
Gadolinium	X	X	X
Samarium	X	X	X
Yttrium	X		
Ytterbium	X	X	X

The author has described the principal requirements for cathodes, anodes, and electrolytes for a fuel cell capable of yielding ultrahigh power with significantly improved performance in terms of efficiency and reliability [6]. The classification of fuel cells is strictly based on the types of electrolytes used, which can be summarized as follows:

■ Alkaline electrolytes
■ Phosphoric electrolytes
■ Molten-carbonate electrolytes
■ Solid-polymer electrolytes
■ Solid-oxide electrolytes

A fuel cell using a solid-oxide electrolyte is called a solid-oxide fuel cell (SOFC) [9].

Fuel cell engineers and scientists have identified the rare earth raw materials capable of producing thin-film electrochemistry with outstanding thermal expansion-matching technology [4]. The deployment of rare earth materials for the fabrication of cathodes, anodes, and electrolytes—the critical elements of a high-power fuel cell with enhanced electrochemical efficiency—have been recommended [4,6]. The anode is the negative, which is considered to be the fuel electrode that gives up the electrons to the external circuit or terminal. The cathode is the positive or oxidizing electrode, which accepts electrons from the external circuit. The output power of a fuel cell is strictly dependent on the cathode's current density and the maximum safe operating temperature. However, the optimum performance of a fuel cell in terms of power output, electrochemical efficiency, safety, and reliability is strictly dependent on the materials used for the cathode, anode, and electrolyte.

2.5.1 Design Requirements for High-Temperature, High-Power Fuel Cells

In the 1960s and 1970s, fuel cell designers were hesitant to pursue extensive research and development activities on high-temperature, high-power cells due to several technological difficulties associated with high-temperature fuel cell technology. The early high-temperature fuel cell design used a semi-solid molten salt electrolyte, which was a mixture of sodium, potassium, and lithium carbonate. The electrode surfaces used a metallic powder of silver and air for the cathode electrode, whereas the anode used a mixture of iron, nickel, and zinc oxide/silver as a fuel. Fuel cell designers found that this particular design could operate at temperatures ranging from 500 to 700°C and would yield a current density close to 100 ampere/cm^2 and a maximum efficiency of 35%.

General Electric (GE) fuel cell scientists used zirconium oxide as a solid electrolyte with no doping [5]. When the scientists doped the ZrO_2 with yttrium oxide (Y_2O_2), significant improvement was noticed in the cell efficiency (from 35% to

48%) and in cell life (from 2,100 hours to 3,600 hours) with no compromise in safety or reliability. The GE scientists claimed that by using higher doping levels of yttrium oxide, one can expect further improvement in cell efficiency. This efficiency improvement was due to electrolyte doping with yttrium oxide. Significant improvement in fuel cell electrical performance and reliability can be expected if appropriate rare earth materials are used in the fabrication of cathode and anode electrodes.

In the early 2000s, American Element's fuel cell scientists concluded that deployment of rare earth materials for the cathode, anode, and electrolyte would produce a highly compatible thin-film electrochemistry with outstanding thermal expansion-matching capability and will yield optimum performance in terms of power output, efficiency, safety, and reliability. High-power fuel cells generally require operating temperatures close to 1,000°C or higher. At such high temperatures, conventional electrolyte materials, such as semi-solid molten electrolytes or aqueous electrolytes, will not meet the output power and reliability requirements. In the early 1960s, several fuel cells were designed, developed, and evaluated including 10 fuel cells using aqueous electrolytes, 6 fuel cells using molten electrolytes, and 3 fuel cells using solid electrolytes. Extensive experimental verification of their performance levels in terms of output power, efficiency, safety, and reliability concluded that the fuel cells using the solid oxide technology for the electrolyte met the above-mentioned performance parameters. The author believes the use of rare earth materials for the cathode, anode, and electrolyte will allow for high-temperature operations with significant improvements in power output, efficiency, operating life, safety, and reliability of the fuel cells. However, deployment of rare earth materials will slightly increase the cost and cell complexity.

2.5.2 Optimum Rare Earth Materials for Anodes, Cathodes, and Electrolytes

American Elements has recommended the rare earth materials that are best suited for the critical elements of a fuel cell. The rare earth materials in the following sections are most ideal for the critical elements.

2.5.2.1 Perovskite Cathode Materials

The recommended perovskite cathode materials are as follows:

- Lanthanum strontium ferrite
- Lanthanum strontium chromate
- Lanthanum strontium magnetite
- Lanthanum strontium cobaltite
- Lanthanum strontium gallate magnetite
- Lanthanum calcium magnetite

These lanthanum rare earth compounds have different doping levels, surface areas, and other parameters needed for optimum performance of the cathode electrode. For example, nickel cermet compositions contain yttria-stabilized zirconia (YSZ) doping and physical levels and controlled particle size distribution. The ionically conductive electrolytes include YSZ, scandium-doped zirconia (SDZ), samarium-doped ceria (SDC), gadolinium-doped ceria (GDC), and yttrium-doped ceria (YDC). These doped materials are best suited for perovskite cathode structures.

These cathode materials are best suited for SOFCs. The above-mentioned rare earth cathode materials are electronically conductive and stable in oxidizing environments. Their perovskite structures allow doping on both sites of the cathode. For example, lanthanum magnetite doped with strontium would be a case of doping on one site, whereas doping with greater strontium levels would lead to significantly enhanced activity. Lanthanum strontium cobaltite ferrite will allow doping on both sites of the structure. When fired in a fuel cell layer of cathodes, nanopowders will partially sinter to form well-defined necks and open gas paths. These open gas paths will allow the simultaneous transfer of gas and electrical energy. As mentioned previously, lanthanum strontium magnetite has an excellent thermal expansion match with YSZ electrolytes. It is highly electronically conductive and offers long-term stability. Lanthanum strontium chromite is available as both a cathode and an interconnect.

2.5.2.2 Rare Earth Materials for Electrolytes

Powders and nanopowders that are suitable for tape casting, air spray, extrusion, and sputtering applications are readily available from commercial suppliers. YSZ (zirconium oxide stabilized with yttrium oxide) makes a robust electrolyte that is purely ionically conductive (which means it has no electronic conductivity) and operates at a wide range of partial pressures. This electrolyte material can operate at temperatures ranging from 900 to 1,000°C with very little or no impact on the material properties or electrolyte performance. GDC (cerium oxide stabilized with gadolinium oxide), YDC (cerium oxide stabilized with yttrium oxide), and SDC (cerium oxide stabilized with samarium oxide) form a class of electrolytes with higher ionic conductivities and operating temperatures 700°C lower than YDC. These electrolyte materials operate at narrow partial pressures and will electronically conduct if operated at lower partial pressures. Doped ceria oxide tends to reduce from the cereus to the ceric state under open circuit conditions. SCZ electrolyte material is approximately three times more ionically conductive than YSZ electrolyte and operates very efficiently at temperatures less than 800°C.

The following two examples each use a solid electrolyte. GE scientists developed a fuel cell using natural gas as the fuel. The gas is enclosed in a heating jacket. The operating temperature of the cell is 1,093°C (2,000°F) when the cell is operating at full capacity. When this temperature is reached, the natural gas is fed directly into the cell. The natural gas decomposes into carbon and hydrogen at this operating temperature. The carbon deposits on the outside of a cylindrical cup made of a solid electrolyte. The electrolyte is solid gas-impregnated zirconia (ZrO_2) doped with calcium oxide (CaO) to provide enough oxide ions to carry the cell current. The oxidant (air or oxygen) is bubbled through the molten silver (Ag) cathode, which is held inside the long cylindrical zirconia cup. The byproducts—namely, the carbon oxide and hydrogen that were formed in the initial fuel decomposition process—are then burned outside the cell to keep the cell at the operating temperature. Hydrogen is not involved in the electrochemical reaction.

Westinghouse engineers designed a fuel cell using a solid electrolyte involving yttrium oxide. The zirconia or zirconium oxide, doped with either 15 mol% of calcium oxide or 10 mol% of yttrium oxide (Y_2O_2), was used as a solid electrolyte. The engineers used two types of fuels, either 7% hydrogen in nitrogen or pure hydrogen only. Air was used as an oxidant. Both the air and fuel electrodes were made from platinum, initially applied to the ZrO_2 as a platinum paint. The ZrO_2 electrolyte contained typically 10 mol% of yttrium oxide. A fuel cell using pure hydrogen as a fuel and air as an oxidant can yield a current density of 50 mA/cm^2 at a temperature of 1,000°C [9].

In this example, a cell conversion efficiency of close to 30% was achieved without doping the zirconium oxide with yttrium oxide (Y_2O_2). After doping the zirconium oxide with yttrium oxide, the cell conversion efficiency jumped to more than 48%. Fuel cell scientists believe that higher doping levels with yttrium oxide can obtain an efficiency that is even better than 58%, but it will add additional cost and complexity. Both fuel cell designs had operating lives exceeding 3,500 hours with no performance degradation. This higher operating life and significantly improved cell efficiency are possible using solid electrolyte and yttrium oxide doping.

Fuel cells using appropriate rare earth materials for cathode and anode electrodes and a solid electrolyte doped with rare earth materials can be summarized as follows:

- A solid electrolyte consisting of zirconium oxide doped with a suitable rare earth material
- Operating temperature exceeding 1,000°C
- Fuel that is hydrogen or reformate
- Oxidant that is air or oxygen

A typical conversion efficiency exceeds 60%, which can be further improved using higher doping levels with rare earth materials. Efficiencies as high as 80% may be possible using appropriate rare earth materials and optimum doping levels. Applications are best suited for large utility companies.

2.5.2.3 Rare Earth Materials for Anodes

SOFC powders and nanopowders for the anode electrode are commercially available. Nickel cermet compositions of nickel oxide and YSZ are used to produce a thin-film layer with optimum doping levels and particle mix. The proportion of nickel to YSZ reflects a tradeoff between the stability of YSZ and conductivity of nickel, which must be balanced to prevent coarsening during operations and to maximize the long-term stability of the fuel cell.

2.6 Performance Improvement of Rechargeable Batteries, Infrared Lasers, and Fuel Cells Using Rare Earth Materials or Their Oxides in Conjunction with Other Metals

This section identifies rare earth elements or their oxides in conjunction with other conventional metals that can be deployed in the design and development of rechargeable batteries, fuel cells, and lasers to enhance their performance levels. Performance improvement in terms of energy level, power output, service life, and efficiency have been observed with the use of rare earth elements and their oxides. Critical properties of rare earth elements and their oxides responsible for the performance of rechargeable batteries, fuel cells, and lasers are summarized in Table 2.13.

Preliminary studies performed by the author on the applications of selected rare earth elements seem to indicate that these elements significantly improve device performance at elevated temperatures due to high electrical conductivity. The room-temperature resistivity and conductivity of conventional metals and rare earth metals are provided in Table 2.14.

CePt, a compound consisting of the rare earth element cerium and the conventional metal platinum, is best suited for fabrication of millimeter-wave micro-electro-mechanical system (MEMS) series and shunt switches requiring very high isolation (as high as 42 dB at 40 GHz). The room temperature electrical conductivity for both cerium and platinum is identical—0.013×10^6 mho/cm. Therefore, CePt provides optimum isolation for the series and shunt radiofrequency (RF) switches using MEMS technology. These MEMS-based switches using cerium offer high isolation for RF switches, which is of critical importance for radar and other RF systems. Similarly, when samarium is doped with the conventional metal cobalt, it forms a compound that is widely deployed in designing powerful permanent magnets for RF TWTAs, electric motors, and generators. This particular permanent magnet continues to retain reliable magnetic performance at operating temperatures as high as 300°C. No other magnetic material could meet or beat this performance level at elevated temperatures.

Rare earth–doped crystals are widely used in the design and development of infrared lasers with high efficiency and excellent beam quality. A classic example of a rare earth compound is the neodymium-based YAG (Nd:YAG) laser, which was

Table 2.13 Characteristics of Rare Earth Elements and Other Metals Used in the Design and Development of Batteries, Fuel Cells, and Lasers

Element	Symbol	Valance	Applications for Performance Improvement
Cadmium	Cd	2	Energy, power, and longevity of batteries
Cerium*	Ce	2, 4	Battery electrodes and power level
Cobalt	Co	2, 3	Electrodes for high-temperature fuel cells
Lanthanum*	La	3	Temperature, power, and energy of cells
Lithium	Li	1	Longevity, efficiency, and electrodes
Manganese	Mn	2, 3, 4, 6	Improved power and energy of batteries
Neodymium*	Nd	3	Improved power and energy in cells
Nickel	Ni	2, 3	Improved electrodes for fuel cells
Praseodymium*	Pr	3	Used to produce mischmetal
Vanadium*	V	3, 5	Improves battery efficiency and reliability
Zirconium	Zr	4	Best suited for high-power fuel cells

* Rare earth material (other materials appearing in the table provide supporting roles when used in conjunction with rare earth materials).

used for optical communication equipment in space in the late 1970s. For example, the Nd:YAG laser used two rare earth elements (neodymium and yttrium), one conventional metal (aluminum), and high-temperature ferrite ceramic. This particular laser system provided three communication functions: space-to-space, space-to-ground, and space-to-aircraft. Both covert as well as conventional communications and data transfer capabilities were provided by this laser using two rare earth materials. Laser systems using various rare earth materials are discussed in great detail in Chapter 7, Section 7.6.

Table 2.14   Room-Temperature Resistivity and Electrical Conductivity of Some Conventional Metals and Rare Earth Elements

Metal	Resistivity (×10⁻⁶ Ω-cm)	Electrical Conductivity (×10⁶ mho/cm)
Aluminum (Al)	2.62	0.380
Cerium (Ce)*	78	0.013
Copper (Cu)	1.72	0.580
Neodymium (Nd)*	79	0.013
Nickel (Ni)	6.9	0.145
Platinum (Pt)	10.5	0.095

* Rare earth element

2.7 Applications for Sputtering Targets, Evaporation Materials, Foils, and Thin Films

This section identifies potential applications for sputtering targets, evaporation materials, foils, and thin films produced from the rare earth elements, alloys, and compounds.

2.7.1 Sputtering Targets

Sputtering targets come in different shapes, sizes, and geometrical configurations. The materials are produced using crystallization forms, solid state, and other exotic purification processes such as sublimation to improve or refine the end product. Custom compositions can be produced for specific commercial and scientific research applications and to develop proprietary technologies involving rare earth materials. Rare earth metals and alloys can be converted into rods, bars, and other shapes. Furthermore, it is possible to convert rare earth elements or their alloys into nanoparticles or nanoquantum dots, which are well suited for the development of MEMS- and nanotechnology-based electro-optic and electronic devices, which have potential applications in the defense, space, and medical fields.

2.7.2 Rotating Sputtering Targets

Rotating sputtering targets are best suited for applications in which large film areas are required, such as photovoltaic and other coatings. The technology allows for reliable installment of printed solar devices, particularly on large surfaces, with minimum cost and complexity. Large targets can be finished to a flatness that is better than ±0.15 inches. Tighter tolerances are possible but at higher costs.

Rotating sputtering targets can be produced from a number of metallic, oxide, or alloy sources. The sputtering decomposition process requires a plasma, which is needed to bombard the target material for a thin film.

2.7.3 Applications of Sputtering Targets and Other Evaporation Materials

Potential commercial, scientific, and medical applications of sputtering targets and other evaporation materials are briefly summarized in the following sections.

2.7.3.1 Electronics and Semiconductors

The first commercial applications of sputtering targets were in the fields of electronics and semiconductors for front-end and back-end packaging, diffusion barriers in the scientific disciplines, phase change memory, integrated circuit interconnects, microcontacts for MEMS circuits, electronic sensors, and MEMS and light-emitting diode (LED) devices.

Sputtering targets and evaporation materials of copper and its alloys, such as copper-nickel and copper-chromium, are best suited for electronic and other packaging applications. Nickel and its alloys, including nickel-platinum, nickel-aluminum, nickel-vanadium, nickel-copper, and nickel-chromium, are widely used in high-performance electronic circuits in which reliability, insertion loss, temperature, and circuit efficiency are the principal design requirements under severe thermal and mechanical operating environments. Titanium (99.99%) can be alloyed in titanium-tungsten. Gold is considered an important decomposition material due to its conductive and solder-wetting properties. Prominent gold alloys include gold-silicon, gold-tin, gold-antimony, gold-copper, and gold-germanium. Germanium-antimony alloyed with silver, indium, telluride, and platinum yields transparent conductive oxides, which are ideal for light-emitting applications such as light sensors and LEDs.

2.7.3.2 Antiabrasive Coatings

Electroplating of cutting and drilling tool surfaces offers adequate protection against wear and extends a tool's life. Decomposition of coating materials is considered to be the most cost-effective technique. The most widely used protective materials using sputtering targets and other evaporation materials include titanium carbide, silicon carbide, boron carbide, and tungsten carbide.

2.7.3.3 Magnetic Materials

High-strength permanent magnets are manufactured using rare earth materials and alloys such as samarium cobalt and neodymium-iron-boron. Samarium-cobalt permanent magnets offer high magnetic induction and field strength at operating temperatures as high as 300°C. Rare earth–based permanent magnets are widely used in electric motors and generators in electric and hybrid electric vehicles, aerospace systems, biomedical imaging equipment, sputtering targets, and defense systems.

2.7.3.4 Optical Coatings and Architectural Glass

Certain rare earth elements have inherent capabilities to absorb and emit at specific wavelengths and to reduce glare due to their high refractive indexes. Therefore, they can play important roles in antireflection coatings and devices. When the glass surface is treated with an antireflective coating, the glass substrate not only cuts down or reduces the glare experienced by the observer, but it also results in the development of sputtering and evaporation materials of rare earth elements, such as neodymium, dysprosium, and other optically active and antireflective materials. Anti-reflective architectural glass is widely used for residential buildings, commercial installations, and office buildings to eliminate glare from the sun. Building designers claim that the use of such glass not only cuts down on the glare but it also keeps the building interior relatively cool, thereby significantly reducing the cost of air-conditioning.

2.7.3.5 Solar Energy Panels Equipped with Photovoltaic Cells

Solar energy technology provides an alternate source of electricity with no adverse environmental effects. Large commercial buildings and office buildings with large surface areas can be covered with a flat plastic substrate printed with solar cell panels [10]. When the sunlight strikes the solar cell panels, direct current electricity is generated by the solar cells. Highly efficient organic solar cells and compound semiconductor cells such as cadmium telluride, copper indium gallium selenide, gallium arsenide, and silicon are available for deployment in solar panels. The compound semiconductor–based solar cells are layered structures that require sputtering targets and suitable evaporation materials at several stages, including transparent conductive oxides such as indium oxide and doped zinc oxide. Molybdenum is used for the back plate to enhance efficiency and provide structural strength for the solar panel [10]. Solar cells made from polycrystalline silicon materials offer high conversion efficiency with moderate cost.

2.7.3.6 Solid Oxide Fuel Cells

As mentioned previously, fuel cells using solid electrolytes will provide high power output, long operating hours, improved reliability, and stable electrical performance [9]. A typical SOFC design includes an electronically conductive low-density cathode, a high-density ionically conductive electrolyte, and an electronically conductive open-air electrode. A new technology is needed for the deposition of various layers in the cell. Sputtering targets play a key role in deposition of the layers, including the perovskite cathode materials, which may contain lanthanum strontium manganite, lanthanum strontium ferrite, lanthanum strontium cobalt-ite ferrite, lanthanum strontium chromite, and lanthanum strontium gallate magnetite. Recommended doping levels and ionically conductive electrolytes include YSZ, scandium-doped zirconia, SDC, and YDC.

2.7.3.7 Data Storage Materials

Sputtering targets and other evaporation materials are essential to the development of coatings for the manufacturing of optical storage devices, namely CDs and DVDs. These provide wear protection and reflectivity during storage periods.

2.7.3.8 Potential Deposition Techniques without the Sputtering Target Method

2.7.3.8.1 High-Power Laser Pulse Technique

The high-power pulsed laser deposition technique offers the most efficient and cost-effective deposition method. Laser pulses from a high-power pulsed laser are used to ablate the target material. High-power laser pulses play a key role in instantly evaporating the material on the target surface, which turns into plasma and returns back to the vapor phase. The ablated material collects and deposits on the top of a correctly placed substrate. This technique has several benefits over other techniques, including preserving the stoichiometry of the target on the film formed and offering higher deposition rates compared to other methods.

2.7.3.8.2 Physical Vapor Deposition Techniques

Physical vapor deposition (PVD) techniques [4] consist of the purely physical formation of a thin film on the top of a substrate. The method does not involve any chemical reaction in the formation of the thin film. In general, PVD is done in low-pressure environments. The evaporation deposition raises the temperature of the thin-film material so that its vapor pressure is reached in a reasonable range. The vapor then moves and deposits on the top of the substrate. Electron beam evaporation is a form of PVD in which the target anode is bombarded with an intense electron beam provided by a charged tungsten filament in a high-vacuum environment. The electron beam causes atoms from the target material to transform into a gaseous phase. The atoms then return to a solid-form coating in the vacuum chamber with a thin film. This can also be used in conjunction with a molecular beam epitaxy process. The EBE technique is widely used in various applications, including medical, metallurgical, telecommunications, micro-electronics, optical coatings, and semiconductor devices using rare earth elements, oxides, and compounds.

As mentioned earlier, the CVD process involves a chemical reaction. The CVD technique generally uses a gas-phase precursor, which is often a hydride of the rare earth element to be deposited. In the case of MOCVD, an organometallic gas is used. The MOCVD technique is currently being used in the manufacture of graphene carbon nanotubes, laser-emitting diodes, multijunction solar cells, MEMS devices, microelectronic circuits, and photo detectors.

2.8 Summary

This chapter was strictly dedicated to the properties and applications of rare earth elements, oxides, alloys, and compounds. Rare earth elements and their properties were briefly summarized, with emphasis on molecular weight, valance, electrical conductivity, and thermal conductivity. All rare earth elements were described, identifying their critical applications in commercial, industrial, and scientific areas. Half-lives of the rare earth elements and their isotopes were provided for the benefit of readers.

Mining, surveying, and processing requirements to obtain rare earth materials were summarized, with an emphasis on cost and complexity. Critical issues such as environmental considerations, geopolitical considerations, and pricing considerations in obtaining rare earth materials from mines located in different parts of the world were discussed in great detail. Important oxides of rare earth materials were identified, along with their applications in various disciplines. Applications of the most important rare earth elements (ytterbium, lanthanum, cerium, neodymium, praseodymium, and samarium) in commercial, industrial, and defense products were briefly described, with an emphasis on their toxic effects.

Samarium was discussed. Samarium and its oxides are widely used in commercial and industrial applications. Samarium-cobalt permanent magnets offer unique magnetic properties that are not matched by any other rare earth material. This particular magnet has an intrinsic coercive force of more than 28,000 oersteds, which is not possible with any other material. Doping a calcium fluoride crystal with samarium will yield infrared lasers with unique characteristics. Its oxide is used as infrared absorption glass. The same oxide is used as a neutron absorber in nuclear power reactors.

Estimated procurement costs, specified in dollars per pound, for rare earth elements were summarized [6]. The applications of some rare earth oxides, alloys, and compounds in commercial, industrial, and scientific disciplines were identified. The atomic mass and availability of various rare earth elements were mentioned. Properties and applications of gadolinium oxide, terbium oxide, ytterbium oxide, and thulium oxide were briefly mentioned.

Various forms of commercially available rare earth materials, such as sputtering targets, pellets, rods, foils, granules, and tablets, were identified. Various chemical solutions—namely, acetates, chlorides, nitrates, sulfides, and bromides—were specified for their applications in conjunction with rare earth materials. Rare earth materials that are best suited for rechargeable batteries and fuel cells were recommended, with emphasis on performance, reliability, and safety. Vanadium is considered to be most ideal for the design and development of rechargeable batteries.

Rare earth elements, oxides, and compounds that are best suited for the development of high-power, high-temperature fuel cells were identified. Rare earth compound materials best suited for high-power fuel cell anodes, cathodes, and electrolytes were identified. The critical roles of cerium, cobalt, lanthanum, neodymium, praseodymium, samarium, and vanadium in the design and development of lasers, batteries, fuel cells, electro-optical sensors, electric and hybrid electric vehicles, and sophisticated electronic components were described, with an emphasis on performance enhancement and reliability improvement. The critical role of zirconium oxide and yttrium oxide in the design of high-temperature fuel cells using solid electrolytes offers significant improvement in electrochemical efficiency and power output. The electrochemical efficiency exceeds 48% with moderate doping levels of yttrium oxide, which is outstanding.

Descriptions and applications of sputtering targets using rare earth materials were summarized, with an emphasis on performance. Applications of sputtering targets and evaporation materials in commercial, industrial, and defense-related products were identified. Unique benefits of rare earth sputtering targets and evaporation materials were highlighted, with reference to antiabrasive coatings, magnetic characteristics, optical coating for laser components, and architectural glass for commercial and office buildings. The use of molybdenum in the design of photovoltaic cells has contributed to a significant improvement in cell performance. The advantages of rare earth materials in the design of CDs and DVDs were highlighted. Potential deposition techniques, such as PVD, PVC, CVD, and MOCVD, for various applications were summarized.

References

1. Wikipedia. *Rare earth element*. http://en.wikipedia.org/wiki/Rare_earth_element.
2. AmericanElements.2014.*Rareearthsinformationcenter*.http://www.americanelements.com/rare-earths.html
3. Glasstone, S. 1955. *Principles of Nuclear Reactor Engineering*. New York: Van Nostrand.
4. Data from American Elements, Technical Director, Los Angeles, CA.
5. Peattie, C.G. 1963. A summary of practical fuel cell technology to 1963. *Proceedings of the IEEE* 51(5):795–806.
6. U.S. Geological Survey. 2002. *Rare earth elements—Critical resources for high technology: Fact Sheet 087-02*. http://pubs.usgs.gov/fs/2002/fs087-02 (accessed December 18, 2013).
7. Weast, R.C. 1971. *Handbook of Chemistry and Physics*, 51st ed. Boca Raton, FL: CRC Press, B7–B8.
8. Jha, A.R. 1992. *Microwave Filters with Sharp Cut-Off Properties*. Cerritos, CA: Jha Technical Consulting Services, 5–8.
9. Jha, A.R. 2012. *Next-Generation Batteries and Fuel Cells for Commercial, Military, and Space Applications*. Boca Raton, FL: CRC Press.
10. Jha, A.R. 2010. *Solar Cell Technology and Applications*. Boca Raton, FL: CRC Press.

Chapter 3

Properties and Applications of Rare Earth–Based Superconductive Magnetic Materials

3.1 Introduction

This chapter describes the properties of the original 17 rare earth materials, as shown in Figure 3.1. The materials that are most suitable for permanent magnets will be identified, with emphasis on their potential applications in various fields. Magnetization and anisotropic properties of potential rare earth materials are summarized at low temperatures. Some rare earth materials suited for high-speed levitated trains will be identified, with emphasis on reliability and speed performance. Rare earth material films, wires, and particles that play critical roles in the design and development of commercial, industrial, and defense products will be mentioned, with emphasis on their cost-effective performance aspects.

The benefits and potential applications of superconducting thin films, such as yttrium barium copper oxide, thallium barium calcium copper oxide, lanthanum barium copper oxide, and lanthanum strontium copper, will be identified, along with their transition temperature limits. The benefits of these films in the design and development of radiofrequency (RF) filters, traveling wave-tube amplifiers

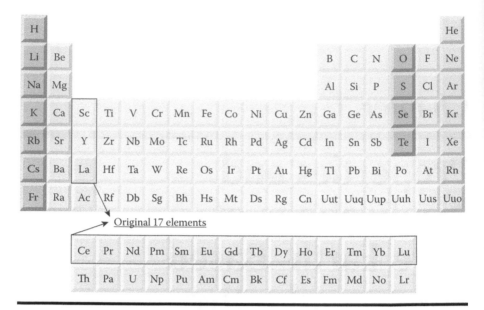

Figure 3.1   Original 17 rare earth elements.

(TWTAs), electromagnetic devices, low-frequency antennas, microwave solid state devices, yttrium iron garnet (YIG)-tuned filters with sharp cutoff characteristics in the stop-band regions, and underwater detecting sensors will be highlighted [1]. Properties of potential superconducting substrates will be summarized. The performance parameters of yttrium aluminum garnet (YAG), YIG, and other superconducting substrates will be summarized. Performance improvement in RF and high-frequency microwave components or devices will be summarized as a function of superconducting temperatures. YAG-tuned transistor design will be discussed, with an emphasis on tuning range and efficiency at higher RF frequencies. Rare earth materials used in the design and development of permanent magnets for various commercial and industrial applications will be identified, with particular emphasis on weight and size reductions, performance improvement, and enhanced reliability while the rare earth magnets are operating under harsh thermal and mechanical environments [1].

It is important to know the behavior and critical functions of various magnetic materials. *Ferromagnetic materials* create a strong magnetic field in the same direction as an externally applied field and can retain some magnetization after the field is withdrawn. This magnet sticks to iron, nickel, cobalt, and refrigerator doors. *Diamagnetic materials* create a magnetic field in opposition to an externally applied magnetic field. This weak opposition field produces a repelling force. Examples of diamagnetic materials include pyrolytic carbon, superconducting materials, and bismuth. Water is a slightly diamagnetic medium. *Paramagnetic materials* create a strong magnetic field in the same direction as an externally applied magnetic field, but they do not retain any magnetization once the field is withdrawn. An example is ferrofluid, which is often classified as a superparamagnet.

3.2 Magnetic Parameters and Glossary of Magnet Terminology

This section defines magnetic parameters and magnet terminology for the benefit of readers. A glossary of rare earth magnet terminology follows [2].

Air gap: The air gap is the distance from one pole of the magnet to the other pole through a nonmagnetic material that could be air.

Anisotropic: Anisotropic magnetic materials have a preferred magnetization direction. Such magnetic materials are typically produced under the influence of strong magnetic fields and can only be magnetized through the preferred axis. For example, neodymium-iron-boron and samarium-cobalt magnets are anisotropic devices.

Coercive force (F_c): The coercive force is the demagnetizing force necessary to reduce observed induction or flux density (B) to zero after the magnet has been brought to a saturation condition. This force is measured in oersteds.

Curie temperature (T_c): At the Curie temperature, magnetic material loses all magnetic properties. Therefore, the operating temperature of a magnet must be much higher than the Curie temperature of the material if all magnetic properties of the material are to be preserved.

Demagnetization curve: This is the force in the second quadrant of the hysteresis loop that describes the behavior of the magnetic characteristics in actual use. It is also known as the *B/H* curve.

Demagnetization force (F_d): The demagnetization force works in the opposite direction of the force used to magnetize in the first place. Environmental parameters such as shock, vibration, and temperature can act as demagnetization forces. In other words, environmental factors can contribute to the overall demagnetization force.

Dimensional tolerance: An allowance given as a permissible range for the overall or nominal dimensions of a manufactured magnet. The principal objective of dimensional tolerance is to specify the allowed variations due to imperfections in the manufacturing process.

Electromagnet: This particular magnet consists of a coil wound around an iron core or solenoid with an iron core. This type of magnet has a magnetic field only during the period in which the electric current flows through the solenoid.

Ferromagnetic material: Ferromagnetic material is either a source of magnetic flux or a conductor of magnetic flux. Any ferromagnetic material must have some component of iron, nickel, cobalt, or another component that can be used to store magnetic energy.

Flux density/magnetic field (*B*/*H* curve): The plot of the magnetic field (*H*) versus the resultant flux density (*B*) describes the qualities of any magnetic material.

Hysteresis loop: The hysteresis loop is essentially the plot of a magnetizing force versus the resultant magnetization of the magnetic material as it is successfully magnetized to a saturation level, demagnetized, magnetized in the opposite direction, and finally remagnetized with full magnetic properties. With continued recycling, this plot will be a closed loop that completely describes the characteristics of the magnetic material. The size and shape of the hysteresis loop is important for both soft and hard magnetic materials. Soft magnetic materials are generally used in alternating current circuits, and the area inside the loop should be as thin as possible, which indicates the measure of magnetic energy loss. Conversely, a large hysteresis loop indicates large magnetic loss. In brief, the magnetic energy loss in the magnetic material is proportional to the area of the hysteresis loop. In hard magnetic materials, flatter loops indicate stronger magnets. The first quadrant of the loop, which covers the +*X* and +*Y* region, is called the magnetization curve. This curve is of great interest because it shows how much magnetizing force must be applied to saturate a magnet. The second quadrant, which covers the +*X* and −*Y* region of the loop, is known as the demagnetization curve [3]. A typical hysteresis loop or curve of a ferromagnetic material can be seen in Figure 3.2.

Maximum energy product (*BH*$_{max}$): *BH*$_{max}$ is defined as the magnetic field strength at the point of maximum energy product of a particular magnetic material. The magnetic field strength of fully saturated magnetic material is measured in megagauss-oersteds (MGOe). The *BH*$_{max}$ product is dependent on residual flux density (*B*$_r$), coercive force (*H*$_c$), intrinsic coercive force (*H*$_{ci}$), and the magnet's physical properties. This product is a function of the above-mentioned magnetic parameters and is different for different rare earth magnetic materials.

Residual induction (*B*$_r$, max): This parameter is also known as *residual flux density*. It is the magnetic induction remaining in a saturated magnet material after the magnetizing field has been removed. This indicates the point at which the hysteresis loop crosses the magnetic flux density (*B*) at zero magnetizing force and represents the maximum flux (product of flux density and area of pole) out from the given magnetic material. By definition, this point occurs at zero air gap and therefore cannot be seen in the practical use of a magnetic material.

It is important to remember that the hysteresis loop is strictly a function of magnetizing force or magnetic field strength (*H*), induction density (*B*), or flux density and the operating temperature of the magnetic material. At temperatures well below the Curie point of the material, all the elementary magnetic moments in a sufficiently

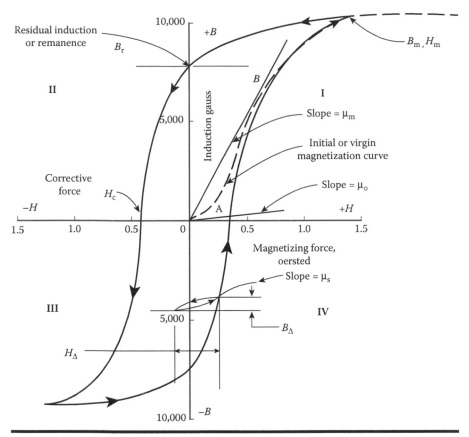

Figure 3.2   Typical hysteresis curve of a ferromagnetic material.

small-diameter particle (which is typically less than 1 μm [10^{-4} cm] for ferrites) will become spontaneously aligned and create an external magnetic field. The energy associated with this field can be decreased if a fraction of the elementary dipoles become aligned in the opposite sense. This will happen if the decrease in the field strength is more than the energy required to form a boundary separating two domains having opposite magnetic moments. Because larger particles can better accommodate domain-wall strain energy, growth will take place with increasing particle size until the external magnetic field disappears. When all the magnetic moments are parallel with the applied field, the magnetic saturation condition is satisfied.

External magnetization can be induced in a magnetic material by applying an external magnetic field that will align the individual domain moments. This induced magnetization (M) is described by the magnetization curve shown in Figure 3.2, which is derived by plotting either parameter M, the magnetic induction, or the flux density B as a function of magnetic field strength (H). The equation for the flux density or the magnetic induction can be written as $B = H + 4\Pi M$.

As mentioned, the ratio of flux density to magnetic field strength (B/H) is known as the permeability (μ) of the magnetic material. Initially, the domains that are most nearly aligned with the magnetic field grow at the expense of the misaligned domains by the movement of domain walls. As the field strength is increased, easy domain-wall movement leads to maximum permeability, after which permeability decreases. Eventually, most of the domain walls disappear, with the remainder reaching stable positions. Note that further increase in the magnetization (M) occurs by the rotation of the elementary ion movements in the direction of the applied magnetic field (H_a). When all these moments are parallel with the applied magnetic field, magnetic saturation is reached.

Intrinsic coercive force (H_{ci}): This force indicates material resistance to the demagnetization force. It is equal to the demagnetization force that reduces the intrinsic induction (B_i) or intrinsic flux density to zero in the magnetic material after magnetizing to saturation. It is measured in oersteds.

Irreversible losses: Partial demagnetization of a magnet occurs due to exposure to high or low temperatures, external fields, shock, vibration, and other environmental factors. These losses are only recoverable by remagnetization. Magnets can be stabilized against irretrievable losses by partial demagnetization induced by the temperature cycles or by external magnetic fields.

Isotropic material: A magnetic material that can be magnetized along any axis or direction is called an isotropic material (also called a magnetically nonorientated material). This is the opposite of an anisotropic magnetic material.

Magnet: A magnet is a device made from certain magnetic materials that create a magnetic field. Every magnet has one north pole and one south pole. The magnetic field lines leave the north pole end and enter the south pole end, which is a classic example of a dipole (two poles). If a magnet is broken into two pieces, each of the smaller portions will then have a north pole and a south pole. If the small piece is further broken into two, then each of those pieces will have a north pole and a south pole as well, although with reduced magnetic field strength [4].

Magnet orientation: This orientation is used to describe the direction of magnetization of the magnetic material. This is the direction in which an anisotropic magnet should be magnetized to achieve maximum magnetic properties and their optimum values.

Magnet pole: The magnet pole is the area where magnetic lines are concentrated.

Magnetic circuit: A magnetic circuit consists of all elements including air gap, nonmagnetic material, and magnetic material.

Magnetic field density or induction (B): A magnetic field has a certain magnetic field density, which is measured in gauss. The presence of field intensity is strictly dependent on the geometrical configuration of the magnet. For example, for an axially magnetized disc or cylinder, it is specified on the surface of the magnet along the central axis of magnetization, as illustrated in Figure 3.2. In magnetic blocks, it is specified on the surface of the magnet as well as, along the center axis of magnetization. In a ring magnet, there are two values of magnetic field intensity—$B_{y, center}$ and $B_{y, ring}$—as illustrated in Figure 3.2. Note the vertical component of the magnetic field on the vertical surface of the ring magnet, midway between the inner and outer diameters of the ring.

Magnetic field strength (H): The magnetizing or demagnetizing force is a measure of the vector magnetic quantity that determines the ability of the electric current, or a magnetic body, to induce a magnetic field at a given point. The magnetic strength is measured in oersteds.

Magnetic flux (φ): Magnetic flux describes the flow of a magnetic field when the magnetic induction or magnetic field density (B) is uniformly distributed and is normal to the area. The magnetic flux φ is a product of area and magnetic induction. This means flux φ is equal to $B \times A$, where A is the area normal to the flux.

Magnetic flux density: Magnetic flux density indicates the lines of flux per unit area, as measured in gauss per the centimeter-gram system. Note one line of flux per square centimeter is recognized as one Maxwell.

Magnetic induction (B): The magnetic field induced by a magnetic field strength H at a given point is known as magnetic induction. It is a vector sum, at each point within the substance, of the magnetic field strength and the resultant intrinsic induction. The magnetic induction indicates the flux per unit area normal to the direction of the magnetic path.

Magnetic line of force: The magnetic line of force defines an imaginary line in a magnetic field on which every point has the direction of the magnetic flux at that point.

Magnetic pole: The magnetic pole is the magnetic area where the lines of flux are concentrated, indicating the locations of south and north poles of the magnet.

Magnetization curve: This is the B/H curve, which appears in the first quadrant of the hysteresis loop for a given magnetic material.

Magnetomotive force: This force is defined as the magnetic potential difference between any two points and is analogous to voltage difference in electrical circuits. The magnetomotive force tends to produce a magnetic field that is equivalent to a magnetic force produced by a current flowing through a coil of wire. The magnetomotive force is measured in Gilberts in the centimeter-gram system and in ampere turns in the international system.

Maximum energy product (BH_{max}): This is the product of magnetic inductance (B) and maximum magnetic field strength (H_{max}). This product represents the magnetic field strength at the point of the maximum energy product of a magnetic material.

Maximum operating temperature (T_{max}): This temperature is also known as the maximum service temperature. At this temperature, the magnet will continue to operate but with adverse effects on long-range stability, structural changes, or magnetic characteristics.

Paramagnetic materials: Paramagnetic materials, such as wood, plastic, aluminum, and other metals, are not attracted to magnetic fields. In most cases, the permeability has two values: initial permeability and maximum permeability. A paramagnetic material will have a permeability slightly greater than 1, whereas magnetic materials will have permeability values ranging from tens to hundreds to thousands, as shown in Table 3.1.

Permanent magnet: A permanent magnet retains its magnetism after it is removed from a magnetic field. It is always on and retains its magnetic properties forever unless its magnetic properties are intentionally destroyed or neutralized. Neodymium-iron-boron and samarium-cobalt are permanent magnets using rare earth materials.

Permeability (μ): Permeability is the ratio of the magnetic induction of a magnetic material to the magnetizing force (B/H). The magnetic permeability of a vacuum medium is denoted by μ_o and has a value of 12.27×10^{-7} N/A^2, where N is the number of turns in the coil and A is the current flowing in the coil. The permeability of various materials along with other critical parameters is summarized in Table 3.1.

Alloy 1040, containing 72% nickel and 14% copper, is best suited for coil applications where high-Q and optimum permeability are of critical importance. Pure

Table 3.1 Widely Used Materials in Electronic Components and Permanent Magnets and Their Critical Magnetic Properties

Metal, Alloy, or Ferrite	Best-Suited Application	Initial Permeability	$B_{residual}$ Maximum	Kilogauss	Curie Temperature (°C)
Silicon iron	Transformer	400	7,000	12	600
Cobalt iron	Saturation	600/800	5,000/10,000	24/24	970/980
Nickel iron	Saturation	400/9,000	2,000/90,000	11/16	300/600
Alloy 1040	72/14 Ni/Cu	49,000	100,000	2.5	290
Ferrites	High-Q coils	1,800	2,640	4.6	170

Source: Modified from Brown, D. 2000. Developments in the processing and properties of NdFeb-type permanent magnets. *Journal of Magnetism and Magnetic Materials* 248:432–440.

cobalt is difficult to find in the earth's surface because it is found in minerals consisting of iron, copper, and other metals. Complex processing is required to find cobalt. Cobalt-60, an artificial isotope, is generally available in the market; its cost varies from $0.50 to $7 per Curie. Cobalt-60 is highly toxic and widely used for medical x-rays. Costs for both nickel and cobalt materials, which are used in the manufacturing of neodymium-iron-boron and samarium-cobalt permanent magnets, are very high.

Permeance (P): Permeance is a measure of the relative ease with which flux passes through a given space or material. It is calculated by dividing magnetic flux by magnetomotive force ($P = \Phi/mmf$). Permeance is the reciprocal of reluctance.

Permeance coefficient (P_c): This coefficient is also known as the operating slope of the B/H curve of a magnet. Essentially, this is the line of the demagnetization curve where a given magnet operates. Its value is dependent on the geometrical shape of the magnet and its surrounding environments. It is a number that defines how hard it is for field lines to go from the north pole to the south pole of a magnet. According to magnet designers, a tall cylindrical magnet will have a high P_c value compared to short and thin magnets.

Pull force (F_{pull}): The pull force is the force required to pull a magnet free from a flat steel plate using force perpendicular to the surface. It is also defined as the limit of the holding power of a magnet. The pull force is a function of magnet pole strength.

Reluctance (R): Reluctance is defined as a measure of the relative resistance of a magnetic material to the passage of flux within the medium. This can be calculated by dividing the magnetomotive force by the magnetic flux ($R = mmf/\Phi$).

Remanence: Remanence is defined as the magnetic induction that remains in a magnetic circuit after the withdrawal of an applied magnetizing force.

Residual flux density ($B_{r,\,max}$): Residual flux density is the magnetic induction remaining in a saturated magnetic material after the removal of the magnetizing field. This represents the point at which the hysteresis loop crosses the B-axis at zero magnetizing force. It also represents the maximum flux output from a given magnetic material.

Residual induction: Residual induction is the residual flux density remaining in a saturated magnetic material after the magnetizing force has been withdrawn. It represents the maximum flux output from the magnet that occurs at zero air gap.

Saturation: Saturation is the state where an increase in the magnetizing force produces no further increase in magnetic induction or flux density (B) in the magnetic material.

Stabilization: Stabilization is defined as the process of exposing the magnet to elevated temperatures or external magnetic fields to demagnetize the magnet to a predetermined level.

Temperature coefficient: The temperature coefficient is a factor used to calculate the decrease in magnetic flux corresponding to an increase in operating temperature. The loss in magnetic flux is recovered when the operating temperature is reduced.

3.3 Measurement Systems and the Units Used

Dimensions of a magnet, magnetic flux, flux density, magnetizing force, and magnetomotive force can be expressed in three distinct measurement systems:

- The centimeter-gram system
- The English system (e.g., inches)
- International System of Units (SI)

Specific details of the measurement systems and commonly used units are summarized in Table 3.2. Conversions between the centimeter-gram system and SI system units are shown in Table 3.3.

Table 3.2 Measurement Systems and Their Units

Units	Centimeter-Gram System	English System	International System
Physical dimensions	Centimeter	Inch	Meter
Magnetic flux (φ)	Maxwell	Maxwell	Weber
Flux density (B)	Gauss	Lines per square inch	Tesla
Magnetizing force	Oersted	Ampere-turn per inch	Ampere-turn per meter
Magnetomotive force	Gilbert	Ampere-turn	Ampere-turn

Table 3.3 Conversion between the System Units

Centimeter-Gram Unit	International Unit
1 oersted	79.62 ampere-turn/meter
10,000 gauss	1 tesla
1 Gilbert	0.7958 ampere-turn
1 Maxwell	1 line = 10^{-8} weber
1 gauss	0.155 lines/inch^2

3.4 Rare Earth–Based Permanent Magnets and Their Applications

This section discusses the rare earth materials that are best suited to the design and development of permanent magnets for potential commercial and industrial applications. Neodymium plays a key role in the development of neodymium-iron-boron (Nd-Fe-B) magnets, whereas samarium-cobalt plays a critical role in the development of samarium-cobalt magnets, which are well suited for defense applications [4]. To explore the suitability of these magnets for operation under harsh environments, the thermal and mechanical properties of various metals and high-temperature ceramics or ferrites involved in the production of these magnets are discussed.

Rare earth hard permanent magnets are made from rare earth elements (Figure 3.1) and three-dimensional transition metals, such as iron and cobalt. The transition metals iron and cobalt offer the highest magnetic moments, exceeding 245 emu/g under ambient temperatures ranging from 280 to 300 K. Hysteresis loop characteristics for a samarium-cobalt ($SmCo_5$) magnet at cryogenic and room temperatures are shown in Figure 3.3. Table 3.4 summarizes the important thermal and mechanical characteristics of potential rare earth materials, metals, and high-temperature ceramics.

Nd-Fe-B and samarium-cobalt ($SmCo_5$ and $SmCo_7$) permanent magnets are widely used in electric motors and generators, magnetic resonance imaging (MRI) scanners, electric generators for wind turbines, and focusing magnets in TWTAs, which are best suited for airborne and shipborne electronic warfare equipment [3]. Original design and development efforts, tests, and evaluations were carried out by General Motors and the Sumitomo Special Metals Division in the 1980s. General Motors and Sumitomo invested a significant amount of time and money in the development of Nd-Fe-B rare earth permanent magnets, which are made from an alloy consisting of neodymium, iron, and boron with a tetragonal crystalline structure.

Neodymium-based magnets are the strongest type of permanent magnets. These permanent magnets are widely used in cordless tools, hard disk drives, and magnetic fasteners. The tetragonal crystal structural of an Nd-Fe-B magnet offers exceptionally high uniaxial magnetocrystalline anisotropy—close to 7 tesla or 70,000 gauss. The potential applications, advantages, and disadvantages of the two widely used rare earth–based permanent magnets will be discussed here. The following sections present an overview of the performance capabilities and limitations of Nd-Fe-B and $SmCo_5$ permanent magnets, with an emphasis on their performance capabilities and limitations.

3.4.1 Neodymium-Iron-Boron Permanent Magnet

As mentioned, the Nd-Fe-B permanent magnet jointly developed by General Motors and Sumitomo in 1982 is widely used among the rare earth magnets. This

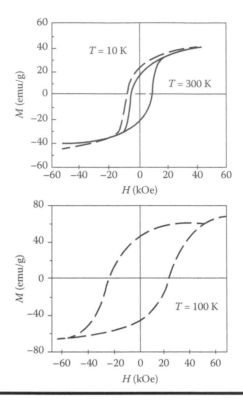

Figure 3.3 **Hysteresis loops for a noncrystalline samarium cobalt (SmCo$_5$) permanent magnet.** M = remnant moment; T = cryogenic temperature (K); H = intrinsic coercivity.

Table 3.4 Important Thermal and Mechanical Properties of Rare Earth Materials, Metals, and High-Temperature Ceramics

Property	Samarium	Neodymium	Iron (Fe)	Boron (B)	Cobalt (Co)
Density (g/cm³)	7.7	7.05	7.87	2.3	8.9
Melting point (°C)	1,072	1,024	1,535	2,077	1,495
Tensile strength (psi)	5,105	10,940	29,315	43,000	32,000
Young's modulus (psi)	5×10^6	4.4×10^6	20×10^6	11×10^6	16×10^6
Curie temperature (°C)	320	800	770	1,250	1,131

Note: All values are estimated and may have 10 to 15% errors.

magnet is the strongest type of magnet and is best suited for applications where strong magnetic properties are the principal requirements [5].

Due to the tetragonal crystal structure of the Nd-Fe-B alloy, the magnet offers high uniaxial magnetocrystalline anisotropy close to 7 teslas or 70,000 gauss [6]. This rare earth magnetic compound possesses high coercivity, which means that this magnet offers high resistance to demagnetization. The compound has a high saturation magnetization exceeding 16,000 gauss. Because the maximum energy density is proportional to saturation magnetization, this magnet offers great storage capability of magnetic energy (BH_{max}) exceeding 64 MGOe, which is significantly more than the samarium-cobalt ($SmCo_5$ or $SmCo_7$) magnets. Because of its high magnetic storage capability, this magnet has potential commercial and industrial applications.

In practice, the magnetic properties of neodymium-based magnets depend strictly on the alloy composition, microstructure manufacturing techniques, and tolerances. When magnets are advertised with strengths rated for 12,000 to 15,000 gauss, the magnets will not actually yield the advertised strength values. For example, a neodymium-iron-boron magnet (N50) that has a residual flux density (B_r) of roughly 14,500 gauss will actually measure from 1,000 to 5,600 gauss for the center of a 2-inch square pyramid that is 1-inch thick and tapers to 1 inch square on the top. Users must ask for magnet ratings from a competent authority at the supply source.

The advantages and disadvantages of this permanent magnet are highlighted as follows:

■ Sintered $Nd-Fe_{14}-B$ tends to be vulnerable to corrosion.
■ It needs a protective coating if used in the manufacturing of commercial products.
■ The protective coating consists of nickel plating or dual-layered copper-nickel plating, which can be provided using a standard method.

3.4.1.1 Nd-Fe-B Magnet Performance Capabilities

The Nd-Fe-B magnet is made from an alloy of neodymium, iron, and boron and is widely used in various commercial and industrial applications. This alloy forms a tetragonal crystalline structure with unique magnetic properties. It is considered to be the strongest permanent magnet and was jointly developed by General Motors and Sumitomo in 1982 [3]. General Motors focused on the development of melt-spun nanocrystalline $Nd_2-Fe_{14}-B$ magnet design, while Sumitomo was perfecting a full-density sintered $Nd_2-Fe_{14}-B$ magnet design configuration. The neodymium manufacturing methods include the classic powder metallurgy or sintered magnet process and the rapid solidification or bonded magnet process, which makes the manufacturing process more cost-effective. The $Nd_2-Fe_{14}-B$ tetragonal crystalline structure results in the strongest type of permanent magnet, which is best suited for motors in cordless tools, hard disk drives, and magnetic fasteners for which strength, safety, and reliability are the principal requirements.

Outstanding characteristics of this magnet include the following:

- The magnet has demonstrated exceptionally high uniaxial magnetocrystalline anisotropy, close to 7 teslas or 70,000 gauss.
- The alloy compound possesses high coercivity, which means high resistance to demagnetization.
- The permanent magnet possesses high-saturation magnetization exceeding 16,000 gauss.
- The unique properties of this magnet offer a maximum energy density (BH_{max}) close to 64 MGOe, which is significantly greater than the samarium-cobalt ($SmCO_5$ or SmCo7) permanent magnets.

The magnetic properties of any rare earth–based permanent magnet are strictly dependent on the alloy composition involving a rare earth element and other appropriate metals, microstructure, manufacturing processes, and quality control techniques.

3.4.1.2 Potential Applications of Nd-Fe-B Magnets

In applications where cost is the most critical design requirement, conventional alloy magnets must be given serious consideration. In the absence of rare earth–based permanent magnets, alnico (an alloy of aluminum, nickel, cobalt and ferrite; hard ceramic with ferromagnetic properties) was widely used in producing permanent magnets. However, these magnets were not able to meet the stringent magnetic properties needed by various electrical devices. For example, neodymium-based permanent magnets are capable of operating at temperatures as high as 80°C (at approximately 75°C, one can expect some degradation in the magnet's performance). Because of greater strength of these Nd-Fe-B magnets, manufacturers were able to produce smaller and lighter magnets with moderate to high strength for the following applications:

- MRI scanners for diagnostic applications.
- Magnetic bearing and couplings.
- Magnetic head actuators for computer hard disks.
- Headphones and loudspeakers.
- Electric motors and generators [6].
- Servo motors.
- Cordless tools for various mechanical applications.
- Compressor motors for various commercial and industrial applications.
- Lifting magnets for removing or lifting heavyweight components.
- Magnetic toys for children.

- Modern sporting events.
- Electric power steering.
- Synchronous electric motors.
- Drive motors for electric and hybrid electric vehicles. (The Toyota Prius requires 2.2 pounds of neodymium, which typically costs $130–$150 per pound depending on the purity of the magnetic material.)
- Electric generators for wind turbines. (Approximately 600 kg [1,323 pounds] of neodymium alloy compound is required per megawatt of electrical energy generated by the wind turbine. Neodymium content is estimated to be approximately 31% of the permanent magnet's weight [4].)

The strength and magnetic field homogeneity of the neodymium magnets have opened new applications in the medical fields with the introduction of open MRI scanners, which are used as an alternative to superconducting MRI magnets (which use a superconducting coil to produce the magnetic field) to image the body in hospital radiology departments. The superconducting magnet MRI equipment makes a loud noise that is disliked by many patients. The magnetic field gradient of the neodymium-based magnets decreases towards the center of their surfaces; thus, there is a magnetic force that attracts the metallic objects to the edges of the magnets.

Nanocrystalline $Nd_2Fe_{14}B$ permanent magnets tend to be vulnerable to corrosion along the grain boundaries, which can cause deterioration of the sintered magnet. However, by providing a protective coating consisting of nickel plating or two-layered copper-nickel plating, corrosion can be minimized or eliminated. Protective coating can be achieved using other metals or polymer and lacquer coatings, although the effectiveness of the latter coating is yet unknown.

Two distinct manufacturing technologies are used:

- The classic powder metallurgy or sintered magnet technology
- The rapid solidification or bonded magnet technology

Sintered neodymium-magnet technology is widely used and is cost-effective. This process requires raw materials to be melted in a furnace, cast into a mold, and cooled to form ingots. The ingots are pulverized and milled into small particles, which undergo a process of liquid-phase sintering whereby the powder is magnetically aligned into dense blocks. These blocks are then heat treated, cut to the desired shape, surface treated, and magnetized to the desired value. Currently, approximately 50,000 to 60,000 tons of sintered neodymium magnets are produced in Japan and mainland China annually.

3.4.2 Samarium-Cobalt Permanent Magnets

The important aspects of samarium-cobalt permanent magnets are summarized in this section. The first commercialized samarium-cobalt permanent magnets were introduced in the early 1970s. These magnets are best suited for applications requiring high magnetic properties. These magnets are used today in applications that involve operating temperatures as high as 300°C [2]. Wide-scale deployment of these magnets in TWTAs in military electronic warfare systems operating in environments close to 300°C is convincing proof of the reliability and safety of samarium-cobalt magnets at elevated operating temperatures.

Samarium-cobalt permanent magnets have magnetic strengths exceeding 32 MGOe. These magnets are very brittle because of poor structural properties; therefore, they can be easily chipped or broken upon slight impact. This shortcoming is due to the extremely low impact resistance and compressive strength of the material used by the magnet. However, unlike neodymium magnets, these magnets have no oxidation problems. The saturation magnetizing field requirement for this material is approximately 50,000 Oe.

This magnet is available in various shapes, including from bars, discs, cylinders, rings, arc segments, and so on. The centric axis of magnetism for some of these magnets is illustrated in Figure 3.4. Machining of these magnets is extremely difficult. These magnets are very powerful and require extreme care in handling to avoid injuries or accidents. This particular magnet can be produced with nanocrystalline grain structures, which have demonstrated great potential for MRI contrast and magnetic fluid hyperthermia.

This material contains cobalt and, the cost of samarium-cobalt is very high, ranging from $22 to $26 per pound for commercial-grade metal. The cost of cobalt is determined by the purity of the metal. The procurement cost of Sm-Co magnets is roughly 20 to 50 times that of ceramic ferrite magnets with the same magnetic performance parameters and three times the cost of neodymium magnets of the same physical size. However, the neodymium magnet is roughly 40% stronger than the SmCo magnet of the same physical size.

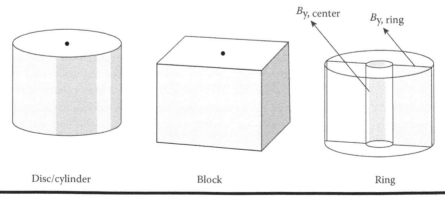

Disc/cylinder   Block   Ring

Figure 3.4 Centric axis of magnetization for an axially magnetized disc/cylinder, block, and ring.

Future high-performance permanent magnets will exploit the properties of the well-known magnetic nanoparticles of ferrites, metals, alloys, and rare earth materials. These magnetic nanoparticles can be superparamagnetic (H_c) or strongly ferromagnetic ($H_c > 1$ tesla or 10,000 gauss) with magnetic moments ranging from those at the ferrite level of about –80 emu/g to the highest alloy of exceeding 200 emu/g. Magnetic nanoparticles have been investigated extensively for their potential applications. These magnetic nanoparticles are classified into two categories—hard magnetic nanoparticles with large coercivity and high magnetic moments and soft magnetic nanoparticles with low coercivity and high magnetic moments; they are best suited for ultrahigh storage applications and electromagnetic devices, respectively.

3.4.3 Conventional Low-Price Magnetic Materials with Acceptable Performance Levels

In certain applications requiring low cost and moderate magnetic properties, conventional magnetic materials could be used, which would save money and manufacturing complexity. These conventional materials include alnico (sintered) and strontium-ferrite (sintered). Magnet performance characteristics of conventional magnetic materials and rare earth materials are summarized in Table 3.5.

Table 3.5 Comparison of Performance Characteristics between Conventional Magnetic Materials and Rare Earth Magnetic Materials

Magnet Type	Magnetic Performance Characteristics			
	Remanence (T)	Coercivity (Oe)	BH_{max} (kJ/m³)	T_c (°C)
Nd-Fe-B (sintered)	1.00–1.40	750–2,000	200–440	310–400
Nd-Fe-B (bonded)	0.60–0.70	600–1,200	60–100	310–400
SmCo$_5$ (sintered)	0.80–1.10	600–2,000	100–200	720
SmCo$_7$ (sintered)	0.90–1.15	459–1,300	150–240	800
Alnico (sintered)	0.60–1.40	275	10–88	700–860
Strontium-ferrite (sintered)	0.20–0.40	100–300	10–40	450

Note: All values are estimated; BH_{max} = maximum energy product; T_C = Curie temperature.

Table 3.6 Comparison between the Physical Properties of Rare Earth Magnets and Commercial Alnico Magnets

Physical Properties	Magnetic Materials		Commercial Materials	
	Nd-Fe-B	Sm_2Co_5	Ferrite (Ceramic)	Alnico
Density (lb./in³)	0.275	0.304	0.180	0.265
Bending stress (psi)	42,000	17,000	20,000	21,000
Compressive strength (psi)	130,000	130,000	130,000	140,000
Resistivity (μΩ/m)	1.4	0.8	100	4.7
Coefficient of expansion	1.7×10^{-6}	9.2×10^{-6}	10×10^{-6}	11.4×10^{-6}
Curie temperature (°C)	655	825	450	825

Neodymium permanent magnets have higher remanence (M_r), much higher coercivity (H_{ci}), and large energy product (BH_{max}). However, they have lower Curie temperatures than other rare earth magnetic materials. Neodymium is alloyed with other rare earth materials (namely, terbium and dysprosium) to preserve its magnetic properties at high temperatures. A comparison between the physical properties of rare earth magnets and commercial magnets is summarized in Table 3.6.

3.4.4 General Comments on Various Magnets

Cobalt alloy is a mixture of cobalt, iron, and other metals. The magnetism and strength of material properties vary as a function of cobalt and iron content in the alloy. For example, alnico 2 (sintered) alloy consists of 10% aluminum, 17% nickel, 12.5% cobalt, 6% copper, and 54.5% iron. This alloy compound has a Curie temperature ranging from 700 to 860°C, which is very impressive, but its other magnetic properties are not close to those of rare earth materials. Iron has a Curie temperature of 770°C, cobalt has a Curie temperature of 1,107°C, and nickel has a Curie temperature of 354°C. Despite high Curie temperatures, these magnetic materials are unable to produce magnetic field strength and flux density comparable to rare earth–based magnets.

For applications where cost is the critical factor and magnetic requirements are moderate, one should look at other conventional magnetic materials, such as alnico cobalt-nickel based alloy. This magnetic alloy compound contains 10% aluminum, 17% nickel, 12.5% cobalt, 6% copper, and the balance of iron. Because the compound contains only 17% cobalt, the magnet's cost will be substantially less than a rare earth–based $SmCo_5$ compound. $SmCo_5$ offers the following magnetic characteristics:

- Curie temperature: 700 to 860 K (427–587°C)
- Maximum magnetic field strength (H_{max}): 1,000 to 2,000 Oe
- Maximum flux density (B_{max}): 12,000 gauss
- Coercivity: 500 to 520 Oe
- Maximum energy product (BH_{max}): 1,430,000 GOe (minimum)

Close examination of the magnetic characteristics of an alnico magnet will reveal that this particular magnetic material is cost-effective, but it may not meet the high operating temperature capabilities of $SmCo_5$ (300°C). However, this magnet will eliminate the brittle characteristic of the samarium-cobalt. In addition, the estimated cost of an alnico magnet will be less than one-fifth of the $SmCo_5$ magnet.

As mentioned, samarium-cobalt permanent magnets have demonstrated satisfactory magnetic performance at operating temperatures as high as 300°C, compared to 80°C for neodymium-iron-boron permanent magnets. Samarium-cobalt magnet suppliers provide an "SH" symbol on the magnet, which means the magnet will retain the specified magnetic properties at the temperature associated with the SH symbol. As the operating temperature increases, the strength of the magnet will decrease. In general, samarium-cobalt magnets are approximately 40 to 60% weaker than neodymium-iron-boron magnets, but they cost approximately 4 to 5 times as much as Nd-Fe-B magnets due to the high cost of cobalt (approximately $23–$26 per pound, depending on the purity of the metal). Sm-Co magnetic material is very brittle because the tensile strength and impact resistance of the alloy compound are not high. The magnetic properties of sintered Nd-Fe-B and Sm-Co permanent magnets are shown in Table 3.7.

The flexural strength and tensile strength of a samarium-cobalt compound are relatively poor compared to an Nd-Fe-B compound, which is why Sm-Co magnets are brittle. Furthermore, the high stiffness (666,000 psi) of boron and high tensile strength of iron (65,000 psi for ductile iron and 40,000 psi for cast iron) mean that the neodymium-iron-boron magnet is free from chipping and brittleness.

3.5 Magnetization and Anisotropic Properties at Cryogenic Temperatures

This section investigates the impact of cryogenic temperatures on the magnetization and anisotropic properties of some metallic rare earth compounds [2]. Hysteresis loops for a samarium-cobalt magnet at various cryogenic temperatures can be seen in Figure 3.3. Preliminary scientific studies undertaken by research scientists at the University of Dayton (Ohio) seem to indicate that magnetization and anisotropic properties can be achieved from crystallographic and magnetic

Table 3.7 Physical Properties of Nd-Fe-B and Sm-Co Permanent Magnets

Properties	Magnet Type	
	Nd-Fe-B	*Sm-Co*
Remanence (tesla)	1.0–1.3	0.82–1.16
Relative permeability	1.05	1.05
Coercivity (MA/m)	0.875–1.992	0.493–1.593
Temperature coefficient		
Of remanence (%1/K)	−0.12	−0.03
Of coercivity (%1/K)	−0.55 to −0.65	−0.15 to −0.30
Curie temperature (°C)	320	720 (SmCo$_5$)
		800 (SmCo$_7$)
Density (g/cm³)	7.3–7.5	8.2–8.4
Coefficient of temperature expansion (1/K)		
Magnetizing direction	5.2×10^{-6}	5.2×10^{-6}
Normal to magnetizing direction	-0.8×10^{-6}	11×10^{-6}
Flexural strength (psi)	36,465	21,879
Compressive strength (psi)	160,446	116,688
Tensile strength (psi)	10,939	5,105
Electrical resistivity (μΩ-cm)	110–170	86
Vickers hardness number	550–650	500–550

Source: Pyrhonem, J., T. Jokinen, and V. Hrabovcova. 2009. *Design of Rotating Electrical Machines.* Hoboken, NJ: Wiley, 232.

data collected from the rare earth metallic compound R_2Co_2, where R stands for a rare earth material (cerium, praseodymium, neodymium, or samarium). In addition, a single crystal must be obtained for praseodymium and neodymium rare earth compounds. Magnetic Curie points, temperature dependence of the magnetization in the temperature range of 77 to 293 K (20°C), and magnetic crystallography are of critical importance in designing the neodymium-based permanent magnets. Rare earth magnetic compounds are most ideal for magnets operating under 30°C crystal temperature and are best suited for space and satellite applications.

A comprehensive research investigation of the anisotropy of neodymium-cobalt revealed the easy-axis magnetic symmetry at room temperature over the easy-cone region between 210 K and 265 K [2]. It is important to mention that the components of R_2Co_7 exist with nearly all rare earth R_2Co_7 materials. (Note that R_2 indicates a rare earth element with a valence of 2, and Co_7 indicates a cobalt metal with a valence of 7.) Furthermore, the rare earth compound material has a hexagonal structure. The compound has known ferromagnetic properties and Curie temperatures as a function of uniaxial anisotropy constants and perpendicular and parallel directions, as shown in Tables 3.8 and 3.9 [2].

From the data presented in Table 3.7, it appears that the anisotropic constant k_1 changes from a negative value to positive value as the Curie temperature approaches room temperature (273 K) from a lower cryogenic temperature of 77 K. Materials scientists believe that compounds of the type R_2Co_7 exist with nearly all rare earth materials. In addition, when R represents lanthanum through samarium, the hexagonal structure modification dominates. The 2–7 compounds, such as Nd_2Co_7

Table 3.8 Uniaxial Anisotropy Constants (k_1 and k_2) of an Nd_2Co_7 Crystal as a Function of Curie Temperature at an Applied Magnetic Field of 15,000 Oe

Curie Temperature (K)	k_1 (erg/cm³)	k_2 (erg/cm³)
100	-92×10^{-6}	35×10^{-6}
200	-18×10^{-6}	8×10^{-6}
300	4×10^{-6}	4×10^{-6}

Table 3.9 Lattice Parameters and Density Values for the Multiphase Samples Investigated

Compound Phase	X-Ray Density (g/cm³)	Lattice Parameters	
		a^0 (A⁰)	c^0 (A⁰)
Ce_2Co_7	8.860	4.949	24.495
Pr_2Co_7	8.445	5.072	24.514
Nd_2Co_7	8.578	5.063	24.456
Sm_2Co_7	8.892	5.041	24.273

Source: Blaettner, H.F. et al. 1997. *Magnetization and Anisotropy of Some R_2Co_7 Compounds at Low Temperatures.* Dayton, OH: University of Dayton School of Engineering.

and Sm_2Co_7, have been known for a long time as ferromagnetic, and their Curie temperatures have been reported in several technical papers.

Scientific investigations indicate that the compound Nd_2Co_7 has easy-plane anisotropy. Domain patterns indicate that the 2–7 compounds of lanthanum, praseodymium, neodymium, samarium, and yttrium have easy-axis magnetic symmetry. Very limited quantitative information has been published on the saturation and crystal anisotropy of the 2–7 phases for light rare earth materials.

Magnetic measurements can be determined from the temperature dependence of the low-field dynamic permeability of coarse powers. Lattice parameters and x-ray density values obtained for the 2–7 compounds are shown in Table 3.9.

An oscillating specimen magnetometer can be used to measure magnetic moment versus applied field and temperature. Magnetization curves must be measured with the field applied parallel and perpendicular to the alignment axis. The temperature dependence of the magnetization is obtained by measuring the moments of the samples at an applied field of 15,000 Oe during cooling using liquid nitrogen (N_2). Fixed temperatures can be established by the slush baths of liquid nitrogen with organic solvents. Temperatures can be varied from 293 to 77 K by controlling the flow of liquid nitrogen.

Single-crystal magnetization curves should be measured with the field parallel and perpendicular to the alignment axis. From the temperature-dependence measurements of easy-axis and hard-axis magnetization at an applied field of 15,000 Oe, one can obtain the saturation magnetization values for the Nd_2Co_7 or Sm_2Co_7 compound samples and the associated anisotropic constants [6]. Saturation magnetization values for a neodymium-cobalt sample for a given applied field of 15,000 Oe as a function of cryogenic temperature are summarized in Table 3.10.

Table 3.10 Saturation Magnetization (M_s) for a Neodymium-Cobalt Compound at a Magnetic Field of 15,000 Oe as a Function of Temperature and Polarization

Temperature (K)	Perpendicular (emu/cm³)	Parallel (emu/cm³)
100 (−173°C)	980	125
150 (−123°C)	955	350
200 (−73°C)	928	562
250 (−23°C)	907	822
300 (27°C)	715	806

Source: Blaettner, H.F. et al. 1997. *Magnetization and Anisotropy of Some R₂Co₇ Compounds at Low Temperatures.* Dayton, OH: University of Dayton, School of Engineering.

3.5.1 Anisotropic Properties

The anisotropic field intensity (H_a) in a rare earth compound is a function of the anisotropic constants k_1 and k_2 and saturation magnetization (M_s). The equation for the anisotropic field intensity can be written as $H_a = [\pm 2(k_1 + 2k_2)/M_s]$.

For the rare earth compound neodymium-cobalt $(NdCo_7)$ single crystal, the temperature-dependent uniaxial anisotropy constants (k_1) and (k_2) in the easy-cone region can be seen in Figure 3.5. The sine of the cone angle φ can be defined as $\sin\varphi = [(-k_1^2)/(2k_2)]^{0.5}$.

It is evident from the temperature versus anisotropic curve that the cone angle changes from 0 degrees at the easy axis at 265 K to 90 degrees at the easy plane at 200 K, with a nearly isotropic situation when the cone angle (φ) is equal to 45 degrees at a cryogenic temperature of approximately 250 K. For specific details on the cone angle versus the cryogenic temperature, see Figure 3.5. The behavior of the Nd_2Co_7 crystal is similar to that of the $NdCo_7$ crystal. From the temperature-dependent uniaxial anisotropy constants versus the cone angle in the easy-cone region, one can state that an easy cone region exists between 200 and 250 K, whereas the easy C-axis region exists approximately between 250 and 293 K. Similar scientific research data on Sm_2Co_7 crystals indicate that the easy C-axis is close to room temperature (293 K) and the easy-cone region appears to be 100 K. The

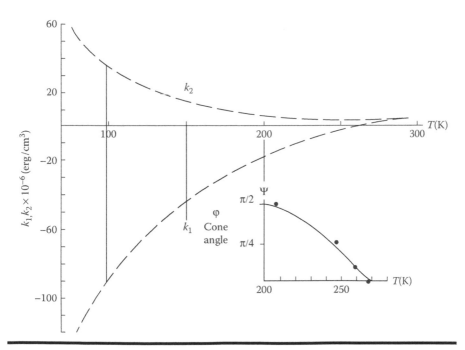

Figure 3.5 Temperature dependence of the uniaxial anisotropy constants and cone angle in the easy-cone region for a neodymium-cobalt crystal.

scientific results indicate that samarium-cobalt permanent magnets can function at lower cryogenic temperatures with compromise in its magnetics properties.

3.5.2 Estimation of Saturation Magnetization and Anisotropy Parameters of Various Rare Earth Crystals

In this section, the k_1 parameter is estimated as a function of cryogenic temperature for $PrCo_7$ and Sm_2Co_7 rare earth crystals. The typical values of saturation magnetization for the Nd_2Co_7 crystal perpendicular and parallel to the crystal axis as a function of applied magnetic field and cryogenic temperature are summarized in Table 3.11. It is evident from the data presented in the table that the values of k_1 are fairly constant for cryogenic temperatures ranging from 100 to 293 K. The parameter values fluctuate from 3.2×10^{-7} to 6.5×10^{-7} under the magnetic field of 15,000 Oe.

It is important to find the impact of cryogenic temperature and applied field strength on the saturation magnetization along the perpendicular and parallel directions of the Nd_2Co_7 crystal. The author has computed the estimated values of the saturation magnetization as a function of temperature and applied field strength (H_a) along the two axes of the neodymium-cobalt crystal, as shown in Table 3.12. Saturation magnetization curves as a function of cryogenic temperature along the perpendicular and parallel directions are displayed in Figure 3.6 for an applied field strength of 15,000 Oe. The easy basal plane, easy cone region, and easy C-axis limit are clearly marked in Figure 3.6.

3.5.3 Low-Temperature Hysteresis Loop from Samarium-Cobalt Magnets

The shapes and areas of hysteresis loops from samarium-cobalt permanent magnets vary as a function of temperature in addition to other magnetic parameters due to the

Table 3.11   Values of Anisotropic Parameter k_1 as a Function of Cryogenic Temperature for Two Widely Used Rare Earth Crystals

Temperature (K)	k_1 (emu/cm³)	
	Pr_2Co_7	Sm_2Co_7
100	10.0×10^{-7}	3.2×10^{-7}
150	10.1×10^{-7}	3.6×10^{-7}
200	10.2×10^{-7}	5.1×10^{-7}
250	10.3×10^{-7}	5.4×10^{-7}
293	10.4×10^{-7}	6.5×10^{-7}

Table 3.12   Computed Values of Saturation Magnetization (M_s) as a Function of Cryogenic Temperature and Applied Field Strength (H_a)

Applied Field Strength (Oe)	Saturation Magnetization (emu/cm³)			
	Parallel to C-axis		Perpendicular to C-axis	
	77 K	300 K	77 K	300 K
5,000	55	692	955	230
10,000	96	714	972	385
15,000	110	722	1,026	513

Note: These saturated magnetization values are strictly dependent on the neodymium-cobalt crystal surface conditions and the quality of the crystal. The author feels that the calculated values are accurate within ±10%; however, emphasis was placed on the trends in magnetization variation, not computational accuracy.

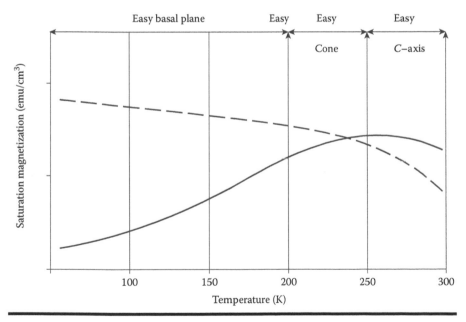

Figure 3.6   Saturation magnetization curves as a function of temperature for parallel and perpendicular directions of an Nd_2CO_7 single crystal.

hexagonal structure of $SmCo_5$ with slight Sm_2Co_7 content. This content has an impact on the intrinsic coercivity and magnetization of the particles, which tend to change from 6,100 Oe and 40 emu/g, respectively, at room temperature (300 K), to 8,500 Oe and 44 emu/g at a cryogenic temperature of 10 K; this is illustrated in Figure 3.5.

Nanocrystalline $SmCo_5$ hard magnets can be prepared by reacting core-/shell-structured Co/Sm_2O_3 with a metallic calcium at a temperature of 900°C in order to reduce Sm_2O_3 and to promote interfacial diffusion between samarium and cobalt. Furthermore, potassium chloride (KCl) can be used as a dispersion medium to accelerate the reduction at lower cryogenic temperatures and to prevent the sintering of the $SmCo_5$ compound into large single-crystals. Rare earth crystal scientists confirmed the hexagonal $SmCo_5$ structure in the annealed product. The coercivity reached close to 24,000 Oe at a cryogenic temperature of 100 K and 8,000 Oe at room temperature (300 K), with remnant moments of approximately 40 to 55 emu/g, which can be verified from the curves shown in Figure 3.3. The phase reaction and high-temperature reduction process can be extended to the synthesis of the exchanged-coupled samarium-cobalt iron ($SmCo_5/Fe_x$) nanocrystalline composites with a maximum coercivity of approximately 11,600 Oe and remnant moments reaching close to 90 emu/g at a room temperature (300 K); this is illustrated by the large hysteresis loop shown in Figure 3.3, which was produced at a cryogenic temperature of 100 K.

Comprehensive chemical synthesis efforts are responsible for the production of various monodisperse magnetic nanoparticles with controlled magnetic properties and chemical stability. The chemical synthesis efforts played a key role in the development of ferrites, metals, alloys, and rare earth magnets. As mentioned earlier, these magnetic nanoparticles could be superparamagnetic with a coercive force (H_c) approaching zero or strongly paramagnetic with H_c greater than 1 tesla (10,000 gauss); they may have magnetic moments ranging from close to 80 emu/g at the ferrite level to exceeding 200 emu/g at the highest alloy levels.

These magnetic nanoparticles have been investigated thoroughly for potential commercial and industrial applications. Hard magnetic nanoparticles with large coercivity characteristics are best suited for ultrahigh-density storage applications, whereas soft magnetic nanoparticles with low coercivity properties and high magnetic moments can be ideal in the design and development of electromagnetic and electro-optical components. Composite nanoparticles containing exchange-coupled magnetically hard and soft phases have potential applications as future permanent magnets with optimum energy product. Superparamagnetic particles are essential for the development of both magnetic imaging contrast enhancement and magnetic fluid hyperthermia treatment.

3.6 Potential Commercial and Industrial Applications of Rare Earth Magnets

In this section, the potential commercial and industrial applications of rare earth magnets are described, with an emphasis on performance, procurement cost, reliability, and safety under harsh thermal and mechanical environments. These magnets are best suited for defense, medical, and other commercial and industrial applications.

Neodymium magnets are a compound composed of neodymium, iron, boron, and some transition metals. This particular magnet offers high remanence, high coercive force, and the highest energy product compared to other rare earth magnets. The magnet also has an exceptionally high performance-to-cost ratio. This material can be easily formed into various shapes and geometrical configurations. These permanent magnets offer high resistance to demagnetization and are best suited for various industrial applications.

Rare earth permanent magnets using embedded rare earth magnetic steel are especially suited for deployment in the design and development of hydroturbines because they offer resistance to turbulence, dynamic impact, and reverse running. Both neodymium and samarium permanent magnets are used in the design of electric motors and generators in electric and hybrid electric vehicles.

3.6.1 Typical Commercial and Industrial Products Using Neodymium and Samarium Magnets

Commercial and industrial applications for neodymium permanent magnets include the following:

- Ceramic magnets
- Disc magnets
- DVDs
- Lifting magnets
- Magnetic hematite jewelry
- Magnets for therapy
- Pyramid magnets
- Radial ring magnets, which can be magnetized from inside to outside
- Lithium batteries
- Rubber-coated supermagnets

Commercial and industrial applications of samarium-cobalt magnets include the following:

- Scientific experimental projects for students
- Polytetrafluoroethylene (Teflon) magnets
- Focusing magnets in TWTAs in electronic warfare equipment
- Laser pointers
- Industrial applications where high magnetic force and high operating temperatures (300°C) are principal requirements
- Ring magnets

This particular magnet is most suitable for electric motors and generators. It is widely used in commercial electric and hybrid electric vehicles, where operating temperatures can reach as high as 300°C. Furthermore, this magnetic material offers minimum weight and size and ultrahigh reliability under harsh operating environments.

3.6.2 Industrial Motors and Generators

Neodymium and samarium-cobalt magnets are often used in electric and hybrid electric vehicles because of their superb magnetic properties at elevated temperatures approaching 300°C. In addition, these magnets offer a significant reduction in weight and size, as well as remarkable improvement in the reliability of these motors and generators. Neodymium magnets are good for continuous operation close to 80°C, whereas the samarium-cobalt magnets are most suitable for operating temperatures close to 300°C.

Neodymium-based magnets have been used in the design and development of minimotors and micromotors [3]. The simplest electric motor using neodymium minimagnets is a homopolar motor, which works without the need for a commutator by rotating along a fixed axis that is parallel to the external magnetic field produced by a permanent magnet. Note that the name homopolar indicates that the electrical polarity of the motor does not change. The first homopolar motor working in principal was demonstrated by Michael Faraday in 1821 at the Royal Institution in London, England. Such motors essentially have a single-turn coil, which restricts their practical applications. These motors operate at low voltages and produce small torques. A simple homopolar motor can be made using a drywall screw, an alkaline battery cell, a wire, and a neodymium disc magnet. The screw and the magnet make contact with the bottom of the battery cell and are held together by the attraction force provided by the neodymium disc magnet.

As far as the operating principle is concerned, the homopolar motor is driven by the Lorentz force. As the motor moves through the external magnetic field produced by the neodymium permanent magnet, the current carriers in the conducting wire experience a push that is perpendicular to both their velocity and the external magnetic field. This force induces a torque around the axis of rotation. Because the axis of rotation is parallel to the external magnetic field, no commutation is required for the conducting wire to keep turning. This motor requires cheap and simple components and is not suitable for most applications. This motor can be made reversible if the conductor is turned mechanically. Under this condition, the device will operate as a generator that is capable of generating direct-current voltage between the two terminals of the conductors. The magnetic material used in the permanent magnet must be conductive.

3.6.3 Medical Applications Involving Rare Earth Magnets

Neodymium and samarium magnets have various medical clinical applications in which high resolution, patient comfort, equipment safety, and reliability are of critical importance. These magnets are best suited for MRI equipment to image either the full body or a specific part of the body. For MRI equipment using high eddy currents through magnetic coils, a nerve-wracking noise is constantly is produced, which can be very annoying and uncomfortable for the patient. Patients may even be given medication to reduce noise-based phobia prior to MRI procedures. In some cases, noise generated by the MRI equipment is so intolerable to the patient that the procedure must be terminated. Alternatively, MRI equipment can use the very high magnetic fields produced by neodymium-based rare earth permanent magnets. These magnets are free from the nerve-wracking vibrations and intolerable noise levels, so the patient will feel very comfortable, even during long durations of clinical treatment. These MRI scanners are also light, portable, and occupy a small amount of space. Furthermore, such scanners can be moved close to the patient's bed.

3.6.4 Applications for Rare Earth Permanent Magnets in Defensive and Offensive Weapon Systems

High-power electronic warfare equipment generally uses TWTAs to counter the threats posed by enemy radars and missiles. These TWTAs are deployed in airborne and shipborne electronic warfare equipment for jamming enemy radars and incoming missiles. The operating temperatures of the collector elements could be as high as 250°C or more, depending on the cooling methods used. Under these conditions, TWTAs using samarium-cobalt permanent magnets to focus the electronic beam are reliable and suitable for temperatures as high as 300°C with no compromise in performance. These rare earth magnets are known as the focusing magnets in TWTAs. No other magnets can meet or beat the performance level and reliability of TWTAs using samarium-cobalt permanent magnets.

Intensive laboratory tests were conducted on TWTAs in 1972 using samarium-cobalt ($SmCo_5$) permanent magnets by the U.S. Army Materials and Mechanics Research Center. The report cited several operational advantages of these magnets, including RF performance and reliability of the TWTAs at temperatures approaching close to 300°C. The tests on these TWTAs were conducted at various RF power levels, higher operating temperatures (ranging from 200–300°C), and various demagnetizing field conditions. No catastrophic failures were reported. The report mentioned no cooling problems in the TWTAs using samarium-cobalt permanent magnets, which is a most common problem observed in TWTAs with conventional magnets. The laboratory test results can be summarized as follows:

- The focusing rare earth magnets displayed great performance superiority over conventional magnets.
- The tests were conducted under varying conditions of applied magnetic fields and temperature. The maximum applied field used was 9,000 Oe and the maximum temperature was close to 300°C.

- No catastrophic failures were observed in the focusing magnets or in the TWTAs when the tests were conducted with temperatures ranging from 200°C to 300°C.
- No deterioration in TWTA performance was observed under conditions of varying demagnetization.
- At operation near 200°C, test engineers noted some time-dependent effects, in addition to the reversible and irreversible effects in the rare earth sintered magnets. However, the overall RF performance of the TWTAs was not affected, except for a slight loss in demagnetization.
- Test results indicated that even 5% magnetization losses at elevated temperatures did not significantly affect TWTA output power.
- Test engineers believed that the time-dependent effects were responsible for the magnetization losses.
- Three TWTA samples, each from three different lots from two different TWTA manufacturers, were used in the tests [4].
- The TWTAs were procured from two different sources (General Electric and Raytheon). They were designed for the same power output over the same bandwidth and frequency band. Each TWTA used samarium-cobalt ($SmCo_5$) focusing magnets.
- Each rare earth permanent magnet was magnetized at room temperature before each run and the applied field intensity level setting was done with a 100,000 Oe pulse field.
- The tests revealed a demagnetization field of 4,000 Oe at 215°C. The test engineers felt that these test parameters were appropriate and reasonable. According the test engineers, a demagnetizing field of 4,000 Oe is close to the point of maximum energy product and is also sufficiently large to give significant irreversible effects.
- Most tests were performed at 225°C and TWTA performance data were recorded under steady and transient conditions. Transient conditions lasted approximately 7 minutes, whereas the steady state condition lasted between 50 and 75 minutes. When the sample was stabilized, it was possible to achieve precision test data.
- Additional tests were performed to determine the effect on magnetization under various stress levels and temperatures. Test data indicate that the effect on the magnetization was less than 0.02% at room temperature, but the effect on the magnetization increased to close to 0.05% when the TWTA temperature increased to 215°C. The test engineers predicted that the effect on the magnetization would be less than 0.1%.

Experimental investigations on TWTAs revealed that the magnetic pole strength of a rare earth permanent magnet decreases with increasing temperatures. To achieve a uniform functional magnetic field, a temperature-compensated alloy is used in the manufacturing of rare earth magnets, which essentially shunts the magnetic lines of force from the poles. To reduce the impact on TWTA performance due to increases in temperature and decreases in magnetic pole strength, the compensating alloy (Sm_2Co_7) must rob more magnetic flux from the functional area. In brief, by proper selection of a compensating alloy, the thickness of the alloy, and the cross-sectional area of the magnet, the functional magnetic field can be controlled to the desired value.

3.6.4.1 Impact of Temperature on Rare Earth Magnet Performance

The use of conventional metals (iron, cobalt, or any other appropriate metal) in rare earth permanent magnets (e.g., neodymium-iron-boron, samarium-cobalt) is intended to improve the thermal and mechanical performance of the magnets. The use of suitable magnetic alloys consisting of appropriate metals can significantly improve the mechanical properties of a magnet, as shown in Table 3.13.

The temperature-compensating characteristics of the alloys may vary with the amount of cold-work stresses induced into the alloy. Cold work decreases the flux carrying capacity, thereby reducing the change in flux density per degree from that level in the annealed condition. Thermal treatments up to 480°C for several hours will help to make the magnetic properties more stable with time and slightly increase the flux-carrying capacity. Therefore, temperature-compensating alloys are preferred for use in permanent magnets for TWTAs, electric motor/generator tachometers, and other microwave devices. Magnetic flux density variations as a function of temperature and alloy content are shown in Table 3.14.

Comprehensive studies undertaken by the author in 1985 on the thermal conductivity of the alloy element indicate that thermal conductivity degrades as the alloy temperature increases, as shown in Table 3.15. As the temperature of the neodymium-iron-boron magnet increases, the ability of the magnet to dissipate heat decreases, thus reducing the flux density of this magnet. This characteristic could affect the performance of the magnet, unless there are higher external magnetization fields.

3.6.4.2 Use of Rare Earth Permanent Magnets in High-Power Microwave Components

Rare earth permanent magnets are well suited for high-power microwave components, such as band pass filters, high-power magnetron sources, programmable

Table 3.13 Improvements in the Mechanical Properties of an Iron-Cobalt Alloy

	Alloy Contents (Volume)	
Mechanical Properties	*49 Co₂, 51 Fe*	*27 Co₂, 73 Fe*
Tensile strength (psi)	170,000–215,000	165,000–200,000
Yield strength (psi)	164,000–205,000	157,000–196,000

Table 3.14 Flux Density Variation as a Function of Alloy Element Content and Temperature

Temperature	Flux Density Variation (Gauss)		
	29.8% Nickel	32.5% Nickel	36.2% Nickel
−40°C	5,650	9,900	14,400
−20°C	4,900	9,350	13,500
0°C	3,920	8,700	13,345
25°C	2,240	7,800	12,700
40°C	1,325	7,255	12,260
50°C	665	6,350	10,700
80°C	350	4,800	10,500

Table 3.15 Thermal Conductivity of an Iron-Containing Alloy at Various Temperatures

Temperature (K)	Thermal Conductivity (W/cm·K)
1	0.750
50	3.722
100	1.325
150	1.043
200	0.942
300 (27°C)	0.803
400 (127°C)	0.694
500 (227°C)	0.613
573 (300°C)	0.545

dual-mode TWTAs for jamming enemy radars and missiles, standard TWTAs for supersonic fighters, electronic countermeasures, directed-energy weapons systems, and other defense products. The deployment of samarium-cobalt and neodymium-iron-boron magnets may improve the efficiency of TWTAs by more than 50%, the efficiency of magnetron by close to 90%, and the efficiency of klystron by more than 70%. The high efficiency of these high-power defense-based products will significantly reduce required input power, thereby improving the system's reliability, weight, and size. The ultrahigh reliability, relatively low cost per watt, higher conversion efficiency, and overall enhanced system performance are possible due to deployment of rare earth magnets.

3.6.4.3 Rare Earth Materials for Filters, Delay Lines, and Limiters

Rare earth–based materials, such as YAG, are widely used in the design and development of microwave filters with sharp skirt selectivity and high attenuation in the stop-band regions, delay lines in side-looking and forward-looking airborne radars for the signal processing needed for clutter rejection, and limiters to protect low-noise front-end receivers. The delay lines use yttrium and play a critical role in designing the signal processing unit capable of rejecting the random fluctuation ground and sea clutter and providing high signal-to-noise ratios. In YAG limiters, yttrium acts as a ferromagnetic material that exhibits major resonance and a subsidiary resonance that occurs for an extremely applied magnetic field strength, which is approximately half of the range required for a major resonance. The subsidiary resonance represents a nonlinear condition for which microwave power is absorbed above a critical value of the microwave field intensity (H_c). With sufficient ferromagnetic material in the transmission line, absorption will increase with microwave power input above the H_c and the output power of the limiter will remain constant; this device is known as a ferrite limiter with a low limiting threshold. These limiters are used in the front of the radar receiver to protect the crystal detector from burn out.

3.6.4.4 Use of Rare Earth Magnets in Electric Cars, Stepper Motors, and Synchronous Motors

Neodymium magnets, samarium magnets, and other rare earth magnets are widely used in various commercial and industrial applications. Preliminary studies undertaken by the author indicate that these permanent magnets are best suited for wind turbines, hydroelectric turbines, electric and hybrid electric cars, and other commercial products. Permanent-magnet synchronous motors (PMSMs) offer efficiencies that are approximately 42% more than the inverter-driven squirrel cage motors. These motors incorporate highly efficient rotor design that uses embedded permanent magnets made from rare earth material. This motor maintains constant speed independent of load conditions. The motors come in ventilated and nonventilated configurations, with power ratings ranging from 0.6 to 15 kW (0.8–20 horsepower).

3.6.4.5 Rare Earth Magnets for High-Speed Trains and Ship Propulsion

In recent decades, several countries, including the United States, Japan, Germany, and France, have used rare earth materials in the design and development of alternating- and direct-current synchronous generators. These machines have demonstrated high efficiency, excellent reliability, and uniform output torque. Electric motors using rare earth permanent magnets are best suited for small, highly maneuverable ships that can provide coastal defense and security. Direct-current synchronous machines are also known as homopolar machines.

3.6.4.6 Application of Rare Earth Crystals for Precision Atomic Clocks

Rare earth crystals (namely caesium and rubidium crystals) have been used in designing atomic clocks. The first caesium atomic clock was built in 1955 at the National Physical Laboratory in the United Kingdom. These clocks provide the most accurate commercially produced time and frequency standards. In caesium atomic clocks, electric transitions between the two hyperfine ground states of caesium-133 atoms are used to control the output frequency. The radiation produced by the transition between the two hyperfine ground states of caesium has a frequency of exactly 9,192,631,770 Hz or 9.192631770 GHz, which provides a time resolution of 0.109 ns. No other measurement device can offer this time resolution. The long-term stability of this clock is 0.107ns after 300 years.

The low-cost rubidium clock and the oven-compensated crystal oscillator were designed strictly for global positioning system (GPS) time, frequency, and synchronous applications. This clock offers short-term stability of 10^{-11} over 10 seconds and long-term stability of 5×10^{-11} over 6 months. The rubidium clock costs approximately \$900; the caesium clock is cheaper. These atomic clocks are best suited for Loran-C and GPS systems. Potential applications for these atomic clocks include use in wireless repeaters, data transmission, covert commercial communication, military communication systems, ethernets, digital network switching, microprocessors, test equipment, telecommunication systems, and industrial controllers.

Low-profile, low-cost rubidium crystal oscillators operating at 1,500 MHz are best suited for applications in which low-phase noise and low spurious levels are the principal requirements. The ovenized compensated crystal oscillator and voltage crystal oscillator are relatively inexpensive sources that can be used where time and frequency resolutions are not stringent. Design reviews of crystal oscillators by the author revealed that high-frequency clocks are essential for achieving high-precision timekeeping, but such clocks draw too much power relative to the sleep mode power of duty-cycled sensors nodes; therefore, they are rarely deployed in low-power settings. Clock designers claim that slow power-proportional operation offers a delay exceeding 1 microsecond, which represents a 10 times improvement in average power and a synchronization accuracy exceeding 1 microsecond at duty cycles below 0.1%.

Note that message time-stamping is the act of associating a specific time to the transmission or reception of a message, whereas timekeeping is the process of maintaining an accurate representation of time between resynchronization attempts. In both cases, atomic clocks offer high accuracy, security, and reliability. However, high resolution and low power are the critical requirements for time synchronization functions.

3.6.4.7 Yttrium-Iron-Garnet Oscillators

Yttrium plays a critical role in microwave components, such as delay lines, filters, and limiters. YIG oscillators have been designed, developed, and tested by various microwave design engineers and scientists. The presence of yttrium in YIG offers significant advantages in the performance of RF oscillators, including stability, improved efficiency, low phase noise at higher microwave frequencies, and high reliability under harsh operating environments. These oscillators have demonstrated wider RF bandwidths, low insertion loss at higher microwave frequencies (less than 2.5 dB at 26 GHz), and ultralow phase noise (approximately –100 dB at 100 kHz offset frequency). The rugged construction allows the oscillator to operate in harsh mechanical and thermal environments with no compromise in the reliability, accuracy, and stability of the device, even at temperatures as high as 95°C. This oscillator has potential applications in radar, electronic jamming equipment, frequency calibration systems, and tunable RF sources.

3.6.4.8 Role of Rare Earth Materials in Superfluid Cooling Devices for Sub-Kelvin Cooling

Researchers at the National Aeronautics and Space Administration Goddard Space Flight Center have developed a supercooling device with no moving parts that is capable of providing sub-Kelvin cooling. The pulse tube refrigerator uses a mixture of ^{3}He and superfluid ^{4}He to cool the device below 300 mK or 0.3 K, while rejecting heat at temperatures up to 1.7 K. This refrigerator is driven by a unique thermodynamically reversible pump that is capable of pumping the ^{3}He–^{4}He mixture without any moving parts [7].

This refrigerator consists of a reversible thermal pump unit, two heat exchangers, a recuperative heat exchanger, two cold heat exchangers, two pulse tubes, and an orifice. The two superfluid tubes operate at 180 degrees out of phase. The reversible thermal pump has been used in superfluid Stirling cycle coolers. This pump unit consists of two canisters packed with pieces of gadolinium-gallium-garnet. This rare earth material has an atomic weight of 157 and atomic number of 64, and it plays an important role in achieving sub-Kelvin temperatures. The canisters are connected by a superleak porous piece of glass with special characteristics. The canisters are surrounded by superconducting magnet coils. This superfluid cooling device is best suited for medical and nanoscience research applications that require sub-Kelvin cooling temperatures [7].

3.6.4.9 Rare Earth Materials for Electronic Applications

Rare earth materials can be used in thin-films for specific electronic applications where reliability, stability, and durability are of critical importance under harsh mechanical and thermal conditions. Rare earth single-crystal films can be made from yttrium or one of the rare earth elements with an atomic number of 62 to 71. The single-crystal films have demonstrated superior magnetic properties, which are considered to be most ideal for ultrahigh frequency resonant insulators, transducers, and surface wave amplifiers widely used for acoustic devices. These films can also be used in logic and storage devices of the domain displacement type. This method permits rapid dissolution of the constituents of the single crystal and consequently results in thin and homogeneous single-crystal films of high quality.

3.6.4.10 Rare Earth Manganese Silicides and Their Applications

The rare earth compound *Ln*MnSi—in which *Ln* is a rare earth element such as gadolinium, lanthanum, or yttrium—can be made by heating stoichiometric amounts of composition of two or more elements. The compositions are magnetic, with Curie temperatures ranging from 275 to 330 K. These compositions can be used in thermally activated magnetic switches with unique performance capabilities. The compounds obtained with this method are lanthanum-manganese-silicate (LaMnSi), gadolinium-manganes-silicate (GdMnSi), and yttrium-manganese-silicate (YMnSi), which are known as ternary compounds and have Curie temperatures of approximately 305, 295, and 275 K, respectively.

The ternary compounds offer greater magnetization that gradually decreases with temperature until a drop at the Curie temperature, which is quite sharp at low external fields (e.g., 75 Oe). This is quite evident from the magnetization values shown in Table 3.16. These composition materials can be used in thermally activated magnetic switches, which can play critical roles in magneto-optic devices. These compositions offer magnetic switching capabilities that are uniquely suited to human convenience in ambient temperatures. The substitution of iron for the manganese in GdMnSi leads to an increase in Curie temperature by as much as 25°C. The new gadolinium-based compound $GdMn_{1-x}Fe_xSi$ will have unique characteristics. In particular, when $x = 0$, this compound has larger magnetization than the lanthanum-based and yttrium-based compounds.

3.6.4.11 Rare Earth Metal Fluoride and Its Applications

High-quality optical materials can be achieved from the powder of an alkaline rare earth metal fluoride or rare earth metal fluoride. The optical body is created by flowing a reactive, reducing gas in combination with hydrogen fluoride gas, directly into the die cavity containing a pressable ionic fluoride powder. These methods will lead to the production of optical coatings. The refractive index of the coating can

Table 3.16 Magnetism of Ternary Compounds as a Function of Temperature

Temperature (K)	Magnetism (emu/g)		
	LaMnSi	*YMnSi*	*GdMnSi*
4.2	8	20	138
40	8	20	134
80	8	20	123
120	8	20	115
160	7.4	20	104
200	7.0	20	92
240	6.4	20	77
280	5.2	20	58
320	3.2	3.5	36
360	0	3.0	17
400	0	2.8	7

Source: McCarthy, G.I., ed. 1980. *Rare Earth Technology and Applications.* Park Ridge, NJ: Noyes Data Company.

be adjusted by heating the film in a controlled environment in a vacuum under a pressure of 10^{-4} torr (1 torr = 1/50 psi) for a suitable time duration.

Typical optical thickness varies from 0.25 to 1 wavelength, which is roughly in the range of 0.1 to 1 μm (10^{-6}) in thickness. The high optical transmission and high refractive index in the visible to near-infrared region, large linear and quadratic electro-optic effect, and the availability of suitable materials with desirable properties make these films extremely useful for electro-optic applications. Lanthanum plays a vital role in producing high-quality electro-optical films that are best suited for laser-based devices.

3.6.4.12 Rare Earth–Based Titanates and Their Applications

Barium titanate, calcium titanate, strontium titanate, and cadmium titanate can be modified using rare earth materials with atomic numbers 57 to 71, which includes lanthanum, cerium, neodymium, terbium, samarium, yttrium, and other elements. These elements are widely deployed in producing ceramic materials for a host of applications, including crystals, electronic components such as ceramic resistance, piezoelectric devices, and semiconductor ceramics by using the liquid-phase reaction.

3.7 Summary

The unique properties and classification of rare earth materials into ferromagnetic materials, diamagnetic materials, and paramagnetic materials were discussed in great detail for the benefit of readers. The important magnetic parameters of permanent magnets and widely used magnetic terms such as induction density, maximum energy product, coercive force, demagnetization force, and magnetomotive force were defined. The importance of the hysteresis loop for magnetic materials was emphasized, as the area of this loop indicates the magnetic loss in the material. Samples of hysteresis loops for two different rare earth magnetic materials were provided for the benefit of readers.

The importance of induction density or flux density, permeability, Curie temperature, and intrinsic coercive force were identified. Important magnetic properties of metals and alloys, such as cobalt-iron, neodymium-iron, and high-temperature ferrite used in the development of permanent magnets, were summarized, with emphasis on Curie temperature, magnetization force, and maximum flux density. Various measurement systems and associated units were used to define the magnetic induction density, magnetization force, magnetomotive force and other magnetic parameters.

The physical and magnetic properties of rare earth permanent magnets, such as neodymium-iron-boron and samarium-cobalt, were summarized in great detail. Performance degradation of permanent magnets under harsh thermal and mechanical environments was discussed. No catastrophic failures under high temperatures exceeding the maximum operating temperatures recommended for the neodymium-iron-boron and samarium-cobalt magnets have been reported. Performance capabilities and limitations of these permanent magnets have been identified while operating under design specifications.

The potential commercial and industrial applications of these permanent magnets were briefly identified, with particular emphasis on electric motors and generators for electric and hybrid electric vehicles, critical parts for wind and water turbines, and large lifting magnets that are widely used in heavy industrial applications. In addition, applications of rare earth magnets for magnetic bearings, magnetic couplings for uranium enrichment plants, and stepper motors were identified. Two distinct manufacturing technologies were briefly described—the classic powder magnet technology and rapid solidification magnet technology. Performance capabilities of conventional magnets, such as alnico and ferrite magnets, were compared with those of rare earth permanent magnets, with an emphasis on cost, complexity, reliability, tensile stress, bending stress, and compressive stress. Cost estimates for the permanent neodymium and samarium magnets were provided for various applications.

Critical characteristics of neodymium and samarium permanent magnets were summarized, with an emphasis on remanence, relative permeance, coercity, temperature, tensile stress, and compressive stress. Magneto-optics and anisotropic properties of rare earth elements and compounds were briefly summarized, with an emphasis on uniaxial anisotropic constants k_1 and k_2 as a function of cryogenic temperatures. Lattice constants and x-ray density parameters for neodymium and samarium magnets were summarized. Computed values of saturation magnetization (M_s) were provided as a function of applied field and cryogenic temperature. Hysteresis loops as a function of room temperature and higher cryogenic temperature were provided.

The potential applications of rare earth elements and ternary compound materials were specified in the design and development of medical devices, MRI scanners, high-power microwave sources, traveling wave tubes, limiters, circulators, duplexers, filters, atomic clocks, ship-propulsion systems, superfluid cooling devices to achieve sub-Kelvin cooling at remote controlled locations, thin-film for micro-miniaturized millimeter-wave and electro-optic components, and high quality magneto-optic films for electro-optic, magneto-optic, and infrared devices requiring high-resolution capability.

References

1. Jha, A.R. 1998. *Superconducting Technology: Applications to Microwave, Electro-optics, Electrical Machines, and Propulsion Systems.* New York: Wiley, 102.
2. Blaettner, H.F., K.J. Stranat, et al. 1997. *Magnetization and Anisotropy of Some R2Co7 Compounds at Low Temperatures.* Dayton, OH: University of Dayton, School of Engineering.
3. K&J Magnetics. 2011/2012. Technical data sheets summarizing the performance capabilities and magnetic properties of materials used in the development of neodymium-iron-boron and samarium-cobalt permanent magnets. Pipersville, PA: K&J Magnetics.
4. Chu, S. 2011. *Critical Materials Strategy.* Washington, DC: U.S. Department of Energy, 2–10.
5. Brown, D. 2000. Developments in the processing and properties of NdFeb-type permanent magnets. *Journal of Magnetism and Magnetic Materials* 248:432–440.
6. Pyrhonem, J., T. Jokinen, and V. Hrabovcova. 2009. *Design of Rotating Electrical Machines.* Hoboken, NJ: Wiley.
7. National Aeronautics and Space Administration. 2012. *NASA Tech Briefs* 36:85.

Chapter 4

Rare Earth Materials and Films Best Suited for Various Applications, Such as Lasers, Electro-Optic Sensors, and Magneto-Optic Devices

4.1 Introduction

This chapter summarizes the rare earth materials, their films, and their properties that are most suitable for the design and development of lasers, electro-optic devices, and magneto-optic devices for commercial, industrial, medical, space, and military applications. Rare earth laser crystals and diode-pump solid-state lasers using rare earth elements are identified for specific infrared wavelength operations and their optical, electrical, mechanical, and thermal properties. The critical electrical performance parameters of each laser are summarized, emphasizing the operating wavelength, continuous wave (CW) or pulsed power output, quantum efficiency,

121

absorption efficiency, conversion efficiency, slope efficiency for normal mode and Q-switched lasers, threshold energy spectral bandwidth, and optical-to-optical efficiency wherever applicable. Laser output, power output, and energy level are identified as a function of temperature. Performance capabilities and limitations of devices are summarized in great detail, with emphasis on performance, reliability, and safety under harsh operating environments.

Diode-pumped rare earth–based lasers, such as hobium:thulim:lutimium-lithium-ferrous (Ho:Tm:LuLiF$_4$), holium:thulium:yttrium-aluminum-garnet (Ho:Tm:YAG), and holium:thulium:yttrium-lithium-ferrous (Ho:Tm:YLiF), operate at room temperature and offer improved optical and slope efficiencies, optimum power output, and higher narrow spectral beams with high reliability (even at slightly higher than room temperature).

Yttrium-iron-garnet (YIG) resonators play a critical role in designing wideband tunable oscillators. For example, an 8- to 18-GHz YIG-tuned field-effect-transistor (FET) oscillator that was designed and developed in the early 1980s demonstrated excellent performance over radiofrequency (RF) bandwidths exceeding 8%. Currently, YIG-tuned FETs have demonstrated wideband performance of more than 16% with no compromise in electrical performance. The power output of wideband oscillators is generally limited to 10 to 20 mW, but a solid-state RF amplifier can be added to the oscillator package to meet higher output power requirements. Significant improvements in the quality of YIG resonators have solved most of the inherent problems, such as fixed frequency resonances, frequency linearity, and power drop at the low end of the frequency range. However, the quality of a YIG resonator is strictly dependent on the purity of the yttrium element, which comes at a higher cost.

Research undertaken by the author revealed that the design and development of an atomic clock is based on hyperfine transitions in hydrogen-1, caesium-133, thallium-205, and rubidium-87. The first caesium-133 atomic clock was designed and developed in 1955 at the National Physical Laboratory in the United Kingdom. Atomic clocks offer optimum accuracy in terms of time, frequency, and synchronization compared to RF crystal-based sources.

In general, rare earth single-crystal films, rare earth manganese silicides, and other rare earth ternary compound materials play important roles in the design and development of electro-optic and magneto-optic devices for commercial, medical, and space applications, as will be discussed in this chapter.

4.2 Theory and Classification of Lasers Using Rare Earth Elements

This section discusses the theory and classifications of laser systems that use rare earth materials. A laser is defined as light amplification by simulated emission of radiation. This particular device uses the natural oscillations of atoms or molecules between energy levels to generate coherent electromagnetic (EM) radiation in the ultraviolet, visible, or infrared regions of the EM spectrum [1]. According to Einstein's theory developed in 1916, a large number of ions are responsible for the laser action in the crystals when they are doped with rare earth elements (REEs) as impurities, which generates divalent or trivalent ions. Many energy levels are involved in laser output from REE ions.

For example, a crystal doped with erbium is one of the most complex laser ions based on the number of different levels where laser action can occur. On the other hand, a crystal doped with yttrium provides the simplest laser action. Regardless of the doping elements, the various REE ions generated have a number of levels that are at the same energy level above the ground state [1]. Therefore, it may be possible to co-dope a host crystal with more than one type of ion, transferring the energy from one species to another (Figure 4.1). This energy transfer can be used to enhance the efficiency of optical pumping.

Let us examine a doping case using a holmium (Ho) host crystal. Research studies performed by laser scientists revealed that Ho, which emits at 2.05 μm, is not considered to be an efficient optically pumped laser. However, sensitizer materials can be deployed to increase the efficiency of the optical pumping. When a host crystal such as yttrium-aluminum-garnet (YAG) is co-doped with another REE such as erbium or holmium, significantly more efficient laser action has been demonstrated due to the energy transfer from one ion to another. However, when neodymium-doped YAG is used, the thermal broadening effect and a Lorentzian line shape in the output of the spectrum of the laser can be noted [1].

The most popular rare earth elements used in the design and development of lasers include neodymium (Nd), yttrium, erbium, holmium, europium, thulium (Tm), and YAG crystal. To illustrate the importance of these lasers, Tm:Ho:YAG lasers emitting at 2.1 μm are best suited for eye-safe coherent radars. Detailed performance parameters and the potential applications of REE-based lasers are identified in the following sections.

4.2.1 Applications of Continuous-Wave and Pulsed Lasers

Lasers can be classified into two categories—continuous-wave (C-wave) and pulsed-wave (P-wave) lasers, which are further divided into subtypes such as gas lasers, chemical lasers, diode-pumped lasers, and flash-pumped lasers. Lasers are sometimes characterized based on their output power, such as low-power lasers and high-power lasers. Low-power lasers are widely deployed in medical applications (optometric and dental procedures), whereas high-power lasers are best suited for the detection and acquisition of military targets and missile illumination. High-power lasers are also widely used in medical, commercial, and industrial applications. The typical applications of lasers can be briefly summarized as follows:

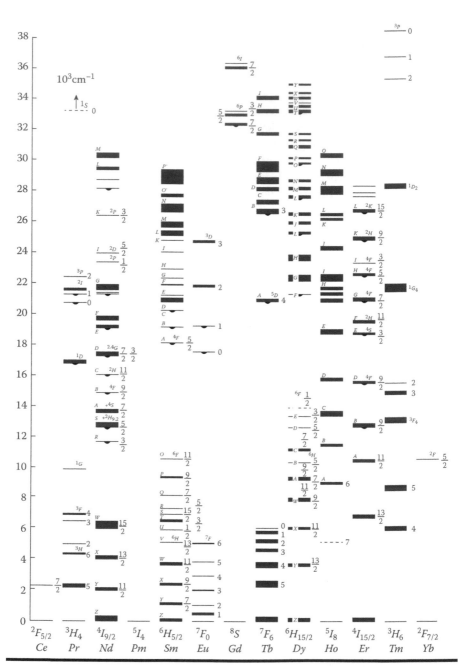

Figure 4.1   Energy levels of rare earth ions. Fluorescence is denoted by a semicircle.

- Optical and dental procedures
- Precision laser-guided bombs
- Laser illuminators
- Infrared (IR) guided missiles
- Detection, tracking, and identification of airborne targets
- Various commercial and industrial laser applications

4.2.2 Classic Problems Associated with High-Power Lasers

This section points out some classic operational problems associated with high-power lasers. In high-power lasers, thermal blooming effects have been observed, which can cause beam spreading, distortion, and bending due to absorption by atmospheric foreign particles and aerosols. Research studies undertaken by the author indicate such problems have been overcome by providing a thin coating of a dielectric material with high thermal conductivity, such as aluminum oxide, silicon nitride, or silicon oxide. This coating will produce a significant reduction in surface recombination by the velocity at the interface. Thermal blooming effects and overheating of the optical elements must be avoided to retain high reliability and safe laser operation.

4.3 Application and Performance Capabilities of Selected REE-Based Lasers

The general characteristics previously mentioned were relevant to lasers made from non-rare-earth-based materials. In the following sections, laser types, system descriptions, performance capabilities, and limitations of laser systems that exclusively use rare earth elements are summarized, with an emphasis on reliability, safety, and optical efficiency. Descriptions and performance parameters of selected laser systems will be provided as briefly as possible.

4.3.1 Nd:YAG and Nd:Yttrium-Lithium-Fluoride Lasers and Their Potential Applications

Nd:YAG lasers offer several applications, including the pumping of dye lasers, early detection and diagnosis of certain life-threatening diseases, and critical surgical procedures. The cutting capability of a 1-kW Nd:YAG-based laser is approximately 2.5 times better than that of a 1-kW Co_2 laser for cutting a 0.5-inch-thick stainless steel plate. YAG is a hard, strong, and durable material that has been widely used as a host material for neodymium-based and other rare earth ions because of its high gain and efficiency characteristics in flash-lamp-pumped, moderate-power lasers operating in the infrared region. The conversion efficiency of a continuous-wave (CW) Nd:YAG laser is approximately 0.15 for a well-designed device. The saturated power density of a high-power CW Nd:YAG laser can be as high as 2,000 W/cm^2. A multistage CW Nd:YAG laser can be designed to a kilowatt level with overall efficiency greater than 2%. The radiation efficiency of xenon flash-lamps is 50 to 60%. However, for large lasers systems, the lamp radiation efficiency could exceed 80% in laboratory environments.

4.3.1.1 Single-Lamp High-Efficiency Nd:YAG Lasers

Most lasers that do not need single-mode or high-brightness performance use a closed-coupled pump geometry. The main cavity member for the storage laser is a block of rare earth material called samarium-doped glass with holes drilled in it for the flash-lamp, Nd-YAG laser rod, and their associated cooling channels. The Sm^{3+}-doped glass acts to suppress the transverse lasing in parasitic modes; it is recognized as an energy-limiting chamber. The coolants could be either distilled water or water-ethylene glycol mixtures, depending on the cooling temperature range requirement for satisfactory laser operation.

4.3.1.2 Nd:YAG Lasers for Space Optical Communication

The first Nd:YAG laser was developed for space optical communications in the early 1980s. Flight tests demonstrated a communications capability of 1,000 Mbits/s with a receiver sensitivity of 20 photoelectrons/bit and a bit error rate of 10^{-6}. The tests further demonstrated an interterminal tracking capability with angle errors less than 1 microradian and station-to-station acquisition capability in less than 6 seconds.

Laser communication channels for space purposes could be used for space-to-space communication, space-to-aircraft communication, and space-to-ground communication. Communication from satellite-to-earth downlinks is much simpler than earth-to-satellite uplinks due to the moment arm involved. Furthermore, atmospheric perturbations to downlink laser signals are not as severe. The Nd:YAG laser receiver is essentially a photon-collecting energy detector and therefore is not dependent on the angle of arrival and the signal phase.

The performance characteristics of an Nd:YAG laser system [2] for asynchronous satellite-to-ground link at a data rate of 1,000 Mbit/s can be summarized as follows:

■ Laser transmitter wavelength: 532 nm
■ Laser beam divergence: 5 microradians
■ Optical system efficiency: 70%
■ Field of view: 100 microradians
■ Bit error rate: 10^{-6}
■ Pointing error: 1 microradian
■ Link margin: 11 dB

- Laser CW power output: 300 mW
- Quantum efficiency: 25% (minimum)

Only laser optical characteristics are summarized here, neglecting the antenna parameters such as size, gain, efficiency, and beam width to limit the discussion. Because the Nd:YAG laser system deploys a direct detection technique, phase coherence of the laser signal is not required at the receiving end. Therefore, a quantum noise limited detector can be employed with moderate cost. Furthermore, the atmospheric effects that disturb the wavefront coherency do not affect the direct detection system. The optical receiver of the Nd:YAG laser does not need to be diffraction limited, thereby allowing low-cost detectors. However, coherent optical receivers require diffraction-limited optics, whose cost and weight could grow in an exponential fashion with size.

The space-based Nd:YAG laser operates at 532 nm, which offers a narrow beam that is a function of optics size [2]. This laser is capable of meeting the angular pointing accuracy and beam-width requirements. Optical multiplexing and demultiplexing of the mode-locked laser techniques are readily applicable for the Nd:YAG laser, which allows data rates exceeding 5,000 Mbits/s on a single laser beam.

This laser system has two major components—the receiver and the transmitter. The receiver package is lighter and occupies minimal space compared to the transmitter package. The transmitter package includes the basic system components, such as the laser subassembly, a modulator capable of providing digital pulse-modulation schemes, imaging optics and telescope, gimbal subassembly acquisition and tracking detectors, communications and signal processing electronics, plus other redundant components needed in an emergency.

Regarding weight and power consumption of the major components, a dedicated 1,000-Mbits/s spacecraft optical transmitter would weigh approximately 250 lbs. (including the weight of lamp-pumped laser elements) and would require an electrical input power between 500 and 650 W. The diffracted-beam receiver would weigh roughly 145 lbs. and would require a power consumption of approximately 275 W using a solar-pumped laser beacon. Both the system size and power requirement increase with higher data rates. Note the laser threshold power requirement and acquisition-and-tracking subsystem power requirement are fairly constant regardless of the data rate requirements.

4.3.1.3 Neodymium:Yttrium-Lithium-Fluoride Lasers

Neodymium:yttrium-lithium-fluoride (Nd:YLF) lasers are widely used to pump appropriate optical crystals in the design and development of optical parametric oscillators. Both Nd:YAG and Nd:YLF lasers are well suited for pumping schemes. These lasers offer compact package dimensions and cost-effective designs with broad tuning capabilities in the mid-IR spectral region. YLF crystal acts like a host for neodymium and other rare earths, even though it is weaker than most commercially developed laser host crystals. Neodymium $^4F_{3/2}$–$^4I_{11/2}$ transition lases at 1.053 μm in one orientation, which closely matches the peak gain on the phosphate glass lasers. This host crystal exhibits near-quasi-thermal behavior. Scientific, research studies indicate that high-quality nonlinear optical crystals are required for phase matching the output of neodymium-based lasers to provide idler output in the 3- to 5-μm spectral range. Furthermore, the nonlinear crystal will allow optical parametric amplifiers to act as continuously tunable infrared lasers using solid-state laser pumping schemes. The frequency schemes are strictly dependent on coherent pump sources, such as 1-μm Nd:YAG or 2-μm Ho:Tm lasers deploying optical crystals with high nonlinear coefficients [3].

When the YLF crystal is doped with holmium and erbium instead of Nd and pumped with a solid-state diode array source, one gets the architectures of Ho:YLF and Er:YLF solid state lasers [3]. These lasers are called diode-pumped solid-state fiber-coupled lasers. Critical components of this laser system include the pump focusing lens, cryogenic cooler, holmium laser rod, output mirror, 780-nm pump source, water cooler with 10°C cooling temperature, collimating optics, Ho:YLF laser rod, optical cavity bending mirror, and Brewster angle polarizer. Two solid-state pump sources involving AlGaAs diode laser arrays are illustrated in Figures 4.2 and 4.3. The energy-level diagram for the Ho:YLF crystal is shown in Figure 4.4. This laser delivers a CW power output of 4 μm and is best suited for commercial and specific medical applications. The block diagram and energy level diagram for the Er:YLF solid-state laser are shown in Figure 4.5 [3].

A YLF crystal doped with erbium is used in the production of an Er:YLF solid-state laser, which emits at 2.8 μm. This single-mode fiber laser system is widely used in clinical research and specific medical applications involving certain diseases. Some Er:YLF lasers have been designed to operate at 3 μm for a particular medical application. The best laser slope efficiency of this particular laser is close to 48%, which can be further improved by using a host crystal with a high degree of purity. However, the interionic conversion efficiency is strictly dependent on the host crystal's geometry, line-width broadening mechanisms, photon energy levels, dopant concentration, and energy transfer levels. Studies performed by various laser scientists on excitation and loss mechanisms of the host crystal geometry and dopant concentration level will determine the optimum pump wave lengths for HO:YLF and Er:YLF fiber-coupled lasers.

The output power rating of these fiber-coupled lasers varies from 12 to 15 W, even when operating at a 1.1-μm wavelength. Higher power levels can be obtained using optical fibers with optimum core diameters and numerical apertures. The selection of these physical parameters of the fiber must not degrade the width and stability of the laser output beam. Potential applications of high-power fiber-lasers include laser spectroscopy, pollution detection, e nvironmental parameter monitoring, specific surgical procedures, remote sensing for deploying aircraft or satellites, and wavelength division multiplexing.

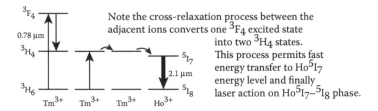

Figure 4.2 **Energy-level diagram and pumping scheme for a dual-doped Tm:Ho:YAG crystal in which $T_m^{3+}\,{}^3F_4$ manifold is pumped by the AlGaAs laser diode at 780 nm.**

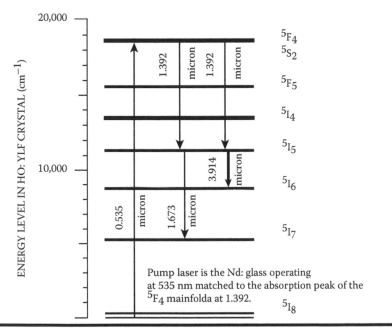

Figure 4.3 **Room-temperature energy-level diagram for a Ho:YLF crystal generating oscillations on S_{I5} and S_{I7} transitions at 1.392 and 1.673 μm and ~5–R6 transitions at 1.392 and 3.914 μm.**

4.3.2 Gadolinium-Scandium-Gallium Oxide Lasers

Gadolinium-scandium-gallium (GSG) oxide host crystals have a large thermo-optic distortion compared to YAG host crystals. Co-doping the crystal with chromium as a sensitizer for neodymium will lead to further scientific investigation to obtain performance characteristics of this host crystal. This is a glass-based laser host crystal. Its properties were extensively evaluated as a part of the laser

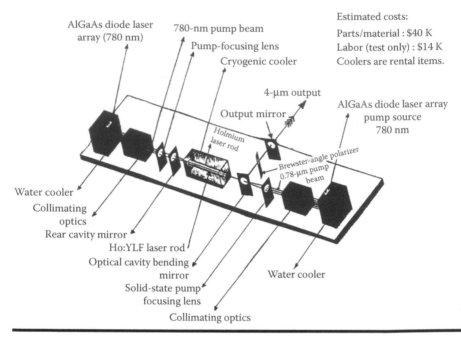

Figure 4.4 **Block diagram for a diode-pumped solid state Ho:YLF laser system (4 μm).**

fusion program at the Lawrence Livermore National Laboratory. In addition, an extensive compilation of the properties of laser glasses was made by Soviet scientists in the late 1970s. Most of the well-developed laser glasses are silicates or phosphates involving glass silicate gorilla glass (GSGG) crystal, which are the leading commercial glasses used in commercial and industrial applications. Note that a diode-pumped solid state laser as shown in Figure 4.4 has limited applications in scientific research areas due to its wide beam width and incoherent laser illumination.

4.3.3 *Performance Capabilities and Advantages of Erbium:YLF Fiber Lasers Emitting at 2.8 μm*

Erbium-doped lithium fluoride single-mode fiber lasers have been widely used in specific medical applications. The fiber laser's efficiency can be significantly improved by concentrating the pump energy from the optical cavity and excited electron population in the optical fiber core. The block diagram for the Er:YLF laser and the energy-level diagram for the 2.8-μm fiber laser are shown in Figure 4.5. Both the excitation and loss mechanisms of the host crystal geometry will yield optimum laser performance if the pumping is done at a wavelength of 970 nm for the Er:YLF fiber laser and 791 nm for the erbium-doped fluorozirconate fiber

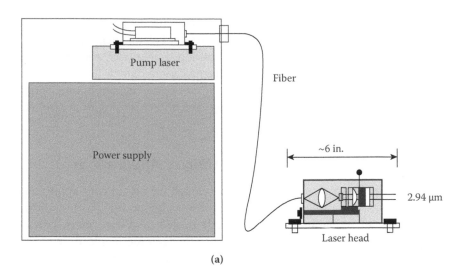

(a)

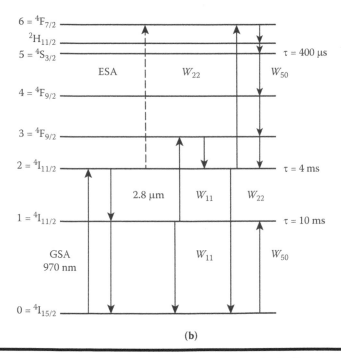

(b)

Figure 4.5 Er:YLF laser emitting at 2.8 μm. (a) Block diagram. (b) Energy-level diagram.

laser. The laser slope efficiency is approximately 40% for the Er:YLF laser and only 23% for the fluorozirconate fiber laser. Researchers at Polaroid demonstrated a CW power output of 35 W from a ytterbium (Yb)-doped fiber laser, with fiber length ranging from a few meters to 10 m. Potential applications of high-power fiber lasers include spectroscopy, pollution detection, environmental data monitoring, surgical procedures, wavelength division multiplexing, and remote sensing from an aircraft or low-orbiting satellite.

4.3.4 Yb-YAG Laser Systems for Military Applications

Low-power Yb:YAG lasers have not been developed for nonmilitary applications. However, it may be possible to design and develop high-power CW and ultrafast Yb:YAG near-diffraction-limited cryogenically cooled solid-state lasers, provided thermal management problems are eliminated. The laser rod temperature must be maintained close to 0°C if optimum laser performance is desired. Such lasers can be deployed as pulsed laser guns and armor carriers in battlefield environments, thereby providing accurate and powerfully effective weapons against enemy tanks and high-speed armor vehicles.

To solve the thermal management problems, some laser scientists have proposed thermoelectric coolers in the form of thins films with high thermal conductivity. Others are exploring the possibility of massive microchannel cooling technology. Laser engineers at the Massachusetts Institute of Technology are exploring the potential of cryogenic technology, which can be used in cryogenic laser systems for military applications. Laser pulses demonstrated the following performance parameters in a cryogenically cooled, near-diffraction-limited Yb:YAG laser system:

- A full-width half-maximum pulse width of 12.4 ps
- An energy-per-pulse capability of 15.2 μJ
- Peak power capability of 1.23 mW
- A highly sustained average or continuous-wave power rating of 750 W (minimum)
- Excellent beam quality
- Impressive beam profile
- Minimum thermal management problems due to cryogenic cooling

A maritime laser demonstrator (MLD) with a 15-kW power output was designed and developed using solid-state technology by Northrop Grumman. Tests were performed jointly by the U.S. Navy and Northrop Grumman engineers at a Pacific Ocean test range off the coast of central California [4]. This MLD tracked and lased land targets, then moved on to remotely piloted, unmanned small boats cutting across the water, successfully disabling the small boats. A 100-kW laser may provide a variety of protection and strike missions, including self-defense against threats from rockets, artillery, mortar rounds, high-speed swarming boats, unmanned aerial vehicles, and aircraft. This laser system will also enable ultrapre-cise strikes from a variety of ground-, sea-, or air-based platforms for prosecution of the enemy targets, while minimizing the risk of collateral damage to property in the vicinity. Testing indicated that the MLD laser system can effectively oper-ate in an open ocean environment, overcoming obstacles including unpredictable atmospheric conditions, waves, and the motions of both the host and target vessels, while providing precision tracking and laser beam delivery requirements for self-defense capability [4].

MLD-type laser systems mounted on large moving platforms, such as destroy-ers and other battleships, targeting missiles, and drones, will provide high-precision strikes with high probability of a kill. In the opinion of the author, no other laser system can provide such precise strikes under the conditions mentioned above.

4.3.5 Laser Systems Deploying Multilevel Doping of Rare Earth Host Crystals

Lasers using single-level doping, such as Re:YLF and Nd:YAG lasers, have limited performance in terms of power output and spectral wavelength. Ho:Tm:LuLiF$_4$ and Ho:Tm:YAG lasers use two-level or dual-level doping of Ho and Tm host crys-tals. The following sections summarize the performance capabilities, limitations, benefits, and potential applications of such laser systems. Lasers using two-level doping require no cooling and offer excellent performance at room temperature, thereby providing maximum economy and reliability.

4.3.5.1 Ho:Tm:LuLiF$_4$ Laser with Two-Level Doping Involving Two Host Crystals

This particular diode-pumped laser system uses an LuLiF$_4$ host crystal, which is doped by two rare earth elements—holmium and thulium. This laser system has achieved an optical efficiency exceeding 9.8% under the normal mode of operation. The Ho:Tm:LuLiF$_4$ laser demonstrated 1.5 times greater optical efficiency than the Ho:Tm:YLiF laser system under identical operating conditions, with impres-sive absorption spectra and lifetimes as a function of pump energy [5]. Studies performed by the author indicate that laser systems using dual-doping technology operate at a 2-μm wavelength at room temperature. The development of room-temperature solid-state lasers in the 2-μm range has received considerable attention because of the lasers' potential applications in radar altimetry, ranging systems, low-altitude wind shear sensors, and remote sensing systems, which includes Doppler radar wind sensing and water vapor profiling. High-power Q-switched holmium is very useful as a pump source for parametric optical oscillators. In addition, there are several medical applications for 2-μm lasers, such as laser angioplasty in the coronary arteries; ophthalmic procedures; and arthroscopic, laparoscopic, chole-cystectomy, and refractive surgeries. The $Y_3Al_5O_{12}$ (YAG) and YLiF$_4$ (YLF) lasers operate at 2-μm wavelengths as well, but their applications are limited because of lower optical efficiency and a lack of quantum mechanical calculations for various garnet hosts and glass-based materials involving different rare earth elements.

Ho:Tm:LuLiF$_4$, Ho:Tm:YLF, and Ho:Tm:YAG have two-level doping using LuLiF$_4$ (LuLF), YLF, and YAG rare earth host crystals, respectively [6], and possess unique laser properties and optical capabilities. These diode-pumped solid-state lasers operate at room temperature (25°C); therefore, the laser system has minimal cost, requires no cooling source, and provides maximum reliability. Threshold energies and slope efficiencies for LuLF and YLF rare earth host crystals are close to 1.3% under the normal mode of operation. Therefore, flash-pumped Ho:Tm:Er:LuLiF$_4$ and Ho:Tm:Er:YLF lasers that employ three-level doping with higher-purity rare earth materials could yield threshold energies and slope efficiencies that are slightly better than 1.35% under the normal mode of operation at room temperature [6]. Research indicates that a fluoride-based LuLF host crystal doped with trivalent Tm and Ho ions has a higher slope efficiency. The studies further indicate that the threshold energy levels for LuLF will be approximately 90% of the YLF crystal thresholds. LuLiF$_4$ is a glass-based isostructure similar to the YLiF$_4$ crystal structure. The smaller ionic radius of lutetium (Lu^{3+}) is 1.11 A° compared to 1.16 A° for yttrium (Y^{3+}).

Higher CW power levels are possible using a multilevel doping technique. The energy levels, transition wavelengths, and lifetimes of critical trivalent rare earth element ions, such as Ho, Er, and Tm, are absolutely essential for achieving the desired power levels and emission wavelengths. Energy-level diagrams for multilevel doping of rare earth–based host crystals demonstrate how the infrared energy is transferred from one wavelength to another wavelength. Trivalent rare–earth doped linear laser crystals tend to have improved laser performance in terms of CW power output, differential quantum efficiency or slope efficiency, electrical-to-optical conversion efficiency, laser beam stability, and beam quality. Nonlinear rare earth–based optical crystals seem to yield lower power output levels compared to linear crystals. However, this power level can be improved by the addition of chromium material to the host laser crystal, which permits strong absorption of visible wavelengths less than 800 nm. Improved laser performance levels require that the appropriate rare earth dopant elements (e.g., erbium, thulium, holmium) are added in the right mole concentrations to the solid-state host crystals (e.g., YAG, YLF) to absorb pump light and transfer it to the host laser crystals. Erbium is the most desirable rare earth material for absorbing the pump light with higher efficiency, whereas thulium is best suited to transfer the light energy from erbium to thulium (Figure 4.5).

4.3.5.2 Advantages and Disadvantages of Lamp Pumping and Solid-State Diode Schemes

A diode-pumped solid-state (DPSS) scheme using indium-gallium-arsenide (InGaAs) laser diode arrays emitting at 970 nm or 805 nm can be used for a pumping mechanism. Each laser diode array can provide a CW pumping capability exceeding 10 W.

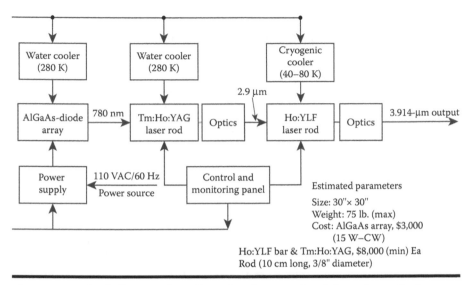

Figure 4.6 Block diagram for a three-level system involving a laser diode array. The Tm:Ho:YAG laser and Ho:YLF laser rods are fully equipped with the necessary accessories.

Laser diode arrays with CW power ratings greater than 25 W require cryogenic cooling, which will increase the system cost, complexity, and mechanical package dimensions. Cryogenic cooling may also affect the overall laser system reliability.

Preliminary calculations indicate that a lamp-pumped Er:Tm:Ho:YAG laser as shown in Figure 4.6 is capable of delivering a CW power output greater than 15 W at 2 μm. However, it requires cryogenic cooling to maintain the needed power output capability and differential quantum efficiency. Implementation of lamp-pumping technology will provide a broad spectral range compared to solid-state diode pumping schemes, which result in high infrared energy wastage and a reduction in energy transfer efficiency.

A CW power level of 2 W is difficult to get from a single laser diode array. However, a laser diode array composed of several diode arrays is capable of delivering a power output greater than 2 W using improved optical conditioning techniques, enhanced optical coupling efficiency, and the right cooling temperature. This was the performance limit for a DPSS scheme. The emission wavelength of the diode-pumping source is selected to match the absorption peak at a specific wavelength to enhance the pumping efficiency. The energy transfer efficiency is much higher in the diode-pumping scheme compared to a lamp-pumping scheme. Furthermore, the diode-pumping scheme generally requires moderate to high currents at low voltage, ranging from 5 to 7 V. Because of the advantages of the diode-pumping technique, DPSS devices are widely deployed in the design of lasers involving rare earth host crystals.

The lamp-based pumping scheme is best suited for applications in which higher CW power is the principal design requirement. Research studies and experimental

Table 4.1 Estimated Values of Power Output (P_{out}) for a 4-μm Laser Diode Array and Differential Quantum Efficiency (DQE) as a Function of Cryogenic Temperature

Laser Diode Array Size	80 K		60 K		40 K	
	P_{out} (mW)	DQE (%)	P_{out} (mW)	DQE (%)	P_{out} (mW)	DQE (%)
500 × 50 μm	25	2.2	35	5.6	50	10
500 × 100 μm	55	4.5	68	7.5	108	22
500 × 200 μm	105	8.9	124	14.6	255	53

data indicate that a Er:Tm:Ho:YAG laser system using a lamp-based pumping scheme is capable of delivering CW power output exceeding 15 W at a spectral wavelength of 2 μm. This solid-state laser crystal requires cryogenic cooling to keep the population of the lower emissions from blocking the laser operation at the desired wavelength. Note that accurate information on the lifetimes and energy levels of the transfer of potential trivalent rare earth ions requires extensive experimental data, which could involve higher costs. The CW power output and differential quantum efficiency (DQE) data for a 4-μm all solid-state laser diode as a function of cryogenic temperature are summarized in Table 4.1. The power levels shown in the table are for a single laser diode array only. Note that the CW power output from a solid-state diode array is strictly dependent on the array size and the cryogenic temperature. The improvement in laser efficiency is due to the use of rare earth materials and the cryogenic temperature. It is very difficult to distinguish between the contributions from these two sources. However, major improvements are likely due to cryogenic cooling, particularly at lower cryogenic temperatures.

Studies undertaken by the author indicate that the performance levels of trivalent rare earth–doped linear laser crystals are not fully optimized in the infrared region, particularly over the 3- to 5-μm spectral region. Trivalent rare-earth linear crystals have demonstrated a CW power output of more than 500 mW at 3-μm wavelength and at room temperature. However, higher CW power levels (much greater than 500 mW) are possible over the 3.9- to 4.1-μm range under cryogenic temperatures using trivalent rare earth–doped linear crystals, such as Ho:YLF, Er:YLF, Tm:YLF, or Tm:Ho:YLF

4.3.5.3 Threshold Energy and Slope Efficiency of Ho:Tm:LuLiF₄ and Ho:Tm:YLiF₄ Laser Systems Using Rare Earth Materials

Threshold energy and slope efficiency are the critical performance parameters of lasers using rare earth materials. Design calculations indicate similar absorption efficiencies for these lasers at a peak wavelength of 0.793 nm, despite the difference in Tm concentration in the LuLiF$_4$ (LuLF) and YLF rare earth–based crystals [6]. The absorption efficiency is 5% in LuLF lasers and 6% in YLF lasers. Both lasers can be designed to emit at 2.055 nm while operating at room temperature (300 K).

The normal-mode laser performance parameters of LuLF and YLF lasers as a function of output-coupling mirror reflectivity are shown in Table 4.2. It appears that an output mirror reflectivity of 0.94 yields the maximum efficiency for both lasers. The maximum optical efficiency was recorded at approximately 9.4% with the reflecting mirror. The corresponding slope efficiency was 19.3%, with a threshold energy of 0.455 J. The maximum laser output energy was 0.0515 J for an input optical energy of 0.84 J, which brings the optical efficiency to 6.13% for the YLF laser. The threshold energies for various output coupling mirrors are approximately 0.10 less for the Ho:Tm:LuLF laser compared to the Ho:Tm:YLF laser, which correlates with the computed values. The laser output wavelengths for various output-coupling mirrors centered around 2.063 nm, with slight variations for the Ho:Tm:YLF solid-state laser. Fluorescence lifetimes for the low and high pump levels were approximately 12 ms and 8 ms, respectively, for the Ho:Tm:LuLF laser and 12.5 ms and 8 ms for the YLF laser system. The reduction in lifetime is due to various conversion processes.

RF laboratory measurements indicate that the slope efficiencies are quite different under the normal mode of operation and the Q-switched mode of operation. Maximum laser output energies of 14.7 mJ in single Q-switched RF pulses were recorded for an input optical energy of 0.978 J, which corresponds to an optical efficiency of 1.5%. The 150-ns-wide RF pulses were used in the measurement of slope efficiencies. The Q-switched laser output wavelength was centered around 2.055 nm for the Ho:Tm:LuLF solid-state laser. In comparison, a maximum Q-switched laser output energy of 11 mJ was observed at the input optical energy of 0.978 J for the Ho:Tm:YLF solid-state laser using the 150-ns-wide pulses. The Q-switched laser output wavelength for the Ho:Tm:YLF laser was centered at 2.052 nm [6].

Table 4.2   Normal-Mode Performance Parameters of Diode-Pumped LuLiF$_4$ and YLiF$_4$ Lasers as a Function of Output-Coupling Mirror Reflectivity

Mirror Reflectivity	*LuLiF$_4$ Laser*		*YLiF$_4$ Laser*	
	Threshold (J)	*Slope Efficiency (%)*	*Threshold (J)*	*Slope Efficiency (%)*
0.98	0.4124	0.1776	0.4727	0.1451
0.94	0.4547	0.1931	0.5162	0.1518
0.90	0.5037	0.1731	0.5653	0.1476
0.86	0.6194	0.1485	0.6637	0.1114

Table 4.3 Normal Mode and Q-Switched Mode Performance Parameters for LuLiF$_4$ and YLiF$_4$ Solid-State Lasers at Room Temperature

Laser	Threshold Energy (J)	Wavelength (nm)	Slope Efficiency (%)	
			Normal Mode	Q-Switched
LuLiF$_4$	0.61	2.055	12.2	4.1
YLiF$_4$	0.69	2.052	11.9	3.9

Laboratory experimental data for the normal mode and Q-switched mode for LuLF and YLF solid-state lasers are summarized in Table 4.3. The optical-to-optical efficiency was 9.4% for the normal mode of operation for the Ho:Tm:LuLF laser at room temperature compared to 6.1% for the Ho:Tm:YLF laser. Optical-to-optical efficiency for the Q-switched mode of operation was 1.51% for the LuLF laser and 1.12% for the YLF laser. The data reveal that the Ho:Tm:LuLF laser has a lower threshold energy and higher slope efficiency compared to the Ho:Tm:YLF laser.

4.4 Quantum-Well Lasers Involving Rare Earth Materials and Their Potential Applications

Quantum-well (QW) lasers are best suited for applications where broadband tuning capability with strong nonparabolicity of the bands is the principal requirement. The effects of band nonparabolicity offer significant increases in gain and current density performance parameters for a laser using rare earth materials, such as a laser using EuSe-PbSe-Te-EuSe incorporating IV–VI semiconductor along with rare earth materials. It is interesting to mention that several kinds of QW heterostructure laser diodes have been designed and developed using materials such as InP–indium gallium arsenide phosphate (InGaAsP). Research in the early 1990s focused on the IV–VI diode lasers involving superlattices and QW heterostructures, which exhibit strong quantum optical effects that significantly affect the gain and current density characteristics of the laser [7].

QW lasers have existed since 1987. Several kinds of QW heterostructure laser diodes have been developed using complex semiconductor materials, such as InP-InGaAsP and AlGaAs-GaAs. Single or multiple QW diode lasers using IV–VI compounds have been fabricated using rare earth materials such as PbEuSeTe-PbTe, PbSnTe-PbSeTe, and PbEu-PbSe compounds. The PbEuSeTe quantum well demonstrated CW operation at 4.4-μm wavelengths under cryogenic cooling at 175 K and pulsed operation at 3.9 μm under cryogenic cooling at 270 K (–23°C). The IV–VI compound laser diodes have demonstrated impressive infrared radiation performance over wavelengths ranging from 3 μm to 30 μm at very cryogenic temperatures. The band gap in these materials is highly dependent on the operating temperatures. Therefore, by controlling the operating temperature, the emission wavelength can have a wide range. Due to this unique capability, a tunable laser source is possible in the infrared region, which offers extensive applications ranging from ultrahigh resolution spectroscopy to local oscillators in heterodyne systems.

There is a serious drawback in the operation of QW lasers: the nonparabolicity of the bands results in an energy-dependent density of the states in the junction plane of the compound structure. The effects of nonparabolicity have been observed in all directions on the gain and current density parameters. Essentially, the gain–current density relationship causes a reduction in the current density for any given gain and an increase in the gain saturation level. Furthermore, an approximately 20% shift in the output lasing energy generates nonparabolicity of the bands, which in turn lowers the modal gain values. This could be a problem in some optical systems.

The IV–VI theoretical two-band model ignores the adverse effects of intraband relaxation processes on the gain spectrum. This model does not take into account the current density and threshold calculations. Although this model is not perfect, it demonstrates the basic physical properties of QW devices and aids in the design of various useful QW structures.

4.4.1 System Performance Parameters and Anisotropy Characteristics

This section describes the performance parameters and anisotropy characteristics of the materials involved in QW lasers. For example, consider a EuSe-PbSe$_x$Te$_{1-x}$-EuSe QW laser device. Here, the barrier material is EuSe and the well material is PbSe$_x$Te$_{1-x}$. The values of laser energy gaps (E_g) can be calculated as follows:

$$E_g(x,T) = (0.17 + 0.057x - 0.095x^2) + (0.00004 + 2.56 \times 10^{-7} \times T^2),$$

where x is the percentage of tellurium material in the well region of the diode. Calculated values of the energy gap as a function of parameter x and cryogenic temperature T are summarized in Table 4.4. The laser structure and the calculated values of E_g are shown in Figure 4.7. As shown in Figure 4.7, the energy gap for the barrier material EuSe is 1.8 eV at the cryogenic temperature of 300 K (27°C), and it is assumed to be independent of the temperature. The well-growth is in the [100] direction, and it is assumed that the discontinuities in the conduction and valence band edges are equal according to the recombination processes.

The anisotropy of the constant energy surfaces can be determined through calculations of the mobility effective masses at the conduction and valence band edges [7]. The carrier mobility effective masses in the [100] direction, which is

Table 4.4 Energy Gap Values as a Function of Parameter x and Cryogenic Temperature for Quantum Well Material

Cryogenic Temperature (K)	Energy Gap Values		
	x = 0.60	*x = 0.70*	*x = 0.78*
77	0.2010	0.2028	0.2002
100	0.2210	0.2144	0.2110
200	0.2814	0.2634	0.2077
300	0.3220	0.3164	0.20567

called the growth direction, can be calculated using $1/m_w = 1/3\,[2/m_t + 1/m_l]$, where m_w is the mobility effective mass at the cryogenic temperature 77 K inside the well material in the [100] direction, m_t is the mobility effective mass in the transverse direction, and m_l is the mobility effective mass in the longitudinal direction. The effective mass outside the well is denoted by the parameter m_b. The mobility effective mass for the well material in the conduction and valence bands is equal to 5% of the electron free mass. The effective mass in the barrier material is considered parabolic and constant. The effects of the barrier material are less important due to the small value of the carrier electron wave-function in the barrier medium. For a well material with parabolic bands in the growth direction or z-direction, the effective masses in the above equation are those of the extreme of the bands and are independent of the energy.

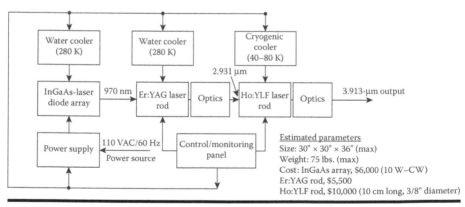

Figure 4.7 Block diagram for a two-level laser including an InGaAs diode array, Erbium-doped YAG laser and, Ho-doped YLF laser along with all accessories. Note that the optics include mirror and output coupler; the optical cavity and mirror will be coated with antireflection coatings; and the water cooler, power supply, and cryogenic cooler will be used as rental items to minimize overall system cost.

Table 4.5 The First and Second Energy Levels for a PbSe-Te Quantum Well

Calculation Method	First Energy Level (eV)	Second Energy Level (eV)
Parabolic equation	0.033	0.149
Luttinger-Kohn equation	0.028	Beyond approximate limit
Energy-dependent mass equation	0.027	0.083

The nonparabolic discrete energy levels can be calculated by the same equations derived for the parbolic bands. The energy level shift due to nonparabolicity effects differ significantly depending on the methods used in the calculations. The maximum energy level can be calculated using $(1/4)E_g = 0.05$ eV. For wells of 50 Aº and 3,100 Aº, the first energy levels lie close to 0.05 eV and 0.15 eV, respectively. Wider wells move the first and second energy levels below 0.05 eV. For thinner wells or higher energy levels, the deviations of the quantized energy levels from those neglecting the nonparabolicity are pronounced. The difference between the level shifts decreases as the well width increases. For larger well widths, the effects of the nonparabolicity on the energy levels decreases significantly.

The first and second energy levels for a 100-Aº $PbSe_{0.78}Te_{0.22}$ quantum well are summarized in Table 4.5. For the first energy levels in the quantum wells with well width of 100 Aº and higher, the small difference between the Luttinger-Kohn equation and the energy-dependent mass equation is neglected. The energy-dependent effective mass is adopted throughout the calculations.

4.4.2 Maximum Gain as a Function of Current Density for Parabolic Quantum Well Systems

The maximum gain of QW laser devices is strictly dependent on the current density and well width [7]. As mentioned, the anisotropy of the bands can be accounted for by using the value of the mobility effective masses at the band edges. The nonparabolicity in the well bands can affect the gain versus injected carrier concentration and gain versus current density relationship. The gain versus current density relationship is similar to that of the maximum gain versus carrier concentration relationship. The gain saturates at lower values as the well width is increased and the emission from the second excited state can be observed, which is evident from the 200 Aº well in the $PbSe_{0.78}Te_{0.22}$ QW laser system. It is interesting to note that the minimum current density decreases with increasing well width.

Maximum gain values as a function of current density and well width for this particular parabolic QW laser system are shown in Table 4.6. The gain decreases with an increase in the well width, while the maximum gain decreases for a given

Table 4.6 Maximum Gain as a Function of Current Density and Well Width

Current Density	Well Width			
	50 A°	100 A°	150 A°	200 A°
50 A°/cm²	1,350	1,120	845	680
100 A°/cm²	1,358	1,205	1,225	1,225
150 A°/cm²	875	916	950	950
200 A°/cm²	705	745	758	760

well width with an increase in current density. Band gap IV–VI semiconductors are highly dependent on operating temperatures; therefore, by controlling variations of temperature one can vary the emission wavelength, which can lead to the design and development of tunable laser sources over an infrared spectrum of 3 μm. In narrow gap semiconductors, the bands are strongly nonparabolic and anisotropic with prolate spheroid constant energy surfaces. These quantum well lasers are best suited for ultrahigh-resolution spectroscopy, scientific research, and local oscillator applications.

4.4.3 Quantum-Well Laser Diodes Using Conventional Semiconductor Materials

So far, the discussion has focused on quantum-well laser diodes using complex group IV–VI materials, such as EuSe-PbSeTe-EuSe. In that particular quantum-well diode, the barrier is made from EuSe material. In this section, the discussion is focused on standard semiconductors materials, such as InGaAs and InGaAsP materials [8]. Studies undertaken by the author reveal that high-performance InGaAs and InGaAsP quantum-well semiconductor laser diodes with heterostructure device technology and strained-layer configuration are capable of meeting vital performance parameters with minimum power consumption, improved reliability, and moderate cooling. Such quantum-well laser diodes offer infrared emissions over a very narrow spectral bandwidth compared to the spectral bandwidth associated with EuSe-PbSeTe-EuSe laser diodes. A semiconductor laser design configuration using InGaAs/InGaAsP quantum-well diodes is shown in Figure 4.8.

This type of quantum-well diode laser offers a low threshold current level, much improved conversion efficiency, higher differential quantum efficiency, and low-cost fabrication. Important operating features of this type of quantum-well laser can be summarized as follows:

- The perpendicular far-field (F-F) is independent of the drive current levels.
- The parallel F-F remains symmetrical at all output power levels.

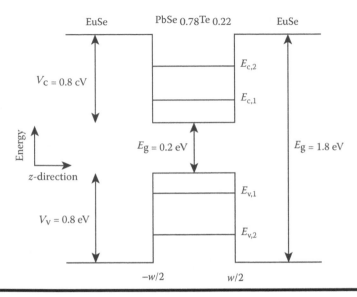

Figure 4.8 Block diagram for the EuScPhcTc quantum-well structure and various energy levels.

- The aspect ratio of 1.5 or less offers a circular beam with high quality and optical stability.
- The variation in laser beam divergence is negligible, with a 3-dB spectral width of 5 nm.

As mentioned, the total DQE is the most important performance parameter for a quantum-well laser. Research studies performed by the author on quantum-well lasers indicated that the total DQE is dependent on diode parameters, mirror reflectivity, optical cavity length, distribution losses, injection efficiency of the diode, and output emission wavelength. The calculated values of total quantum well efficiency (TQWE) as a function of the critical variables are summarized in Table 4.7. An injection efficiency of 73% has been assumed for these calculations. It is evident from these calculated values that the TQWE decreases with an increase in cavity length, but it improves at lower distribution losses for a given reflectivity. In some cases, the TQWE drops between 20 and 25.4% as the cavity length is increased from 1 mm to 2 mm. However, at cryogenic temperatures, both the TQWE and the power output increase. Note that cryogenic cooling of QW diodes provides higher differential efficiency, higher power output, and lower threshold current, in addition to improved reliability and longevity.

Experimental investigations by leading scientists revealed that the use of the binary compound InP in the InGaAs/InGaAsP heterostructure of a QW diode offers improved thermal performance and high reliability. Furthermore, the lower

Table 4.7 Calculated Values of Quantum Well Efficiency of a Quantum-Well Laser Diode

	Cavity Length			
	$L_d = 7.5$		$L_d = 6.5$	
Mirror Reflectivity	*1 mm*	*2 mm*	*1 mm*	*2 mm*
0.5	70%	46%	75%	50%
0.6	58%	37%	64%	41%
0.7	46%	28%	51%	32%

Note: L_d is the distribution loss per centimeter in decibels.

series resistance of the QW diode yields ultralow power dissipation in the junction, leading to significantly improved reliability and ultralong service life.

4.5 Applications of Rare Earth Elements in Atomic Clocks

The idea of using atomic transitions to measure time was first suggested by Lord Kelvin in 1879. In 1945, Isadora Rabi suggested that the atomic beam magnetic resonance could be used as the basis of atomic clocks. The first atomic clock was an ammonia maser device built in 1949 at the U.S. National Bureau of Standards, but the device was less accurate than the existing quartz clocks. In modern quartz clocks, the quartz crystal vibrates at a frequency of 32,768 Hz (32.768 kHz), which has a vibration frequency equal to 10^{15} cycles per second. The quartz crystal oscillator has become the backbone of quartz analog watches. The fundamental frequency (f) of the vibration of a cantilever of a quartz crystal is given as $f = [(1.875^2/2\pi) (a/l^2) (E/12\rho)]$, where a is the thickness of the cantilever, l is the length of the cantilever, E is the Young's modulus of the cantilever material, and ρ is the density of the cantilever material.

4.5.1 Performance Capabilities of Atomic Clocks Using Rare Earth Elements

Today, the frequency accuracy and time accuracy of atomic clocks exceed all other clocks. The first accurate atomic clock was based on a certain transition of the rare earth element caesium-133 atom; it was built by Louis Essen in 1955 at the National Physical Laboratory in the United Kingdom. Calibration of the caesium standard atomic clock was carried out using the astronomical time scale. This led to the internationally agreed definition of the second international (SI) being based on atomic time. The SI second has been verified to be within 1 part in 10^{10}.

Since the beginning of their development in the early 1950s, atomic clocks have been based on the hyperfine transitions in hydrogen-1, caesium-133, and rubidium-87. The first commercial atomic clock was manufactured by the National Company, and more than 50 units were sold between 1956 and 1960. This bulky and expensive instrument was replaced in 1964 by a much smaller rack-mounted device made by Hewlett-Packard, which was based on the caesium frequency standard. The 13th General Conference on Weights and Measures in 1967 first defined the International System unit of time, the second, in terms of atomic time rather than the motion of the earth. At this conference, a second was defined as the duration of 9,192,631,770 cycles of microwave light absorbed or emitted by the hyperfine transition of caesium-133 atoms in their ground state undisturbed by the external fields.

4.5.2 Performance Capabilities of Caesium Atomic Clocks

Caesium atomic clocks offer dependable, reliable, accurate, and stable frequency standards for navigation, communications, and timing operations. These atomic clocks provide the following performance parameters:

- Accuracy of 1 part in 10^{12}
- Frequency stability of 8 parts in 10^{14}
- Time domain stability of less than 2 parts in 10^{14}, with an averaging time of 5 days

Recent improvements in caesium clock technology include the replacement of the state-selection magnets with laser beams, which can select and detect the required transition with greater efficiency and minimum motion, thereby generating minimum noise from the radiating atoms. In addition, atomic clock performance is further improved through laser cooling, trapping of atoms, and deployment of high-finesse Fabry-Perot optical cavities.

4.5.3 Performance Capabilities of Rubidium Atomic Clocks

Rubidium is another rare earth element widely used in atomic clocks. A rubidium atomic clock uses a frequency standard in which a specified hyperfine transition of electrons in rubidium-87 atoms is used to control the output frequency. A rubidium-based atomic clock is the most inexpensive, compact, and widely used atomic clock. It is used to control the frequency of television stations, cell base stations, heavy test equipment, and global navigation satellite systems. Rubidium atomic clocks are less accurate than caesium atomic clocks, which serve as primary frequency standards. Essentially, the rubidium (Rb) clock is a secondary frequency standard device. Note as far as stability is concerned, recently designed rubidium clocks are considered more stable than caesium clocks.

4.6 Applications of Superconducting Compounds Using Rare Earth Elements

This section discusses the potential applications of rare earth elements in low-temperature and high-temperature superconducting compounds. Low-temperature superconducting compounds include vanadium silicon, lanthanum barium copper oxide, lanthanum strontium copper oxide, yttrium barium copper oxide (YBCO), and thallium barium calcium copper oxide (TBCCO). The High-temperature superconducting compounds include thallium barium copper oxide (221), TBCCO(2223), TBCCO(2212), YBCO(124), YBCO(123), bismuth strontium calcium copper oxide (2212), and BSCCO(2223).

The applications of high-temperature compounds are relatively easy and low cost. The various phases of these compounds strictly depend on the number of copper oxide layers and the valence of the chemical elements present in the compound. Conduction results from the chemical reaction between the copper and oxygen atoms. Specific details on the high-temperature superconducting compounds are summarized in Table 4.8.

4.6.1 Potential Applications of Some Prominent Rare Earth Compounds

Thin films of YBCO, TBCCO, and $LaAlO_3$ compounds are widely used in the design and development of microwave and millimeter (mm)-wave components, electro-optics devices, quantum-well lasers, high-performance optical crystals, IR

Table 4.8 Specific Details on Common High-Temperature Superconducting Compounds

Compound	Transition Temperature (K)	Number of Copper Oxide Layers
TBCO(221)	90	1
TBCCO(2223)	125	3
TBCCO(2212)	110	2
YBCO(124)	81	4
YBCO(123)	95	2

Note: TBCO, thallium barium copper oxide; TBCCO, thallium barium calcium copper oxide; YBCO, yttrium barium copper oxide.

devices, high-performance bolometers, mm-wave phase shifters, medical diagnostic equipment (e.g., magnetic resonance imaging [MRI], computed tomography [CT]), and a host of other components for defense applications. Erbium-doped optical fibers and optical amplifiers demonstrate high gains over wide spectral bandwidths.

4.6.1.1 Applications of Neodymium and Yttrium

A particular rare earth element may have several applications. For example, neodymium is used for permanent magnets, which are widely deployed in hybrid electric vehicles. It is also used in Nd:YAG lasers, which are well suited for space optical communications. Furthermore, yttrium elements are used in YIG filters and also for YAG material.

High-quality thin films are made from these compounds for use in applications requiring excellent surface conditions. Yttrium is widely deployed in the development of microwave YIG filters, in which sharp skirt performance is the principal design requirement. Neodymium is widely used in the neodymium-based Nd:YAG laser, which has demonstrated state-of-the art performance for space-based optical communication application in unpredictable space environments.

4.6.1.2 Applications of Erbium

Erbium-doped fibers are widely deployed in the design and development of fiber-optic amplifiers, which demonstrate higher gain over wide spectral bandwidths. These amplifiers offer optimum gain and low noise over wide spectral bandwidths under harsh operating environments. Erbium-based optical crystals are widely used in the design of dual-valence and trivalent laser systems, such as Er:YLF (dual-valence) and Er:Ho:YLF (trivalent). Scientific studies indicate that the trivalent rare-earth-doped linear crystals with lamp-pumping schemes have demonstrated CW power outputs more than 500 mW over the 3.9- to 4.1-μm range when operating under cryogenic conditions. These laser systems have demonstrated higher differential quantum efficiency, enhanced electrical-to-optical efficiency, and significantly improved beam stability with minimum cost and complexity. More output power levels have been observed under the lamp-pumping scheme than under the solid-state diode-pumping scheme. Note that rare earth nonlinear crystals yield extremely low conversion efficiencies.

4.6.1.3 Microwave Filters Using Yttrium

Yttrium has been widely used in the development of YIG microwave filters where higher skirt selectivity and low insertion loss in the pass-band region are of prime importance. It has played a critical role in the development of neodymium-based YAG laser systems, which have demonstrated excellent performance in two-way space communications between aircraft and satellites because of interterminal tracking, with angle errors less than 1 microradian (1^{-9} radian) and station-to-station acquisition capability in less than 6 s. This laser system is capable of operating at ranges of nearly 40,000 km with no degradation in reliability and electrical performance.

4.6.1.3.1 Microbolometer Operations Based on Rare Earth Materials

Another critical application for yttrium has been observed in the design of direct detection in a superconductor YBCO microbolometer [9]. These devices have demonstrated a responsivity of the spiral antenna-integrated microbolometer close to 190 V/W across the range of 330 to 1630 GHz and a response time better than 300 ps at a cryogenic temperature of 77 K. For YBCO high-quality thin films on bulk substrates, a bolometer response time of slightly greater than 1 ps can be achieved. Recent advances in nanoscience and microtechnology allow the cryogenically cooled direct-detector YBCO microbolometers to reach a photon noise-limited noise equivalent power (NEP) in the vicinity of 3 pW/Hz$^{0.5}$, which is much lower than what is available from conventional wideband terahertz detectors, such as pyroelectric detectors. Note that pyroelectric detectors typically have an NEP capability close to 9 pW/Hz$^{0.5}$ [10]. It is important to point out that the detector's responsivity and noise is strictly dependent on the cryogenic temperature. These two performance parameters can be optimized at a specified cryogenic temperature using the trial-and-error method. Terahertz direct detection seems to only be possible using a specific rare earth material—and that material happens to be YBCO, in which yttrium plays a critical role.

Responsivity is dependent on the measurement frequency and bolometer size. During responsivity measurements, bolometers must be biased at constant current to the resistive state (i.e., exceeding the critical current level). At each temperature, the responsivity maxima must be observed at approximately 40 mV for the 1.5-μm and 2.0-μm wide bolometers. The direct-current resistance at this point is approximately one-third of the normal state value. One can now summarize the responsivity measurements at 77 K as a function of frequency. During the measurements, the responsivity variation from 330 GHz to 1,630 GHz appears to be small for the small-size bolometer. For the larger-size bolometer (4 × 1 μm), the responsivity is reduced, scaling approximately as the inverse of the bolometer area. The responsivity degrades as the bolometer temperature approaches the transition or critical temperature (T_c).

The system noise V_n is strongly dominated by the noise that is coming from the readout device. At low modulation frequencies ranging from 18 to 200 Hz, the bolometer noise exceeds the lock-in amplifier noise. This low-frequency noise decreases as the temperature approaches the transition or critical temperature, which happens to be 90 K for YBCO oxide. At modulation frequencies above

500 Hz, the lock-in amplifier noise is dominating. However, the temperature dependence in this case is negligible. From 1 kHz to 100 kHz, which happens to be the higher frequency limit of the lock-in amplifier, the noise remained constant at around 3 nV/Hz$^{0.55}$. Studies performed on such bolometers seem to reveal that both the sensitivity of the temperature coefficient of the resistance and the responsivity of the bolometer improve as the operating temperature of the bolometer is reduced.

4.6.1.3.2 Theory of Superconducting Bolometers Using Rare Earth Elements

Niobium elements and YBCO compound film are best suited for designing a sensitive bolometer. The transition or critical temperatures for these two materials are 9.5 K and 90 K, respectively. Because the cooling cost for a niobium-based bolometer will be approximately three times the cost of cooling the YBCO, the bolometer discussion is limited to the device using the YBCO rare earth oxide.

The bolometer effect indicates the change in the electrical resistance of the responding element due to a change in temperature from the absorption of the incident IR radiation. The absorption of incident IR radiation will cause an increase in temperature, which will make the measuring bridge unbalanced. The change in the bolometer is dependent on the temperature coefficient of the bolometer resistance and the detector temperature. The temperature coefficient is defined as $\sigma_r = (1/R_d) \, [dR_d/dT_d]$, where R_d is the detector resistance and T_d is the detector temperature. The change in detector resistance is given by $\eth R_d = [dR_d/dT_d] \, [\eth T_d]$, where $\eth T_d$ is the change in detector temperature.

Superconducting bolometers are best suited to measure IR radiation levels from visible wavelengths to microwave frequencies. These bolometers can also be used for measuring absolute power levels. Superconducting bolometers use the temperature dependence of a rare earth element resistance in the bolometer near the critical temperature of the superconducting element deployed in the bolometer. The noise equivalent power (NEP) indicates the performance level of the bolometer. The bolometer measures the photon noise component, which consists of a photon noise component present in the incident IR radiation. The first two NEP components provide a fundamental limit to wideband bolometer sensitivity as measured at the critical temperature and the ambient temperature T_a.

4.6.1.3.3 A Reliable Method to Obtain Bolometer Measurements

Because deployment of imaging sensors using terahertz technology is increasing at a rapid rate, an accurate and reliable method is necessary to obtain bolometer measurements. One reliable and accurate method [10] uses spiral antenna-integrated superconducting bolometers operating at the cryogenic temperature of 90 K. Most compact bolometers are available with a cooling capacity down to

77 K, which makes high-T_c (90 K) bolometers more attractive for a wide range of applications.

At the critical temperature, the sensitivity of junction detectors at terahertz frequencies decreases. Even at liquid helium temperatures, the sensitivity or NEP is approximately 20 pW/Hz$^{0.5}$ at 600 GHz, which is significantly less than the sensitivity of YBCO bolometers. Therefore, as direct detectors, YBCO bolometers have demonstrated a photon-limited NEP better than 3 pW/Hz$^{0.5}$, which is much lower than for other wideband detectors, such as pyroelectric bolometers. For many applications, detectors with even higher response rates may be required, such as for observations of short electromagnetic pulses, fast scan spectroscopy, and active imaging. Bolometers using YBCO thin films on bulk substrates are capable of meeting a bolometer response time on the order of 1 ns. For optimum sensor performance, both the detector responsivity and the noise level must be optimized versus the bolometer operating temperature.

4.6.1.4 Applications of Superconducting Ceramic Compounds Involving Rare Earth Elements

Ceramic compound substrates and thin films composed of rare earth materials are widely used in the development of high-performance substrates that are best suited for microwave, mm-wave, and terahertz devices operating at cryogenic temperatures. These devices offer low insertion loss and high reliability, with compact size. They have potential applications in radars, lasers, electro-optics, and other defense and space applications where compact size, low power consumption, and ultrahigh reliability are the principal design requirements. The following sections identify the applications of superconducting substrates and thin films using rare earth elements along with their performance capabilities.

4.6.1.5 Unique Characteristics of Thin Films Using Rare Earth Elements and Their Applications in Microwave and Millimeter-Wave Devices

The unique characteristics of thin films using rare earth materials include anisotropic behavior, London penetration depth, atomic orbitals, conduction layers, electrical doping, transport properties, and microstructure format. To minimize costs, high-temperature superconducting (HTSC) technology should be used, as low-temperature superconducting devices are more expensive. Rare earth–based thin films can be deposited on various substrates, including lanthanum aluminate, magnesium oxide, strontium titanate, ultrastabilized zirconia, and alumina ceramic. The principal features of HTSC films and their potential applications are identified in Table 4.9.

The use of rare earth materials provides high-quality films with excellent surface conditions, greater smoothness, and an enhanced quality of Q factors, which

Table 4.9   Principal Features of High-Temperature Superconducting Films and Their Applications

Symbol	Transition Temperature (K)	Copper Oxide Layers	Potential Device Applications
TBCO(221)	90	1	Microwave and millimeter-wave components
TBCCO(2223)	125	3	Millimeter-wave devices
TBCCO(2212)	110	2	Cryocoolers
YBCO(124)	81	2	Microwave filters, antennas, motor coils
YBCO(123)	95	2	Industrial motors, flux-pinning devices
BSCCO(2212)	90	2	High current millimeter-wave devices
BSCCO(2223)	110	3	Superconducting tapes and wires

Source: Jha, A.R. 1998. *Superconductor Technology: Applications to Microwave, Electro-Optics, Electrical Machines, and Propulsion Systems.* New York: Wiley.

Note: TBCO, thallium barium copper oxide; TBCCO, thallium barium calcium copper oxide; YBCO, yttrium barium copper oxide.

essentially yields minimum insertion loss at higher RF frequencies and mm-wave frequencies. The use of mm-wave frequencies is strictly dependent on surface conditions and the uniformity of the superconducting properties of the HTSC metallic surfaces. For example, a microwave phase shifter using thin films of YBCO on lanthanum aluminate demonstrated impressive performance in terms of insertion loss and phase linearity on the entire RF bandwidth over 6 to 10 GHz. The same device demonstrated precise phase control at microwave frequencies with ultralow insertion loss, fine tuning capability, and fast tuning speeds while operating at a cryogenic temperature of 77 K [9].

A quasi-optical mm-wave band pass filter using thin films of YBCO on lanthanum aluminate substrate yielded remarkable filter performance over the entire operating range of 75 to 110 GHz. When fabricated using thin films of TBCCO on the lanthanum aluminate substrate, the same filter demonstrated much improved filter performance. The significantly lower insertion was due to the ultrasmooth surface, uniformity of superconducting properties of the films, and a lower cryogenic temperature of 15 K. Insertion loss in an RF device is dependent on the rare earth film, film properties, substrate material, and cryogenic temperature.

4.6.1.6 Microwave and Millimeter-Wave Components Using Thin Films Made from Rare Earth Elements

The following sections discuss the performance capabilities and limitations of microwave substrate materials and thin-film technology involving rare earth elements, as well as the performance capabilities and limitations of microwave and mm-wave components and their applications. Performance parameters are provided for several microwave and mm-wave devices, such as filters, phase shifters, printed circuit antennas, delay lines, and a host of other components using rare earth superconducting thin films and appropriate substrate materials. Insertion loss in a microwave device is the most critical performance parameter.

4.6.1.6.1 Delay Lines Using Thin Films Made from Rare Earth Materials

Uncooled delay lines suffer from high insertion losses. Superconducting thin films using rare earth elements play a critical role in the significant reduction of insertion losses. The insertion loss parameter is the most important performance parameter for microwave delay lines. Table 4.10 provides the performance parameters of a 100-ns superconducting delay line as a function of frequency using various substrates and rare earth elements.

4.6.1.6.2 Performance of Various Substrates Involving Rare Earth Elements

Lanthanum aluminate, strontium titanate, and yttria-stabilized zirconia substrates are well suited for designing filters with sharp cutoff characteristics, microstrip RF

Table 4.10 Insertion Loss (in dB) for Various 100-ns Delay Lines Using Yttrium Barium Copper Oxide Films and Various Superconducting Substrates as a Function of Frequency at a Temperature of 77 K

Frequency (MHz)	Delay Line Substrate Material		
	Magnesium Oxide	Sapphire	Lanthanum Aluminate
1,000	0.55	0.02	0.0002
2,000	0.72	0.83	0.005
4,000	0.85	0.28	0.030
6,000	1.81	0.74	0.085
8,000	2.12	0.92	0.23
10,000	3.24	1.64	0.32

components, resonators, delay lines, and mm-wave phase shifters with excellent phase stability. Insertion loss, phase stability, band-pass loss, stop-band attenuation, and current density are strictly dependent on London penetration depth in the niobium, YBCO, and TBCCO films and the unloaded quality factor (Q_{un}) of rare earth–based substrates. Q_{un} is strictly dependent on the film's characteristics and the substrate's physical and electric parameters. London penetration depth is a function of film thickness and cryogenic temperature. The current density is dependent on the current carried by the superconducting surfaces adjacent to the dielectric substrates.

4.6.1.6.3 Unloaded Quality Factors of the Substrate and Current Density of the Film

The unloaded quality factors of the substrate depend on the cryogenic operating temperature, surface condition, and electric and physical parameters of the substrate. The current capacity of the film is a function of London penetration depth, cryogenic temperature, film thickness, RF surface resistance of the film, and the electric conductivity of the film. The surface resistance of the film is a critical performance parameter that is strictly dependent on the superconducting film, RF frequency, cryogenic temperature, and film material.

The microstrip attenuation performance, which is a function of Q_{un} of the strip, is of paramount importance because it indicates insertion loss in the pass-band and stop-band regions in the microwave and mm-wave filters and other microstrip devices operating at microwave and mm-wave frequencies. Typical Q_{un} values of YBCO thin films on MgO and LaAlO$_3$ substrates as a function of frequency and cryogenic temperature are summarized in Table 4.11 [9]. Transition temperature, critical current density, quality factors, surface resistance, London penetration depth, and the Meissner effect are the most important characteristics of superconducting films.

4.6.2 Use of Rare Earth Materials in the Design of Ring Resonators and Radiofrequency Oscillators

Significant improvements in the performance of RF oscillators and ring resonators have been observed due to deployment of thin films of YBCO and TBCCO. HTSC-sapphire ring resonators using YBCO thin films provide a Q_{un} value that is better than 10^6 at a cryogenic temperature of 77 K and 10^7 at 4.2 K. This configuration offers the best overall performance with minimum weight and size. Microstrip ring resonators have demonstrated intrinsic quality factors in excess of 7,500 and 20,000 at cryogenic temperatures of 77 K and 25 K, respectively, at X-band frequencies. The lowest insertion loss in ring resonators can be realized using lower cryogenic temperatures and high-quality thin films of YBCO.

Table 4.11 Unloaded Quality Factors of Yttrium Barium Copper Oxide Thin Films Deposited on MgO and LaAlO$_3$ Substrates as a Function of Cryogenic Temperature and Radiofrequency

Frequency (MHz)	MgO Substrate (e_r = 10)		LaAlO$_3$ Substrate (e_r = 24.5)	
	77 K	4.2 K	77 K	4.2 K
1,000	7,800	19,800	23,100	36,150
3,000	2,600	6,600	7,700	12,000
6,000	1,300	3,300	3,850	6,000
9,000	870	2,200	2,580	4,150
18,000	435	1,100	1,290	2,400

Ring resonators are used in the design of RF oscillators where frequency stability and frequency drift are of critical importance. An X-band superconducting ring resonator-oscillator using a high-quality YBCO thin film on lanthanum aluminate substrate has demonstrated a residual frequency better than 0.1%. A significant improvement in phase noise and frequency stability of mm-wave oscillators is possible with ring resonator-oscillators with unloaded Qs better than 10^7 at a cryogenic temperature of 40 K. These improvements are due to a decrease in London penetration depth at lower cryogenic temperatures and the use of rare earth material films with optimum thickness and high surface quality. A fractional frequency change in an RF oscillator is strictly dependent on the cryogenic temperature and uniform thickness of the YBCO films. A fractional frequency change in an X-band superconducting ring resonator–RF oscillator is near 0% at a cryogenic temperature of 20 K, increasing to 0.5% at 77 K.

4.6.2.1 Performance of a Hybrid RF Oscillator Using Planar Superconductive Microwave Integrated Circuit Technology and YBCO Thin Films

The author was deeply involved in the design and development of a 10-GHz hybrid RF oscillator using planar superconductive microwave integrated circuit (PSMIC) technology. This oscillator design uses a ring resonator made from YBCO thin film deposited on a lanthanum aluminate substrate, a 10-GHz GaAs MESFET oscillator, and a transmission line section fabricated from the thin film using YBCO material. This particular RF oscillator demonstrated a minimum RF power output of 11 dBm and phase noise better than –68 dBc/Hz at an offset of 10 kHz from the carrier when cooled down to a cryogenic temperature of 77 K. The same RF oscillator using copper thin films demonstrated a phase noise of –43dBc/Hz under the same operating temperature, offset conditions, and output frequency. This clearly illustrates the benefit of YBCO thin films.

4.6.2.2 Performance Capability of a Surface Acoustic Wave Compressive Receiver Using Superconducting YBCO Thin Films

Electronic system measurement (ESM) is a critical component of an electronic warfare system. A cryogenically cooled ESM receiver offers significant performance improvement over an uncooled ESM receiver in terms of noise, sensitivity, and instantaneous bandwidth. The cooled version offers a reduction in both system weight and size compared to an uncooled version. A performance comparison between cooled and uncooled ESM receivers is presented in Table 4.12.

In a surface acoustic wave (SAW) compressive receiver, the frequency resolution as a function of receiver bandwidth can be summarized as follows:

- Resolution is 7.8 MHz with an instantaneous bandwidth of 1000 MHz
- Frequency resolution is 0.073 MHz (73 kHz) with an instantaneous bandwidth of 300 MHz
- Frequency resolution is 0.024 MHz (24 kHz) with an instantaneous bandwidth of 50 MHz

From these data, it is evident that frequency resolution improves when the compressive receiver's bandwidth is reduced. This frequency resolution is possible with compressive receivers using SAW filter technology. The compressive receiver architecture illustrated in Figure 4.9 offers improved overall receiver performance.

Table 4.12 Performance Comparison between Cooled and Uncooled Electronic System Measurement Receivers

Performance Parameter	Cooled (77 K)	Uncooled (300 K)
Frequency range (GHz)	2–20	2–16
Instantaneous bandwidth (GHz)	1.5	0.5
Maximum sensitivity (dBm)	90	60
Relative weight reduction (%)	45	45
Relative size reduction (%)	55	55

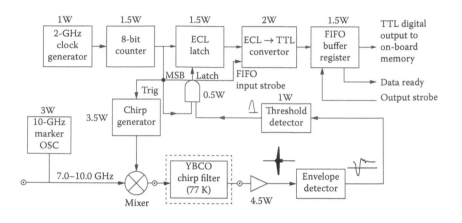

Figure 4.9   Architecture of a compressive receiver using SAW-filter technology.

4.6.2.3 *Cryogenically Cooled Wideband Compressive Receivers Using Chirp Filters Made from YBCO Thin Films*

This section describes cryogenically cooled wideband compressive receivers using chirp filter technology and incorporating YCBO thin films. This particular compressive receiver is widely deployed in electronic warfare systems in which improved frequency resolution over instantaneous bandwidth and high sensitivity are the principal design requirements. These chirp filters offer ultralow side lobe levels that, eliminate unwanted electronic signals; therefore, they provide high signal-to-noise ratios, leading to very high signal detection capabilities under noisy signal environments.

It is possible to detect ultralow-level electronic signals using this particular compressive receiver in combat and battlefield environments. However, the probability of signal intercept is extremely high under heavy signal and noise environments. The compressive receiver design configuration using a chirp filter incorporating YCBO film technology (Figure 4.8) offers improved receiver sensitivity, accurate amplitude measurement capability, high frequency resolution, instantaneous time-of-arrival capability, short-pulse measurement capability, and a wide dynamic range compared to a compressive receiver using SAW filter technology. The chirp filter provides dispersive time delays as long as 50 ns, which provide instantaneous bandwidth as wide as 20,000 MHz. It is possible to design chirp filters with time delays as low as 20 ns, but this introduces undesirable pulse jitter, which means it will be extremely difficult for the spectrum analyzer to measure narrow pulses and an accurate time of arrival.

4.7 Applications of Rare Earth Materials for Heavy Electrical Equipment

This section discusses applications of rare earth materials deployed in the design of heavy electrical equipment, such as levitated high-speed trains, high-power electrical motors, synchronous generators, and other electrical systems.

4.7.1 Application of Rare Earth Materials in the Design of High-Speed Levitated Trains

Advanced research and development programs carried out by German scientists and engineers indicate that electrodynamic levitation technology is most ideal for high-speed trains. In 1979, Siemens Research Center performed extensive levitation tests using linear synchronous motors (LSMs) with three-phase armature windings consisting of rare earth elements at the bottom of the track.

4.7.1.1 Operating Principals and Performance Capabilities of Electrodynamic Levitation Trains

An LSM motor is the major component of the electrodynamic levitation system. An LSM motor can generate thrust in excess of 4,500 lbs., which is sufficient to provide a maximum train speed of 150 km/h. Higher thrust can be achieved through other means readily available to design engineers. The Japanese National Railways (JNR) carried out extensive tests in 1978 on such a system, demonstrating a top speed of 517 km/h over a 7-km track using a 40-foot long, 10-ton test vehicle [9]. This particular levitated train was equipped with eight rare earth superconducting magnets (four magnets for levitation and four magnets for propulsion and guidance). The cooling of high-power superconducting magnets is accomplished by sealed helium-bath cryostats that are refilled between the test runs. The JNR introduced the bullet trains with top speeds exceeding 300 km/h using the superconducting electrodynamic technology. Later, JNR introduced bullet trains with top speeds greater than 550 km/h (395 mph). The levitated trains provide fast, efficient, reliable, and smoke-free transportation. Highlights of these trains can be summarized as follows [9]:

- Maximum number of cars in the bullet train: 16
- Top speed: 550 km/h (395 mph)
- Typical lift speed: 100 km/h
- Effective levitation height between coil centers: 25 cm
- Time of continuous operation without helium supply: 18 hours (equivalent to six trips between Tokyo and Osaka)
- Number of superconducting magnets per car: 8
- Heat load per car: 50 W at a cryogenic temperature of 4.2 K
  - Weight of one car: 30 to 33 tons (60,000–66,000 lbs.), including the weight of the car body, superconducting magnets and cryostats, helium recovery tank, and miscellaneous equipment and maintenance components
- Passenger weight: 6 tons (12,000 lbs.), which is included in the previous weight; this means that only 100 persons each weighing approximately 120 lbs. can travel in one car

4.7.1.2 Rare Earth Superconducting Magnet Requirements

Both neodymium-iron-boron and samarium-cobalt superconducting magnets seem to be suited for levitated train applications. However, samarium-cobalt magnets are brittle and therefore are not suited for adverse mechanical operating environments. This means only Nd-Fe-B superconducting magnets are suited for levitated train applications.

4.7.2 Important Characteristics of Nd-Fe-B Superconducting Magnets

Nd-Fe-B superconducting magnets offer high uniaxial magnetocrystalline anisotropy— close to 7 T or 70,000 G flux density. This rare earth compound has high coercivity, which means that this magnet offers high resistance to demagnetization. Furthermore, this compound has a saturation magnetization capability exceeding 16,000 Gauss. Because the maximum energy density is proportional to saturation magnetization, this magnet offers maximum storage capacity for magnetic energy, which is a product of maximum magnetic field intensity and flux density (BH_{max}) and exceeds 64 MG-Oe. Thus, the maximum storage capacity of the Nd-Fe-B superconducting magnet is significantly higher than that of samarium-cobalt.

Sintered Nd-Fe-B material tends to be vulnerable to corrosion, so a protective coating must be provided for high lifting or pulling purposes. This protective coating consists of nickel plating or dual-layered copper-nickel plating, which does not require a large investment. The Nd_2-Fe_{14}-B tetragonal crystalline structure is thought to be the strongest type of superconducting magnet. It is best suited for high-power motors, large lifting magnets, levitated trains, and magnetic fasteners in which strength, safety, and reliability under harsh mechanical environments are the principal requirements. Chapter 3 provides specific detailed performance capabilities and vital information on the lifting capacity of Nd-Fe-B magnets.

4.8 Applications of Rare Earth Compounds in Commercial Equipment and Medical Diagnostic Systems

This section discusses the potential applications of rare earth materials in industrial and commercial equipment, with an emphasis on cost, reliability, and longevity. Rare earth–based magnets are widely used in the design of electric motors, synchronous generators, heavy lifting equipment, and other industrial equipment.

4.8.1 Use of Rare Earth Compounds in Automobile Components

As mentioned in Chapter 2, neodymium-iron-boron permanent magnets are used by automobile manufacturers in electric motors and generators, which are considered the most critical elements in electric and hybrid electric vehicles. This particular magnet provides a significant reduction in the weight and size of motors and generators, in addition to significant improvements in reliability, electrical performance, and longevity. Of course, these benefits come with a higher cost of the automobile.

4.8.2 Role of Rare Earth Materials in the Medical Field

Neodymium permanent magnets offer a maximum energy product (BH_{max}) that is best suited for medical diagnostic equipment, such as MRI and CT, for which high resolution and accuracy are of critical importance. Furthermore, the presence of vibrations, irritating high noise levels, and discomfort is significantly reduced when using rare earth magnets compared with conventional diagnostic equipment.

Vanadium-based batteries are widely used in implantable pacemakers and other medical devices. Vanadium-based batteries offer high reliability, continuous sources of electrical energy, and longevity. These batteries provide continuous and reliable operation over extended periods exceeding 10 to 15 years.

4.8.3 Rare Earth–Based Magnets for Offensive and Defensive Weapon Systems

Samarium-cobalt magnets are widely used in the design and development of offensive and defensive weapon system components, such as radars and electronic warfare equipment. These permanent magnets retain all magnetic properties at operating temperatures close to 300°C. No other magnet can operate at this temperature. High-power RF sources, such as magnetrons or klystrons used by radars and traveling wave tubes (TWTs) in warfare equipment, can experience operating temperatures ranging from 100 to 250°C during combat operations. These rare earth magnets will ensure high reliability, improved clutter rejection, and enhanced frequency stability for the radars with coherent RF operation, which is vital for clutter rejection. For electronic warfare, TWTs using samarium-cobalt magnets will provide high RF power while maintaining uniform RF beams over the entire length of the helix structure of the TWT. Samarium-cobalt magnets will maintain their magnetic performance under harsh thermal and mechanical environments, which is essential during combat operations.

4.8.4 Ubitron Amplifiers

Ubitron amplifiers operate like TWT amplifiers but behave like high-power mm-wave amplifiers and are capable delivering 1.6 mW and 150 kW output power at Ku-band (16 GHz) and V-band (54 GHz), respectively [11]. Ubitron's operating principal is based on a periodic beam interacting with an unmodulating electron beam with the transverse of an unloaded waveguide section. The amplifier system consists of a magnetron injunction gun, an input coupler rectangular to circular waveguide, a samarium-cobalt permanent magnet, an electron beam section, a circular ceramic output window, and a laminated steel and copper interaction circuit. This ceramic window is capable of handling more power. The interaction area is approximately 9 inches long. The V-band amplifier requires a beam voltage of 70 kV at 35 A°. The permanent magnet or solenoid provides a magnetic flux density of 10,000 G. The tube requires water cooling to increase the average power output level to 250 W. The size and heat-handling capabilities of the waveguide allow extremely high power levels to be generated and amplified at higher frequencies compared to conventional TWTs. A flux density much greater than 10,000 G is possible with a samarium-cobalt magnet. As the frequency increases, a higher flux density is required. As mentioned, samarium-cobalt will maintain all magnetic properties, even up to 300°C, and therefore can meet both high CW and pulsed power capabilities at mm-wave frequencies. Samarium-cobalt permanent magnets are best suited for dual-mode mm-wave TWT amplifiers in electronic warfare (EW) systems.

4.8.5 Platinum in the Design of Jet Engines, High-Speed Gas Turbines, Scientific Equipment, and Nuclear Reactors

The rare earth element platinum has a melting temperature of 1,772°C and has a coefficient of expansion almost equal to that of soda-lime-silica glass, which makes it most ideal for sealed electrodes in glass systems. It is widely used for electrical contacts, which is the principal requirement for corrosion-resistant terminals needed in medical devices and other commercial components. Platinum wire is best suited for high-temperature electric furnaces needed for scientific and commercial research applications. Fuel nozzles in the jet engines of fighter aircraft and supersonic bombers continuously experience high temperatures as well as harsh thermal and mechanical environments. Rare earth element coatings, such as platinum coatings, are essential for preserving the structural integrity and performance of the nozzles under extreme thermal and mechanical operating conditions.

Platinum coatings are extremely important for steam turbine blades maintaining their structural integrity and longevity under harsh thermal and mechanical environments. Steam turbine blades continuously experience superconducting steam temperature, pressure, and entropy while the turbine rotor is rotating under high speed conditions. When coated with platinum thin films, steam turbine blades will maintain structural integrity and retain reliable turbine performance under superheated steam pressure, high temperatures, fluctuating entropy, and critical rotor vibrations.

Thin-film coatings of platinum are widely used to preserve and protect the structural integrity and surface conditions of missile nose cones, which experience high surface temperatures and high speeds in space environments. The safety and integrity of the missile cone is absolutely necessary for preserving the reliable operations of various electrical, electro-optical, and mechanical sensors housed in the missile cone during the missile launch, cruise, and terminal phases. The thickness of the platinum coating is strictly dependent on the atmospheric temperature gradients, inclination angle of the orbit, and the cruising speed at which the missile is traveling in the various regions of space.

4.9 Deployment of Rare Earth Materials in the Design of Electronic, Electro-Optic, Piezoelectric, Magnetoelastic, Ultrasonic, Surface Acoustic Wave, and Ferroelectric Components

In this section, the various types of electronic, electro-optic, piezoelectric, magnetoelastic, ultrasonic, surface acoustic wave, and ferroelectric devices using rare earth elements and compounds are described, with emphasis on their unique performance capabilities and potential applications. Devices such as solid-state delay lines, dispersive delay lines, ultrasonic delay lines, piezoelectric transducers, ultrasonic transducers, ubitron amplifiers, and high power millimeter-wave TWTs are described, as well as single crystals such as YIGs and YIG-doped erbium and terbium garnets.

4.9.1 Acoustic Materials for Room-Temperature Propagation of Microwave Energy in an Acoustic Form

Certain materials, such as sapphire (Al_2O_2), quartz (SiO_2), YIG, and YAG, are well suited for acoustic delay lines, SAW filters, SAW delay lines, and pulse-compression devices. The first two materials can be used at room temperature, whereas the remaining two require cryogenic cooling temperatures at 77 K. Ferroelectric lithium-niobate ($LiNbO_2$) may have transmission properties similar to those of YIG. Therefore, one of these materials can be used in the above-mentioned devices.

4.9.1.1 Pulse Compression Filters Requiring Dispersive Delay Lines

Pulse compression filters can be designed using dispersive delay lines, which use surface acoustic wave technology or RF pulse-forming networks. YIG filters can be designed as dispersive delay lines, which are compression filters. Note that if the time delay of the dispersive delay line is too short, the dispersion will not be complete; if too long, diversification of the compressed pulse will occur. To avoid these situations, the matched filter forms the basis for pulse compression-filter design. The pulse compression technique is widely used in forwarding-looking radars and side-looking radars to obtain high-range resolution, which is vital for target acquisition and identification purposes. Pulse compression ratios as high as 100:1 have been achieved by radar designers. The most critical design requirement for the dispersive delay line is a low insertion loss, with a dispersion curve of delay versus frequency accurately matching the design curve.

4.9.1.2 Solid-State Microwave Delay Lines

Solid-state microwave delay lines are suited for low-frequency microwave applications. Low-frequency ultrasonic delay lines meet the performance requirements of such a delay line. Low-frequency ultrasonic delay lines are most attractive for radars operating from 10 to 2,000 MHz. These delay lines are also known as acoustic delay lines; they are widely used by radars for which weight and size requirements are a few ounces and a few cubic inches. These delay lines are most suitable for electronic countermeasure systems and airborne radars, including moving-target indicators, pulse Doppler, and pulse-compression airborne radars.

4.9.1.3 Materials for Transducer Applications

The usual transducer materials are Y-cut quartz crystals, lead-zirconate-titanate (PZT), and sodium-potassium-niobate. The properties of these materials for the thickness-shear mode are summarized in Table 4.13.

Sodium-potassium-niobate contains traces of niobium. This transducer material is a hot pressing material and is widely used in the design of high-frequency transducers operating over the 10- to 40-MHz frequency range. This material is relatively expensive but highly recommended for the manufacture of thin high-frequency transducers. This material offers high acoustic velocity and has an advantage over PZT material in high-frequency thickness extensional and thickness shear transducers because this material allows greater thickness and therefore lower capacitance. Only sodium-potassium-niobate belongs to the rare earth material category. This material has a lower dielectric constant compared to PZT-7 material and therefore would see lower insertion loss at microwave frequencies.

4.10 Deployment of Rare Earth Materials in Nuclear Power Reactors

Some rare earth elements play critical roles in nuclear power reactors. A nuclear power reactor is a complex apparatus in which a chain reaction of fissionable material, such as thorium, uranium, or plutonium, is initiated and controlled with extraordinary safety. Orbiting atoms undergoing fission are ejected and many orbits of the atoms are released as fission products. The range of the fission products in various materials is very important for the nuclear design and for the safety of nuclear reactor operators. It is absolutely necessary to prevent their escape from the fuel elements into the surrounding medium. The fission products produced in a nuclear reactor are shown in Table 4.14, with an emphasis on their half-lives and

Table 4.13 Transducer Materials and Their Properties for Thickness-Shear Mode

Material	Dielectric Constant	Coupling Coefficient	Thickness-Shear (mm)
Y-cut crystal	4.58	0.137	1.87
Lead-zirconate-titanate	840	0.68	1.00
$Na_{0.5}K_{0.5}NbO_{0.2}$	545	0.65	1.54

Table 4.14 Fission Products Produced in a Nuclear Reactor during the Fission Process

Fission Product	Half-Life	Fission Yield (%)
Cerium	290 d	5.3
Europium	15.4 d	0.013
Lanthanum	40 h	6.1
Rubidium	17.8 min	3.1
Rhodium	30 s	0.46
Terbium	30 h	0.45

fission yields. Some rare earth materials can be extracted from mines as well as from nuclear reactors.

4.11 Summary

This chapter was dedicated to the potential applications of rare earth elements and their compounds in the design and development of electronic, electro-optic, medical diagnostic systems, and a host of defense-related sensors. Rare earth elements and their alloys and compounds have been deployed in the design and development of lasers operating at various infrared wavelengths. These lasers are widely used in various industries, scientific research, space acquisition and tracking sensors, airborne missiles, antisatellite systems, and medical applications. Erbium-doped YAG and holmium-doped YAG lasers using optical pumping schemes were discussed in great detail, with particular emphasis on their applications in defense sensors, medical diagnostic equipment, and product reliability verification under harsh operating environments. Performance parameters of CW and pulsed lasers were discussed, along with their applications in precision laser-guided bombs, CW target illumination, IR-guided missiles, and delicate surgical procedures where high resolution, precision optical focusing, and high beam stability are of critical importance. Performance parameters of Nd:YAG and Ho:YAG using flash-lamp pumping technology were briefly described, with emphasis on cost, complexity, and reliability.

An Nd:YAG laser using a flash-lamp scheme is most ideal for space optical communication applications. This Nd:YAG laser has demonstrated excellent beam quality, enhanced quantum efficiency, and low pointing errors. The Nd:YAG laser is also suited for applications in optical parametric amplifiers, optical communication components, spectroscopic measurements, pollution detection, and surgical procedures, where beam quality, optical stability, and improved optical parameters are the principal performance requirements. Cryogenically-cooled YAG lasers offer near-diffraction-limited optical performance and are most ideal for operating in oceans and unpredictable atmospheric environments. This particular laser system is suitable for coast guard and naval operations, where reliability and visibility are of critical importance.

A laser design using a lamp-pumping scheme and multilevel-doping incorporating rare earth host crystals, such as Tm:Ho:YAG, offers unique optical performance, including coherent CW laser output, higher conversion efficiency, and improved energy transfer efficiency at operating wavelengths around 2 μm. Such a laser system has demonstrated significantly improved optical performance at room temperature (300 K), which has potential applications in laser angioplasty procedures, arthroscopic surgery, laparoscopic procedures, and refractive surgery. The advantages and disadvantages of lamp-pumping and diode-pumping schemes were summarized in detail with major emphasis on output power, differential quantum efficiency, and distribution loss as a function of optical mirror reflectivity.

The operating principles and critical performance parameters of rare earth–based atomic clocks using caesium and rubidium were summarized, with particular emphasis on time resolution and frequency resolution. The major applications of atomic clocks include their use in global positioning systems and time signal radio transmitters that are strictly based on atomic time. Time accuracy better than 1 part in 10^{10} and frequency accuracy better than 8 parts in 10^{14} have been demonstrated by atomic clocks.

Potential applications of Nd-Fe-B and Sm_2Co_7 permanent magnets were briefly discussed, with particular emphasis on optimum flux density and maximum energy product (BH_{max}). These rare-earth permanent magnets have been used in the design and development of motors and generators in electric and hybrid electric vehicles. Manufacturers of these vehicles are impressed with the significant reduction in the weight and size of these electric components, in addition to a remarkable improvement in reliability under harsh thermal operating environments—all due to use of rare-earth permanent magnets.

Samarium-cobalt permanent magnets are best suited for applications where operating temperatures approach close to 300ºC. The helix structure in TWTs, which are widely deployed by radars and high-power electronic jamming equipment, experiences temperatures as high as 300°C in the collector region of the TWT. Thus, samarium-cobalt periodic permanent magnets are widely used in TWTs, where optimum flux density, maximum energy product, high reliability under harsh operating environments, and operating temperatures as high as 300°C are of critical importance. Neodymium-based permanent magnets are best suited for applications where operating temperatures could exceed 80°C. Applications of rare earth–based magnets in heavy-duty lifting cranes and levitated high-speed trains were briefly mentioned.

The deployment of rare earth elements in magnetoelastic, piezoelectric, ferro-electric, and ceramic ferrite should generate a new class of devices and components, such as microwave filters with minimum insertion loss in the pass-band region and maximum attenuation in the stop-band region, pulse compression filters with low time lobes and high compression ratios (ideal for pulsed and side-looking radars), transducers, dispersive delay lines that are best suited for pulse compression fil-ters, and other critical microwave components required by offensive and defensive weapons systems. Note that a high pulse-compression ratio is required to achieve high-range resolution and enhanced pulsed-power output.

Bolometer detectors using YBCO thin films and niobium wire elements have demonstrated significantly improved detector performance at infrared wavelengths. These detectors are best suited for terahertz systems or component RF performance measurements. Optimum performance of the bolometer is possible at liquid helium temperature (4.2 K). This detector has demonstrated a photon-limited noise equiv-alent power (NEP) better than 3 $pW/Hz^{0.5}$.

Ceramic substrates using thin films of YBCO and TBCCO materials have demonstrated improved RF performance in the low-loss microwave and mm-wave devices and optical components, which are best for possible integration in IR lasers, electro-optical sensors, mm-wave radars, airborne ECM systems, space tracking and reconnaissance sensors, and IR missiles. Components using ceramic substrates will provide longevity, safety, and reliable performance under unpredict-able vibration, thermal, and space radiation environments. Potential applications of components or devices using ceramic substrates and rare-earth thin films include thin-film miniaturized mm-wave filters, electro-optical devices, high Q-resonator oscillators, mm-wave phase shifters, and dispersive delay lines with high compres-sion ratios and suppressed time lobes.

High-quality hybrid resonators made of YBCO and TBCCO thin films have demonstrated significant improvements in RF performance, exceptionally high quality factors, and impressive reductions in component weight and size. These components are most ideal for airborne radars, space sensors, and airborne elec-tronic warfare systems where minimum weight and size and high reliability under space radiation environments are the principal design requirements.

Rare earth materials play an important role in the design and development of heavy commercial applications, such as electrodynamic levitation trains traveling at speeds exceeding 550 km/hour. These high-speed trains are currently running in Japan and China; they provide clean, safe, and reliable cross-country surface transportation. Cryogenically-cooled thin films of lanthanum can be installed at the bottom of the cars but above the track (less than 2 inches) to provide the necessary electrodynamic forces to produce high surface speeds. Rare earth elements also play a critical role in the design of high-power nuclear power reactors that are vital for electric power generation, wind turbine blades, and steam turbine rotor blades. In addition, rare earth elements are widely used in the design of jet engine fuel nozzles, gas turbine rotor blades, missile nose cones, and synchronous generators that are widely used in hydroelectric power plants. Rare earth elements are also widely deployed in the design of nuclear power reactors. Nuclear fuel rods of uranium and plutonium are used in the reactors to generate electrical energy exceeding 1,000 mW. These fuel rods can be reprocessed and used multiple times, as long as they are capable of sustaining the nuclear reactor's rated power capacity.

In summary, rare earth materials such as cerium, platinum, lanthanum, and other elements are widely employed in the design of fiber-optic amplifiers, magnetoelectric devices, and ultrasonic and surface acoustic wave components for space, defense, and commercial applications.

References

1. Chang, K. 1990. *Handbook of Microwave and Optical Components*. New York: Wiley.
2. Ross, M., P. Freedman, J. Abernathy, G. Matassov, J. Wolf, and J.D. Barry. 1978. Space optical communications with the Nd:YAG laser. *Proceedings of the IEEE* 66:320–323.
3. Jha, A.R. 1998. *Infrared Technology: Applications to Electro-Optics, Photonic Devices, and Sensors*. New York: Wiley.
4. Editor, *Photonics Spectra*, January 2012, p. 101.
5. Jha, A.R. 1995. *Infrared Continuous Wave Lasers*. Cerritos, CA: Jha Technical Consulting Services, 4–11.
6. Jani, M.G., N. Barnes, K.E. Murray, D.W. Hart, G.J. Quarles, and V.K. Castillo. 1997. Diode-pumped Ho:Tm:LUliF$_4$ laser at room temperature. *IEEE Journal of Quantum Electronics* 33:112–114.
7. Khodk, N.F., P.J. McCann, and B.A. Mason. 1996. Effects of band nonparabolicity on the gain and current density in EuSe-PbSe$_{0.78}$Te$_{0.22}$-EuSe IV-VI semiconductor quantum-well lasers. *IEEE Journal of Quantum Electronics* 32:236–239.
8. Jha, A.R. 1995. *SBIR Proposal on the Development of High-Power 1.5–1.8 Micron Semiconductor Quantum Well Lasers*. Cerritos, CA: Jha Technical Consulting Services, 2–7.
9. Jha, A.R. 1998. *Superconductor Technology: Applications to Microwave, Electro-Optics, Electrical Machines, and Propulsion Systems*. New York: Wiley.
10. Hammar, A., S. Cherednichenko, S. Bevilacqua, V. Drakinskiy, and J. Stake. 2011. Terahertz direct detection in YBa$_2$Cu$_3$O$_7$ microbolometers. *IEEE Transactions on Terahertz Science and Technology* 1:390.
11. Enderby, C.E. and R.M. Phillips. 1965. The ubitron amplifier—A high-power millimeter-wave TWT. *Proceedings of the IEEE* 53:1648.

Chapter 5

Critical Roles of Rare Earth Elements in the Manufacturing of Iron, Steel, and Other Alloys Suited for Industrial Applications

5.1 Introduction

This chapter investigates the role of rare earth elements in the production of various types of iron and steel, with particular emphasis on the metallurgical aspects and mechanical properties of the materials. Iron and steel materials are widely deployed in construction, industrial and commercial equipment, transportation, and all sorts of machines. The following sections discuss the critical roles of rare earth elements in manufacturing both ferrous and nonferrous materials.

5.2 Forms of Iron

Both gray iron and nodular iron are iron-carbon-silicon alloys. Gray iron is one of the oldest ferrous alloys, and it is widely used in various commercial

167

and industrial applications. The massive use of iron casting was the driving force behind the Industrial Revolution. Nodular iron a newer alloy that is used as an engineering material in the metal casting industry. It is an outgrowth from gray iron.

Gray iron contains 2 to 4% carbon and 1 to 4% silicon by weight, whereas typical nodular iron contains approximately 3 to 4% carbon and 2 to 4% silicon by weight. Therefore, carbon and silicon are the main ingredients in iron, regardless of the type, because of their effects on the microstructure and physical properties of the castings and any metallic bodies.

The overall properties of all cast irons are significantly affected by the presence of minute concentrations of many other elements, in addition to carbon and silicon. The presence of sulfur, even in concentrations from 0.01 to 0.1% (by weight), significantly affects the graphite morphology in solidified cast iron [1]. The presence of rare earth elements or even magnesium in amounts ranging from 0.02 to 0.1% (by weight) in iron that has a low (0.01%) sulfur concentration can change the entire growth pattern of the graphite. Certain minor elements are intentionally introduced for adverse effects, such as in the treatment of base irons with magnesium or rare earth elements (e.g., lanthanum, cerium, praseodymium, neodymium) to produce nodular iron, which has undefined and irregular surface-geometrical configurations. Other trace elements, such as sulfur, are sometimes introduced intentionally or carried along in the raw materials that are used to produce the iron. Elements such as magnesium, phosphorous, and titanium may be innocently picked up during the scrap melting process or melting operations.

The microstructures of cast irons are also dramatically influenced by the cooling rates. For example, if the cooling is rapid, no graphite can precipitate. Rather, the alloy solidifies in the metastable $Fe-Fe_3$ compound. In this particular process, the carbon is combined with the iron carbides, which has a fractured hard surface and hence is not suitable for machining operations. The fractured surface of the carbide cast iron is white and is very hard to machine, drill, or cut. Carbide iron castings are sometimes used for special applications where the dimensional accuracy, quality of surface conditions, and mechanical strength requirements are not critical. Carbide iron is a binary compound of carbon and a more electropositive element such as iron.

Graphite precipitates only if the liquid is allowed to solidify and cool as a thermodynamically stable iron-graphite assembly or a structure with recognizable geometry. In the case of gray iron, the fractured surface is gray because the carbon has precipitated as graphite and the graphite is visible on the fractured surface. Depending upon the heat transfer, the graphite morphology is altered somewhat. Characteristic forms of graphite in gray iron are therefore a function of the cooling rate of the metal. From a practical standpoint, little can be done to control the cooling rate of iron in a given casting assembly. However, nucleating agents can be added to the iron, which will minimize the constitutional undercooling associated with the formation of the iron-carbide phase. This can be accomplished by the appropriate selection of nodulizing and inoculating alloys that promote a very high degree of nucleation. The degree of nucleation is strictly measured by suitable counts in the nodular iron or chill depth in the gray iron assembly.

Graphite grows upon a prism, having faces of a hexagonal crystal structure, as illustrated in Figure 5.1 along with a polished specimen of gray iron. This growth is not uniform and is affected by the mechanisms that result in the irregular structures. The graphite forms a nearly continuous network within the numerous solidification cells with uniform geometrical aspects. To say that the gray iron is a steel matrix interrupted by large graphite inclusions is an oversimplified explanation. More accurately, when a sample of gray iron is polished and examined using a microscope, the two-dimensional cut through the graphite causes it to appear as isolated flakes.

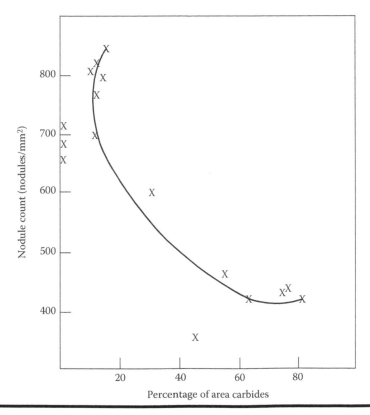

Figure 5.1 The inverse relationship between the nodule count and percentage area of carbides.

5.2.1 Interface of Iron-Graphite

The morphology of the flaked graphite at the shaped edges is generally thought to serve as an internal notch in the material itself. However, the cracks can readily nucleate at the ends and edges of the graphite material, then propagate through the graphite or along the iron-graphite interface. Gray iron generally fractures readily under mechanical stress. It is characterized by low tensile strength (ranging from 21,755–43,510 psi), negligible ductility, and low impact strength. The material is soft and can be readily machined with minimum time and effort.

Machinability is significantly enhanced by the lubricated effect normally observed on the graphite surface. Yields in the molds produced from gray iron are relatively high because of eutectic (an alloy showing the lowest melting point) graphite solidification. Note that the melting point of this material is relatively low compared to other ferrous alloys, which means that the energy input required during production is also low compared to some other ferrous materials. In spite of the limitations of its physical properties, gray iron is considered to be a useful engineering material, with exceptionally high casting tonnage shipped to various parts of the industrial world.

5.2.2 Important Mechanical Properties of Gray Irons and Their Close Alloys

The mechanical properties of gray irons are summarized in other textbooks in great detail. These mechanical properties can be altered using alloying elements such as copper, tin, manganese, chromium, and nickel. The alloys affect the mechanical properties of the matrix more than the morphology of the graphite. The physical properties of the iron are primarily affected by the graphite morphology, regardless of the matrix structure or module configuration. The structure configuration or matrix structure that results from alloying introduces limitations on the modification of physical properties.

If the graphite morphology can be altered, certain physical properties of the cast irons can be improved. The malleable iron industry has long made use of annealing procedures applied to cast irons that had initially solidified the metastable white iron system (Fe-Fe_3C). Heat treatment procedures applied to white cast iron castings cause the iron carbide to break down because white iron has poor physical properties. The carbon diffused through the matrix results in a breakdown, leading to the generation of some form of spheroids rather than flakes. Irons produced in this manner seem to demonstrate higher tensile strengths than gray iron, ranging from approximately 52,925 to 101,500 psi. In addition, these irons show appreciable elongation, ranging from approximately 2 to 18%. This illustrates that the improvements in mechanical properties are strictly due to the change in graphite morphology from flakes to spheroids. Therefore, changes in graphite morphology are essential if improvements in mechanical properties are desired.

5.2.3 Properties of Malleable Irons

Two distinct problems are associated with malleable irons: 1) heat treatment procedures are costly, and 2) it is essential that such irons solidify in a metastable state. The requirement associated with the second problem makes the production of heavier, slow-cooling sections impossible because they would most likely contain free graphite elements. However, the higher tensile strengths and ductility of malleable irons are well suited for some industrial applications.

Therefore, the problem confronting the foundry industry is how to produce spheroid graphite without heat treatments or constraints upon the selection of material thickness. Because of these challenges, nodular iron was developed, and rare earth elements have assumed a critical role in this process. Henton Morrogh of the British Cast Iron Research Associates undertook research experiments that added a variety of alloying elements based upon three assumptions: 1) the nodulizing rare earth material needs to be an iron carbide–stabilizing agent; 2) it must be capable of desulfurizing the iron; and 3) the rare earth element should be readily dissolved in the iron. These research efforts resulted in the discovery that cerium causes spheroidal graphite structures in nickel-carbon alloys. Continued experiments revealed that spheroidal graphite could be created under laboratory conditions, which will produce irons with 0.02% cerium introduced as mischmetal (Figure 5.2). The mischmetal was introduced into the hypereutectic iron-carbon silicon alloys containing less than 0.06% sulfur in the base material or untreated iron. The phosphorous level must be restricted to less than 0.1% in iron-carbon-silicon alloys to the retain improved mechanical properties of the alloy.

At the 1948 International Foundrymen's Society Convention, the International Nickel Company announced that the company had been successful in producing nodular iron in production environments [1]. In this particular production process, magnesium was added as a nickel-manganese alloy. Both hypoeutectic and hypereutectic base irons have been successfully treated using the nickel-manganese alloy. This was the birth of the nodular iron industry.

The acceptance of nodular iron as an engineering material was controversial from the beginning. Further studies were undertaken to explain the mechanisms of graphite growth. However, even after much time, the phenomenon of producing spheroidal graphite was not explained clearly. It was speculated that this effect was a result of the interrelated roles played by the nodulizing elements, such as calcium, magnesium, and other rare earth elements. Other researchers believed that sulfur and oxygen could be removed from the molten product due to the formation of oxides, sulfides, or oxysulfides, although the graphite growth would be affected. Finally, it was recognized that spheroidal graphite is characterized by growth on the basal faces, as seen in gray iron. Note that numerous fibers of the graphite grow radially from a given nucleation site.

Laboratory experiments have revealed some of the spheroids of the graphite in nodular iron test samples. Using transmission electron microscopy, these inclusions

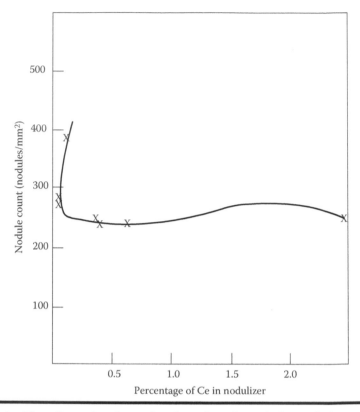

Figure 5.2   The effect of cerium when introduced as mischmetal along with the treatment alloy as a function of the cerium content of the alloy.

can be analyzed as being oxides of the rare earth elements cerium, lanthanum, and magnesium. The electron microscope has allowed the identification of the sulfides of cerium, lanthanum, and magnesium as well as calcium-magnesium sulfides at the centers of the nodules or exposed surfaces. Such conclusions demonstrate the heterogeneous substrates on which some of the graphite can readily nucleate, leading to a stony substance.

The role of the nodulizing elements is important. These elements clean the melt of other elements, such as sulfur or oxygen, which prohibits the growth of spheroidal graphite. The compounds formed in this way provide heterogeneous substrates for graphite nucleation. It should be remembered that the spheroids of the graphite no longer serve as internal parts in the matrix. As a result, the material produced is not as easily fractured as gray iron. Therefore, the tensile strengths, yield strengths, and percentages of elongations in castings containing spheroidal graphite are dramatically increased over those properties in gray iron. Nodular iron materials have demonstrated tensile strengths ranging from 43,500 to 58,000 psi, with percentages of elongations from 6 to 18%; these parameters are on the order of those observed in malleable iron materials.

The physical properties can be modified by alloying techniques or using appropriate heat treatment procedures for gray iron materials. The machinability of the nodular iron castings is as good as gray iron castings. However, the material yield is somewhat lower because of the altered solidification process—approximately 50% compared to 60 to 70% for the gray iron material. The impressive physical properties of nodular iron coupled with its good founding qualities have established its status among high-quality engineering materials.

There is some confusion within the foundry industry regarding the roles of some rare earth materials using mischmetal. In many cases, only the cerium level was reported in rare earth material tests and the presence of other rare earth materials was ignored. Even today, the rare earth elements often mentioned are the first four lanthanides: cerium, lanthanum, neodymium, and praseodymium. Even though their roles are critical, the effects of the other rare earth elements in the series will not be similar to these four elements. The roles of the other elements have not been investigated individually in the same way. Keep in mind that commercially used rare earth materials may contain small quantities of those other rare earth elements.

The use of mischmetal led to a great demand for bastnäsite ore, which contains various alloys with some content of rare earth materials. In these alloys, roughly 50% of the rare earth materials are cerium-based. In the mid-1960s, a more economical source for cerium was introduced, which was in essence a concentrate from which the lanthanum element was removed. This particular material allowed for the production of alloys with a rare earth concentration of approximately 90% cerium. These alloys are considered earth-bearing materials, and their approximate analyses are summarized in Table 5.1. It is evident that cerium plays a critical role in both the low-cerium rare earth materials (or mischmetal) and high-cerium rare earth materials [1]. These materials are widely used in commercial applications.

Table 5.1 Analyses of Appropriate Rare Earth Sources

	Low-Cerium Sources or Mischmetal (%)	High-Cerium Sources (%)
Cerium	50	90
Lanthanum	33	5
Neodymium	12	2
Praseodymium	4	1
Other rare earths	1	2

Various rare earth materials are widely used in the foundry industry as silicides, in which the rare earth content is roughly 30%. Other alloys are used as well in which the level of rare earth materials is approximately 10%. It is estimated that approximately 10% is cerium and 2% is other rare earth materials, mainly silicon and iron. Rare earth elements are present in amounts ranging from approximately 0.1 to 1.0%. These alloys are used in various commercial applications.

Rare earth elements play three distinct roles in the production of iron materials: nodulizing elements, graphite modifiers to enhance the nodule count, and controllers of deleterious elements. The use of rare earth elements for each of these purposes is described in great detail in the following sections.

5.3 Role of Rare Earth Materials as Nodulizers in the Production of Nodular Iron

Simply stated, modern foundry processes for producing nodular iron cores use a sophisticated treatment of a base iron material containing 3 to 4% carbon, 1 to 2% silicon, 0.0005 to 0.01% sulfur, and very little phosphorus (<0.05%). The industrial treatment is carried out by the introduction of an appropriate nodulizer into this base iron. Sometimes, the inadequate addition of a nodulizer could result in an incomplete spheroidization process. On the other hand, excessive concentrations of nodulizers promote the formation of iron carbides.

Nodulizing elements can include rare earth elements, such as magnesium, yttrium, and calcium. However, the latter two elements are rarely used because of economic and technical problems. Furthermore, the range of efficacy for a given nodulizer is quite restricted because the content of magnesium and other rare earth elements is 0.02 to 0.1%, with 0.01% sulfur in the iron. Therefore, nodular iron treatment procedures require the addition of other treatments. This step is required to ensure that the matrix solidifies completely free of iron carbides and to provide for adequate nucleation and the growth of the spherodial graphite.

For economical nodular iron production, the sulfur content in the untreated iron must be between 0.004 and 0.05%. However, the sulfur content is usually held around 0.01% to minimize the necessary addition of nodulizing rare earth elements. Note that a minimum sulfur content is required in the untreated iron to facilitate adequate nucleation to maintain the quality of the material produced.

Base iron materials or untreated iron materials are melted in electric furnaces. The sulfur levels in the iron melted in electric furnaces are generally around 0.05% in the base materials. The use of coke-fired cupola with silica lining could result in the production of iron materials having sulfur content of approximately 0.1%. A major portion of the total tonnage of cast iron production uses these acid-lined cupolas for higher yield. Basic slag cupolas are generally used for the production of base iron materials with a sulfur content that does not exceed 0.01%, as measured at the cupola spout; however, they are rarely used. The basic cupolas are more expensive to reline and more difficult to control than acid-lined cupolas.

Because most untreated iron materials have a sulfur content greater than 0.01%, a desulfurizing procedure that drives the sulfur level to approximately 0.01% is highly desirable. Rare earth elements can be used to desulfurize the iron. However, because of the valence state of rare earth elements and their high mass weights, a significant mass input of rare earth materials would be required to remove the sulfur from the iron; thus, the cost of such a process could be unacceptable for economic reasons. Modern methods, such as the use of calcium-carbide desulfurization, make this procedure more economical than the use of rare earth materials. The cost of rare earth elements are exceptionally high compared to other materials (except for precious metals, such as gold or platinum). The typical costs of rare earth elements are shown in Table 5.2. Note that the prices quoted in the table are approximately 40 years old as current prices are not readily available. However, it is estimated that the current prices would be at least 20 to 25% more.

Table 5.2   Typical Cost of Various Rare Earth Elements

Rare Earth Element	Cost ($/lb)
Cerium (Ce)	70
Dysprosium (Dy)	320
Erbium (Er)	500
Europium (Eu)	4,000
Gadolinium (Gd)	250
Holmium (Ho)	500
Lanthanum (La)	70
Lutetium (Lu)	8,000
Neodymium (Nd)	10
Praseodymium (Pr)	225
Samarium (Sm)	255
Terbium (Tb)	750
Thulium (Tm)	1,200
Yttrium (Y)	200
Ytterbium (Yb)	300

5.3.1 Most Common Nodulizing Materials in the Production of Nodular Iron

Nodulizers are generally introduced into desulfurized base iron materials. Magnesium or magnesium-based alloys are often used as primary nodulizer materials. Economic concerns led to the deployment of the magnesium and magnesium-based alloys in commercial applications rather than rare earth elements. In the early days of nodular iron production, the sulfur content in the iron was high and cost-effective desulfurization means were not available; it was too costly to make nodular iron using rare earth elements. Magnesium became an economically feasible alternative method. Generally, the magnesium residual required to produce spheroidal graphite varies with the thickness of the casting. In the case of a nodular iron pipe with very thin sections, residuals may be as low as 0.02% or less. In heavy section castings, the residual magnesium may be as high as 0.1%.

However, the use of magnesium as a primary nodulizer is not free from problems. Elemental magnesium vaporizes at 1,103°C under 1 atm or 14.7 psi, which is lower than the temperature of the molten metal into which the magnesium is introduced (1,500°C). The alloys introduce the magnesium as either magnesium silicate (Mg_2Si) or as a nickel magnesium (Ni-Mg) phase, and the dissolution of these phases ultimately results in nascent magnesium (which is present in the melt). Therefore, the magnesium tends to vaporize even under unsuitable pressures. To avoid problems, the magnesium sources used (except some nickel-magnesium sources) are currently commercially available sources that are less dense than the liquid base iron. This allows the materials to float close to the surface of the bath, where combustion can take place efficiently. Magnesium oxidation takes place rapidly because of the dissolved oxygen in the melt. Because of these factors, magnesium recoveries range from 30 to 50% when approximately 5% magnesium-ferrosilicon alloys are used.

Because of poor magnesium recoveries and the present availability of desulfurized base iron, significant efforts have been undertaken to establish rare earth materials as primary nodulizers, regardless of the cost of the rare earth materials. Nodular iron was originally produced using mischmetal. It is important to mention that the rare earth becomes dense because of higher material density (6.6 g/cc for cerium element) than the liquid-based iron material (6.2 g/cc) for gray iron. Gray iron materials are liquid at iron foundry temperatures. Recoveries of the rare earth materials present in gray iron have been found to be exceptionally high. However, the efforts to use rare earth materials as primary nodulizers have not been fully recognized in commercial applications for unknown reasons. Rare earth materials are considered to be strong carbide stabilizers, but there is a propensity for the formation of carbides in hypoeutectic iron mixtures if the level of the rare earth materials becomes too high for the nodular section thickness of the casting.

Industrial sources have reported the use of several nodulizers in the treatment alloy. During these tests, rare earth materials—namely yttrium and magnesium—were used to produce nodular iron materials. It was concluded that the recovery of the nodulizing elements was improved. Furthermore, the loss of potency over time was reduced regardless of what element was initially added. It was further reported that the best nodule efficiency was obtained when some magnesium was added. Based on these results, a proprietary alloy that contains approximately 2% rare earth elements was produced. This particular alloy has the advantage of higher magnesium recovery and includes roughly 5% magnesium-ferrosilicon alloys, which will not affect the effective utilization of this product for various commercial applications. This alloy has been used for various applications with no compromise in mechanical and thermal characteristics.

The potential applications of some individual rare earth elements as nodulizers have also been explored. Researchers were able to produce nodular iron that retained adequate physical properties without an excessive presence of iron carbides. This was consistently observed for four rare earth materials—cerium, lanthanum, neodymium, and praseodymium—when they were evaluated as nodulizers. Approximately 1.5 times more nodulizer material was required for neodymium or praseodymium and 2.5 times more for lanthanum or cerium to yield equivalent results. No reasons were given for the variations in the behaviors of these rare earth materials.

5.4 Role of Rare Earth Materials as Nucleating Agents

Material research scientists observe a relationship between the nodule shape, size, physical properties, and structural distribution of the iron developed as a result of the incorporation of a rare earth material as a nucleating agent. Research has attempted to relate these morphological properties to the physical properties. Material scientists believe that quantitative metallography is related to impact resistance and graphite morphology. These research efforts are still in the early stages. However, it is generally accepted that the nodule count (in terms of nodules per square millimeter) and propensity for the formation of primary iron carbides are inversely related, as shown in Figure 5.2.

In nodular iron, the goal is a carbide-free matrix: hence, a high nodule count that implies adequate nucleation is generally considered the most desirable. The nodule count seems to increase with the addition of rare earth materials along with magnesium compared to the addition of magnesium alone. These improvements in nodularity percentage with the addition of rare earth materials are especially observed when the magnesium residual is marginal.

5.4.1 Elimination of Iron Carbon

The elimination of iron carbide contents could cause some deterioration in the physical properties of the iron. A scientific investigation demonstrated that the introduction of cerium as mischmetal in appropriate amounts is effective in eliminating iron carbides that cause deterioration in the physical properties of the material. The elimination of iron carbides in thin sections by the proper use of rare earth materials is a major contribution to the iron industry. Material scientists seem to agree that there is an optimum percentage for the added rare earth material—cerium ranging from 0.01 to 0.02% and other rare earth elements ranging from 0.02 to 0.04%. These amounts of rare earth elements increase the nodule count and control iron carbides when used in conjunction with magnesium-based nodulizers (Figure 5.2). This particular scientific work demonstrated the importance of adding appropriate rare earth elements with a magnesium-ferro-silicon alloy and also made available the ferroalloys containing specified amounts of rare earth elements in the open market at reasonable prices. Most of the ferroalloys currently available in the United States employ cerium-based hydrates. Essentially, they add the cerium element only with magnesium-ferro-silicon.

Material scientists claim that increases in nodule counts are more readily available (under laboratory thermal environments) if the ratio of lanthanum to cerium is higher than the cerium-rich magnesium ferroalloys. Because of this, alloys using refined bastnäsite ores as the rare earth source are also available and have found some industrial acceptance. Until further research is undertaken, both sources of rare earth materials will continue to be used. Variations in the consumption of a particular rare earth element might reflect changes in marketplace, especially by the individual producers of rare earth materials.

The final step in the nodular treatment process is postinoculation. This procedure accelerates the elimination of iron carbides and promotes the enhanced nucleation and proper growth of graphite spheroids, as illustrated in Figure 5.1. This step is generally accomplished by the introduction of silicon or a ferrosilicon alloy along with the calcium, possibly with some magnesium or an appropriate rare earth material. The benefits of rare earth additions are not affected by the moment in the process when they were added. For example, the elimination of iron carbides by rare earth materials is possible whether the rare earth materials are introduced in the process along with the primary nodulizer or with the postinoculants or insert. Note that both the primary nodulizers and ferrosilicon inoculants contain at least 1% calcium. Material research has demonstrated that some calcium addition is required for proper graphite nucleation, which is the process of forming a nucleus. This particular effect is observed even if the cerium is added during the treatment process.

5.4.2 Role of Rare Earth Materials in Controlling Deleterious Elements

Rare earth materials such as cerium, lanthanum, neodymium, and praseodymium can control the effects of certain harmful elements that are present in very minute quantities, with concentrations ranging from 0.01 to 0.05% in the melt (i.e., the material in the molten stage). The effectiveness of rare earth materials in controlling these subversive elements is strictly responsible for the success of the magnesium treatment process. They made the production of nodular iron more economical, leading to the commercial success of the nodular iron industry.

Harmful elements include lead, bismuth, antimony, and titanium. Other harmful elements have been identified, including tellurium, indium, and thallium. However, the first four elements listed are of most concern to the foundry industry. These deleterious elements are generally carried along in pig iron and steel scrap melted by the foundry. If the foundry is reserved for gray iron, slight concentrations of some of these harmful elements will not be catastrophic. However, if a foundry converts some of its capacity to nodular iron, the adverse effects of these elements will be observed. For example, many malleable iron foundries use bismuth and tellurium as carbide stabilizers in their normal operations to avoid primary graphitization. If these foundries convert to the production of nodular iron, possible sources for two deleterious elements already may be in the foundry. In another example, lead is generally used as a metal. Therefore, some lead-containing materials may be inadvertently charged into the melting furnace. Lead paint, which may be used on castings, can have some adverse effects, even if the lead is volatilized to some extent due to the iron foundry temperatures.

Lead also adversely affects flake iron morphology. Lead concentrations of just a few parts per million can result in Widmanstätten graphite, which has strange physical properties. Sufficient quantities of these harmful elements can result in the degeneration of graphite spheroids to vermicular (resembling a worm) or flake graphite. This causes deterioration in the physical properties of the casting. The tensile strengths and percentage elongation are similar to those of gray iron. However, the addition of appropriate rare earth elements can neutralize the adverse effects of these harmful elements, such as lead paint. For example, consider the graphite in a sample containing the minute content of 0.012% lead. After the addition of 0.017% cerium as mischmetal, the total addition of rare earth elements comes to approximately 0.03%, which will restore the sphericity of the graphite to a maximum level, as illustrated in Figure 5.2.

Adverse effects due to small concentrations of bismuth and antimony have also been observed in various alloys that are widely deployed in commercial and industrial applications. These effects have been overcome by the additions of small amounts of rare earth elements. It is estimated that approximately 0.01% cerium will neutralize the adverse effects of harmful elements present in the materials. Note that mischmetal contains roughly 50% cerium and 50% other rare earth elements, such as lanthanum, neodymium, and praseodymium. After the complete neutralization of adverse effects, the production of high-quality nodular iron has been realized, while allowing the use of commercially available steel scrap as a raw material.

Some researchers have suggested that the rare earth elements must be combined with the deleterious elements to form innocuous insoluble intermetallic compounds. The particles of such compounds can be observed using an electron microscope only when the concentrations of both the rare earth elements and deleterious elements are well above those levels usually found in commercial practices. Even then, the compositions of the particular phases are not determined accurately. Note the effectiveness of the level of cerium at which the beneficial effects can be observed suggest that the mechanism may or may not be simply compound formation as mentioned above.

Thermodynamic studies concerning rare earth compounds formed with bismuth, lead, and antimony indicate that the formation of some intermetallic compounds is possible. It is unclear if the amounts in a bath that also contains approximately 0.01% sulfur and 0.01% oxygen are sufficient to control the deleterious elements by forming intermetallic compounds. Furthermore, it is unclear if the rare earth elements are used up completely in other chemical reactions. It is well known that arsenic can be alloyed with nodular iron to form a paralytic matrix. Content levels of 0.2% arsenic and 0.03% manganese have not had significant effects on graphite morphology, even when cerium-free alloys are used as nodulizers. In this particular case, arsenic does not behave like other elements of that family in the periodic table.

Other than titanium, deleterious elements can be used to stabilize the iron-carbide phase. Titanium is the only so-called deleterious element that appears to be controllable by the addition of more magnesium instead of one of the rare earth elements commonly used for this process. The use of rare earth elements allows titanium to be controlled more efficiently. Titanium does not form iron metallic compounds with rare earth elements either. Perhaps the rare earth materials act as active elements on the growing graphite surface or somehow affect the chemical activity of the carbon. It is possible that some of the intermetallics are due to graphite growth by formation of some other compounds. There is a difference in the chemical properties of the various iron intermetallics, which could permit the use of arsenic to stabilize various impurities that are present in the compound. However, more research studies must be undertaken to improve the effectiveness of rare earth elements in overcoming and controlling adverse effects due to deleterious elements. Existing experimental data seem to indicate that cerium is the most effective element for controlling these adverse effects. Cerium is widely available in the market at reasonable prices; however, its potency has not been addressed in great detail.

Because of their effectiveness in overcoming deleterious elements, some rare earth materials have been added to proprietary magnesium-ferrosilicon alloys produced by major manufacturers in highly industrialized countries, such as the United States. A foundry can now easily obtain a magnesium-ferrosilicon alloy containing a specific percentage of rare earth elements (typically ranging from 0.3–1.0%) for its microstructural benefits. If a foundry specifies the "regular" grade of magnesium-ferrosilicon alloy, there is hardly 0.1% of rare earth material present in the alloy. A foundry may add 1.5 to 2.0% of a 5% magnesium-ferrosilicon alloy using the primary nodulizer. In that case, enough rare earth element (ranging from 0.015 to 0.12%) may be added with the regular magnesium-ferrosilicon alloy

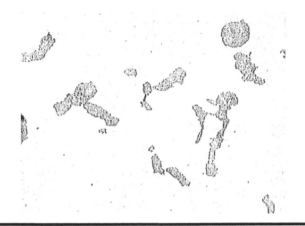

Figure 5.3 Typical microstructure of compacted graphite iron.

to control the harmful elements found in their usual concentrations in the treated iron material.

A new variety of cast iron—compacted graphite iron—is becoming popular for various commercial applications. It is an intermediate form of graphite known as a vermicular (between a flake and spheroidal). It has the same appearance as the graphite in iron-based materials with deleterious elements. Such iron metals can be manufactured by deliberate undertreatment of the base iron with small percentages of magnesium and rare earth elements. There is another method of producing this iron that involves the introduction of enough titanium into the magnesium-ferrosilicon plus an appropriate rare earth alloy to produce a completely spheroidal graphite iron material that is most ideal for casting sections. It is anticipated that rare earth elements will play an important role in the production of these compacted graphite iron compound materials. The typical microstructure of compacted graphite iron is shown in Figure 5.3. If compacted iron compounds are used in large quantities, the demand for rare earth elements may significantly increase.

5.4.3 Future Role of Rare Earth Elements in the Commercialization of Nodular Iron

It is extremely difficult to estimate the tonnage requirements of rare earth elements in the nodular iron industry. According to the best estimates, the future looks very bright. For example, the production of nodular iron castings in 1959 was less than 170,000 tons. However, by 1978, approximately 2.9 million tons of nodular iron castings were manufactured—essentially a seventeen-fold increase over a period of 20 years [1]. Assuming a typical 50% yield of the iron casting in 1978, there would have been approximately 6 million tons produced for nodular iron castings. From this production capability, it appears that the outlook for nodular iron production is not highly optimistic and the foundry industry is experiencing a temporary shortage. Casting shipments of nodular iron were off by about 10% in 1979 compared to 1978.

5.5 The Role of Rare Earth Materials in the Steel Industry

This section discusses the role of rare earth materials in the steel industry and steel-related materials widely used in various industrial applications. Critical roles of specific rare earth elements in steel-making efforts are highlighted, with particular emphasis on the physical properties of the steel-related materials.

5.5.1 Rare Earth Alloys Currently Used in the Steelmaking Industry

Mischmetal, a well-known alloy, plays a critical role in the steelmaking process and is widely used in the production of different types of steel. Mischmetal is produced by fused chloride electrolysis of light lanthanide elements. It constitutes more than 90% of rare earth metals (REMs) consumed for steelmaking in industrialized countries. Approximately 3,000 metric tons of mischmetal alloy, worth $35 million, are added annually to the liquid steel production quota.

It is estimated that the mischmetal used in steel making contains 95 to 98% REM, 0.5 to 5% iron to lower the melting point towards the eutectic, and 0.1 to 1% residual impurities such as aluminum, calcium, hydrogen, nitrogen, oxygen, magnesium, silicon, and others that have no impact on the quality of steel produced. Note that *eutectic* is defined as an alloy with the lowest melting temperature possible.

The relative concentrations of the four light lanthanides and three heavy lanthanides can vary significantly depending on the origin of the ores deployed. Cerium content may vary from approximately 45 to 58%, lanthanum content from 17 to 30%, neodymium content from 11 to 20%, and praseodymium content from 4 to 8%. The "heavies" include only three elements—gadolinium, samarium, and yttrium; their contents are less than 0.1 to 2%, approximately. The effects of these individual elements on the physical properties of steel are not identical. However, the cost of chemical separation prior to electrolysis cannot be justified in terms of any improved physical properties, at least for high-tonnage carbon steels. Note that the lanthanum content is specified in a superalloy and its properties have been accepted at face value.

The price of mischmetal was approximately $6.50 per pound during most of the 1950s. It drifted down to a low of $2.50 per pound from 1967 to 1973 and rose back to $5.30 per pound after adjusting for inflation [2]. For large-scale steel-making and deployment, there is plenty of ore and excess steel production capacity worldwide. Estimated mischmetal output produced by various nations during 1960 and 1975 can be described as follows:

United States: Reactive Metals and Alloys Corporation and Ronson Metals Corporation were capable of producing mischmetal with a total capacity of 2,500 metric tons per year, with their plants working close to full capacity. Note that 85 to 90% of the output alloy was consumed by steel mills.

West Germany: Treibacher Chemische Werke was capable of producing close to 1,000 tons of mischmetal per year.

United Kingdom: Ronson produced more than 150 tons per year.

France: Rhone Poulenc–Pechiney produced close to 250 tons per year. It is important to mention that both the British and French plants were working at less than 50% of plant capacity because of a substitution issue beginning in 1975 in the steel industry.

Brazil: Corona, Colibri, and Fluminense were capable of producing a combined alloy output of more than 1,000 tons per year. Almost all mischmetal produced was exclusively for export.

Japan: Santoku Corporation was producing approximately 200 tons of mischmetal per year, but very limited alloy was deployed in steelmaking industry. This corporation was the only producer capable of converting entirely to the oxide-fluoride electrolysis process in 1979, while increasing the output capacity of the plant.

Based on these production capacities, it is estimated that the worldwide mischmetal production capacity is approximately 6,500 metric tons per year from all sources. However, currently less than 50% of this mischmetal alloy is used mainly due to a substitution clause added to limit the production quota by the European steel industry. U.S. manufacturers are now in effect producing more than 50% of the world's total mischmetal.

5.5.2 Physical Properties of Mischmetal

This section describes the important physical properties of mischmetal. The first property defines the total solubility in all liquid steels. A cerium-iron phase diagram shows complete solubility in the liquid phase and zero solubility in the solid phase. This property also indicates one single low-melting point eutectic at 92.5% cerium [2].

The second property is of critical importance because it indicates the low vapor pressure shown by the mischmetal at liquid steel temperatures, as illustrated in Table 5.3. In this table, the vapor pressures of the main elements present in mischmetal are compared to the common alloying elements that are widely used in the steelmaking process.

Table 5.3 Typical Vapor Pressures of Alloying Elements in the Steelmaking Process at 1,800 K

Alloying Element	Vapor Pressure (µatm)
Cerium (Ce)	0.682
Lanthanum (La)	0.569
Silicon (Si)	2.15
Praseodymium (Pr)	14.9
Neodymium (Nd)	139
Aluminum (Al)	808
Manganese (Mn)	28,100
Calcium (Ca)	1,290,000
Magnesium (Mg)	12,600,000

The third property considers the density of the alloy to be introduced in the liquid steel. Data presented in Table 5.4 compares the densities of the alloying elements to those of liquid carbon steel at room temperature. It is evident from the tabulated values that mischmetal will have little buoyancy in liquid steel and compares favorably with most other additives.

Table 5.4 Comparison of Densities in Liquid Steel

Alloying Element	Density (g/cm³)
Manganese	7.3
Liquid steel	7.2
Mischmetal	6.7
Zirconium	6.5
Vanadium	6.0
Rare earth silicide	5.7
Aluminum	2.7
Silicon	2.3
Magnesium	1.7
Calcium	1.5

The fourth property of mischmetal concerns the melting point of the alloying element as compared to the liquid steel bath. Mischmetal compares favorably with the liquid bath temperature. In addition, mischmetal compares most favorably with other alloying metals used in the steelmaking process, which is evident from the data presented in Table 5.5.

Finally, the fifth property indicates the negligible solubility of all rare earth elements in solid iron, which can only be seen on the binary-phase diagram developed for cerium-iron showing complete solubility in the liquid phase.

By the first four physical properties of mischmetal, which is a complex alloy of rare earth metals, it can be seen that mischmetal is a very favorable steel additive. The combination of these properties makes mischmetal the most desirable alloy among all ferroalloys commonly introduced in liquid steel [2]. While adding mischmetal to a liquid steel ingot after interruption of teeming at the hot-top junction, no visible flare develops, the alloy disappears quickly, and the distribution

Table 5.5   Comparison of Melting Points of the Alloying Elements

Alloying Element	Melting Point	
	°C	K
Molybdenum	2,617	2,890
Chromium	1,867	2,140
Zirconium	1,852	2,125
Titanium	1,675	1,948
Low-carbon steel	1,520	1,793
Silicon	1,416	1,689
Manganese	1,244	1,517
Neodymium	1,024	1,297
Praseodymium	935	1,208
Lanthanum	920	1,193
Calcium	842	1,115
Mischmetal (4% iron)	816	1,089
Cerium	795	1,068
Aluminum	660	933
Magnesium	651	924

of the REMs throughout the ingot body is easily ensured by the resumption of teeming to fill the hot-top volume. (An ingot is a mold in which a metal is cast.) No other metal can be added to steel using that simple practice without risking nonhomogeneous distribution in the steel ingot.

The fifth property of the mischmetal presents a serious limitation for the role of lanthanide elements in the steel-making process because it prohibits any solid solution alloying. Therefore, a metallurgist is forced to carefully avoid excess mischmetal addition for fear of making the steel "hot-short." A low-melting point eutectic alloy develops with any excess metallic REM content in the steel towards the end of solidification, concentrating at the "as-cast" grain boundaries. This grain-boundary film is liquid at all hot-rolling temperatures, thereby making the steel highly "hot-short," as seen in an industrial accident. The same grain-boundary eutectic is often interpreted as "dirt" or oxide inclusions. During the polishing of the samples, the eutectic, which is highly concentrated in the REMs (ranging from 90 to 95%), oxidizes at a rapid rate and requires special precautions to be observed without any alteration.

5.5.3 Rare Earth Silicide Alloy

Rare earth silicide alloy was developed by Vanadium Corporation of America (which became part of Foote Mineral Company) and Molybdenum Corporation of America around 1965. Both then and still today, this alloy is cheaper than mischmetal. The alloy was sold in the market at $0.40 per pound in 1968, which is approximately 42% of the price of electrolytic mischmetal per unit of REM. Its attractive price played a key role in the development of high-strength low-alloy (HSLA) steel. Development of HSLA steel started at the Jones and Laughlin Steel Corporation around 1968 and at Telco in 1970.

Rare earth silicide alloy is produced in a submerged arc furnace by direct reduction of purified bastnäsite ore, quartz, iron, and carbonaceous recusants. This alloy is composed of one-third mischmetal, one-third silicon, and one-third iron. The disadvantages of this alloy are strictly due to its inferior physical properties compared to those of mischmetal. Its lower density, higher melting point, and "chill effect," which requires sensible heat from the steel to melt three times more material than for mischmetal, have contributed to alloying problems, particularly for the popular mold additions. It is highly impractical to plunge the rare earth silicate in a steel ladle due to excessive buoyancy and chilling effects. In mold additions, the rare earth silicide is more conductive to surface and subsurface defects as compared to mischmetal; in addition, it cannot be added later at the hot-top junction.

Even though the ferroalloy enjoyed high consumption in the United States, capturing close to 90% of the total rare earth metal market in steel (with a peak share of approximately 1,800 tons of the rare earth metal equivalent in 1972), its current market share is down to slightly less than 15% of the REM market, which is strictly due to unfavorable alloy ability [2]. Attempts were made for reaching higher REM concentrations (40–50%) by pushing the submerged arc surface closer to its thermodynamic limits, but this led to an unstable alloy exhibiting spontaneous decrepitating or leading cracking during storage. The production costs have increased for the submerged arc process and the electrolytic cell in recent years, at least for rare earth metal production, with the price ratio of rare earth silicide to REM moving progressively from $0.42 to $0.65 per pound; thus, the attractiveness of rare earth silicide alloy for the steelmaking business has reduced.

However, this alloy is favored over mischmetal for various ladle practices, as well as for a combined deoxidation-desulfurization-sulfide shape control effect, deploying large ladle additions in the production of critical pile-line qualities. (A ladle is a deep-bowled long-handled spoon that is specially used for dipping and conveying liquids.) It is important to point out that the pipeline application is threatened or underestimated by substitution with calcium injections. Furthermore, this particular application is based on bastnäsite material, which was exclusively an American ore at that time. Due to its metallurgical limitations, rare earth silicide never regained popularity in industrial nations such as Japan and European countries. However, the recent emergence of a large Chinese bastnäsite deposit may prompt renewed interest in a directly reduced rare earth metal alloy.

5.5.4   Various Rare Earth Metal Alloys

This section discusses the properties and applications of other REM alloys. Few other REM-containing alloys are deployed in the steelmaking process. A Russian "ferrocerium" alloy is very similar to the mischmetal, although very limited information is available about it. The only other rare earth metal compound currently used by a few U.S. factories is known by its trade name of "T-compound" [2]. The T-compound is a mixture of rare earth metal oxides and fluorides that was widely used during World War II as a fluxing agent for nonmetallic inclusions, in the production of alloy steel in electric arc furnaces, and in the design and development of aircraft deployed during World War II. Adolf Hitler ordered aircraft designers in July 1944 to use the T-compound material in the design of jet fighters. These jet fighters were used by the German Air Force against the British with limited success, as the war was already coming to an end at that time.

5.5.5   Physical Chemistry and Ferrous Metallurgy Aspects of Rare Earth Metals

To focus on the essential, temporarily irreplaceable properties of REMs in the steelmaking process, one has to highlight their affinity for sulfur. Since the Iron Age, sulfur has had deleterious effects in the steelmaking process. A new steelmaking process was adopted in 1960, which did not use sulfur. Essentially, they were able to stop the sulfur from making the steel "hot-short" during the rolling process; instead, manganese was included in the process. Soon after, 99% of steel was made using manganese as an essential ingredient in the steelmaking process. However, it was later found that hot-shortness is caused by iron sulfide, which precipitates at cast grain boundaries of the steel ingot because it is liquid at hot rolling temperatures.

Because the affinity of manganese for sulfur is substantially higher than that of iron for sulfur, all sulfur will precipitate as manganese sulfur rather than ferrous sulfide, provided that a high enough manganese-to-sulfur ratio is achieved to counterbalance the large iron-to-manganese ratio. Manganese sulfide is a rather plastic solid at hot rolling temperatures, which eliminates hot-shortness. However, there are limitations to the manganese sulfide approach, which are discussed in the following sections.

5.5.5.1 Some Limitations of the Manganese Sulfide Approach

Manganese solution was used by the steel industry for more than eight decades. Over that timeframe, early steel materials were not thoroughly deoxidized and the plasticity was drastically reduced by the presence of oxygen in the sulfide. Hot plasticity of sulfide inclusions was absorbed during the hot rolling of steel as a function of oxygen, manganese-to-sulfide ratio, rolling temperatures, and substitution with stronger sulfide formers. Hot plasticity can be measured in terms of the length-to-width ratio of elongation after hot rolling in the longitudinal sections of the steel.

However, with the growing deployment of aluminum as a deoxidizer instead of silicon, particularly in hot-rolled steels in which inclusions were not broken by the cold-rolling process, it was observed that that the manganese-to-sulfide inclusions were extremely elongated and the mechanical properties were seriously degraded in the directions perpendicular to the rolling axis. Some relief was obtained with cross-rolling of the plates, but this was not applicable to the continuous hot-rolling process. Some relief may be obtained with lower sulfur content, but there are limits due to the plasticity of the manganese sulfide, which increases with the manganese-to-sulfide ratio.

In more recent decades, manganese has been assigned other metallurgical functions that are capable of strengthening and phase control. Therefore, most HSLA steels containing 1 to 2% manganese may significantly aggravate the plasticity of the manganese sulfides during the hot-rolling process. In conventional aluminum-deoxidized steels, the extra-low dissolved oxygen content will maintain the sulfur in solution form in the liquid steel until the end of the solidification process. Because sulfur shows no solubility in solid steel, a manganese-sulfur-iron eutectic suddenly precipitates at the cast-iron grain boundaries. These cast-iron grain boundaries are known as type II structures. The formation of grain boundaries weakens the cast structure and also results in elongated manganese inclusions in several components, such as hot-rolled steel plates, sheets, bars, coils, wires, and so on.

5.5.5.2 Impact due to Substitution of Manganese Sulfide

Researchers were curious to see what would happen when manganese was substituted in sulfide, which combines a greater affinity for sulfur than manganese with a nonplastic sulfide during the hot-rolling process. The substitution of manganese sulfide generates increased negativity of the free energies of formation, which display a large choice of alternatives, including titanium, zirconium, calcium, and the four lanthanide elements. The free energy of formation of some sulfides and rare earth material oxysulfides, along with their increasing negativity, are shown in Figure 5.4.

Lanthanides not only outperform most other substitutes for sulfur stability, but they also exhibit unique features such as the formation of extremely stable oxysulfide of the lanthanum oxide sulfide (La_2O_2S). The free energy of formation is extremely negative and close to that of the most stable oxides. When rare earth metals are added to liquid steel predoxidized with aluminum, no other oxides are formed.

Research studies to be performed on potential substitutes require considerations other than just sulfide stability, such as the possible interference of the newly introduced elements with other physical properties of the steel, the plasticity of the new sulfides, the physical alloy-ability of the additive, and the cost-effectiveness of the additive. For example, zirconium and titanium interfere with other properties

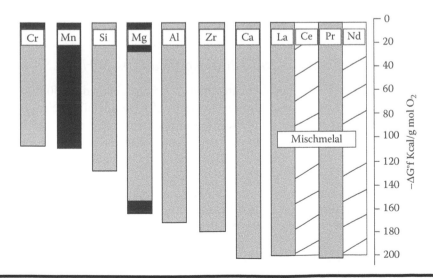

Figure 5.4 **Free energy formation of rare earth metals with increased negativity.**

of the steel because of the excessive stability of their nitrides and carbides. Despite this, these two elements have played critical roles in sulfide substitution. More than 500 metric tons of nuclear zircascrap were consumed in 1974. Their role will fade away, however, because of the low-temperature impact properties of the steel materials treated with zirconium and titanium.

Both calcium and magnesium exhibit excessive vapor pressure at steelmaking temperatures of 1,600 to 1,900ºC. However, the alloy-ability problem has been alleviated by submerging the introduction of calcium alloy under 10 feet of liquid steel. Regardless of vapor pressure, the low solubility of calcium limits the effective substitution of manganese sulfide by calcium sulfide to extremely low sulfur levels, particularly when the manganese content is high (2%).

5.5.5.3 Replacement of Manganese Sulfide with Rare Earth Sulfides

Laboratory tests have demonstrated that the predictability and control of the morphology of sulfide inclusions can be achieved using small quantities of mischmetal and rare earth silicides. This control cannot be achieved using calcium alloys in multiple additions. Manganese sulfide inclusions can be totally eliminated and replaced with small, undeformable, round REM oxysulfides and sulfides, if a ratio of rare earth metals to sulfides in the range of 3:1 and 6:1 by weight is retained. Results demonstrated the complete elimination of cracking of the test plates when the plates were bent with a sharp radius in the transverse direction, as well as more than doubling the transverse shelf energy. These tests also showed progressive improvements as a function of incremental additions of rare earth metals.

5.5.5.4 Commercial and Industrial Applications of HSLA Steel

Mechanical tests on HSLA steel indicate a tensile strength exceeding 80,000 psi. Tests have confirmed the approach used in terms of reproducibility, cleanliness of the steel, superior physical properties, and a reasonable cost of $2 per ingot ton using rare earth silicide inclusions. Typical practice includes an extra-sample mold addition of 5 pounds rare earth silicide per ingot ton during the first half of teeming of each ingot. This particular teeming technique is known as "sulfide shape control." With it rare earth metals spread at lightning speed across various steelmaking plants in the United States, Canada, and Europe. Despite this unique development, mischmetal rapidly displaced rare earth silicides in the steel market.

HSLA steel was widely used in early automobile applications. Wide deployment of mischmetal, however, was confined to steel pipelines; the market for this particular application was very active from 1970 to 1971 due to the great demand for steel pipes. These applications forced the steelmaker to explore other alternatives to control the manganese sulfide inclusions in order to increase steel sales for other commercial and industrial applications. Steelmakers focused their attention on a new application involving the node of an offshore platform with a complex welding structure of tubular shapes, which requires a relatively large quantity of steel. There was a resistance to spelling during the cold-punching of high-carbon plow wheels for agricultural equipment, as well as a resistance to lamellar tearing of welded structures, in which the elimination of elongated manganese sulfide inclusions is highly essential. These resistance activities had a tremendous impact on the sales of HSLA steel.

After 1976, an application requiring large tonnage was developed in the United States—high-strength welded steel tubes for deep oil well casting and drilling, which competed with more conventional seamless steel tubes manufactured using the Mannesmann process. According to market analysis, roughly 1,500 tons of mischmetal are consumed annually for this particular application. The total consumption of steel was expected to increase at least until 1983, when complete predesulfurization of the steel would reduce the rare earth metal consumption per ton of steel by more than 50%. Steel users predict that today the average consumption per ton of steel treated is roughly 1.75 pounds of mischmetal or equivalent in rare earth silicides. Steel market experts predicted that it would drop close to 0.75 pound per ton by the year 1990.

5.5.5.5 Steelmaking Practices Used by Steel Producers

Four vital techniques are used for deploying mischmetal to liquid steel—two involving the ladle and two involving ingot molds. Specific details on these two techniques can be found in the journal *Foundry Management and Technology*. Simply stated, the ladle technique involves the plunging of a canister containing anywhere from 100 to 1,000 pounds of cast mischmetal, alloyed with a little magnesium to promote agitation and mixing. The ingot mold technique (Figure 5.5) involves an introducing a ladle degasser through the alloy feeding system after the degassing process and aluminum trimming additions have been completed.

These two mold practices require hand-feeding of preweighted bags during the early part of the mold-filling process needed to promote the most efficient mixing, as well as the previously described delayed mold-addition practice. Other practices, which are used less extensively in Japan and Europe, feed a wire containing the alloy into the mold of slab casters, hang the mischmetal bar in the ingot during bottom pouring, and inject the rare earth metal alloy powders with calcium silicon to improve sulfide shape control after the desulfurization process with calcium [2]. Recovery of the rare earth metal elements varies from as low as 25% with the second ladle practice to as high as 80% with the delayed mold addition and 95% by wire feeding in the slab casting mold.

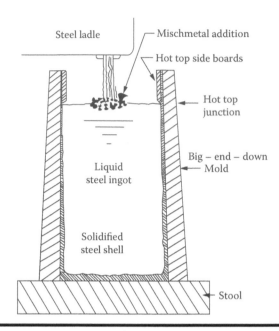

Steel ladle
Mischmetal addition
Hot top side boards
Hot top junction
Big – end – down Mold
Liquid steel ingot
Solidified steel shell
Stool

Figure 5.5   Ingot mold technique involving the alloy wire-feeding procedure.

5.6 Problems Associated with Rare Earth Metals

Despite their many advantages, there are unique problems associated with REMs. This section identifies these unique problems. Note that the precipitation of RE_2O_2S and RE_xS_y as solid particles in the liquid steel medium immediately after the addition is conductive to inclusion clustering, leading to the formation of spatial networks of small nonmetallic particles adhering to each other by virtue of surface tension.

5.6.1 Mechanical Problems from Clustering

Small spatial network dimensions vary from 1 to 10 μm. In addition to the forma-tion of spatial networks, the slow melting and dissolution of rare earth silicides lead to oxidation of the alloy during mold addition, which ultimately will result in surface defects on the finished rolled products. This can lead to quality control problems, particularly in precision steel products such as the bearings deployed in high-speed steam turbines. The delayed mold-addition technique may eliminate such problems, and the use of rare earth silicide in the mold would not be required.

5.6.2 Serious Operating Problems, Such as Nozzle Blockage

The clustering mechanism can create serious operating problems that may not be related to the use of rare earth metals but rather to any additive conductive to solid inclusions in the liquid steel. In addition, deoxidation with zirconium, titanium, aluminum, and high additions of calcium could lead to accumulation in the throat section of the nozzle, which will ultimately stop the flow of liquid steel completely. The feeding of rare earth metals in the form of a wire could solve this particle problem. However, the slab casting mold has the greates tendency of developing this problem. The most effective solution is to minimize the volume of the inclusions passing through the nozzles by upgrading the ladle refractory and performing an efficient stirring process after the rare earth metal treatment, which gives the inclusion clusters maximum time to separate out of the melt by adherence to the ladle slag. This particular problem can only be resolved when the initial sulfur content is very low (0.007–0.012%).

5.6.3 Problems with Bottom Cone Segregation

Bottom cone segregation is another serious operational problem. Layered accumulations of rare earth sulfides and oxysulfides near the bottom of the ingot are observed in all large ingots with sulfur content exceeding 0.01% that are treated with rare earth metals in the mold. This particular condition degrades the through-thickness properties of the steel plates due to the large concentration of small inclusions acting as notches.

Researchers at Kawasaki Steel Chiba Works have performed an exhaustive research study addressing this particular problem. They found that, with the required ratio of rare earths to sulfides (2.7:1) for total manganese sulfide substitution, the controlling factor in the severity of segregation is equal to the product of REMs and sulfides [2]. Furthermore, for a product of REMs and sulfur—known as parameter K_{RE}, with a value either below or equal to 13×10^{-5}—the bottom semimetal zone leaves faintly "enriched A and V segregations" and less top segregation. In turn, this is equal to sulfur content with a maximum of 0.007%. However, a maximum sulfur content of 0.005% with a corresponding K_{RE} product of 10×10^{-5} is preferred. At this value of K_{RE}, all segregations disappear and the resulting steel approaches its maximum properties, with total shape control and no side effects, as shown in Figure 5.5.

5.6.4 Cost Projection of Rare Earth Metal Alloy

Steel cost is strictly dependent on the recovery rate and mischmetal content. A minimum retained REM content of 0.045% is necessary to ensure full manganese sulfide substitution, which corresponds to a minimum REM-to-sulfide ratio of 3:1. Using the delayed mold-addition practice with a minimum recovery rate and market value of mischmetal per pound, the current cost of a typical treatment can be calculated using the following empirical formula:

Current cost of typical treatment per ton = (mischmetal cost/lb.) × (% REM content)/(recovery rate %) × 2,000

Assuming a recovery rate of 75%, a mischmetal cost of $5.20 per pound, and an REM content of 0.04%, the cost per ton for the treatment of ingot steel is $5.55. However, this cost can be much higher if ladle plunging or additions through a degassing unit are required. Under these conditions, the treatment cost may be as high as $12 per ton for guaranteed sulfide shape control and maximum transverse physical properties, which typically require an REM-to-sulfur ratio of 4:1 instead of 3:1.

The most direct and simplest solution to this cost is to reduce the sulfur content prior to mischmetal addition. This is a necessary and sufficient condition for reducing the cost of the treatment of ingot steel. For every 0.001% reduction in sulfur, there may be a minimum cost savings of $0.42 per ton using the delayed mold-addition concept; the savings potential increases with other less efficient techniques. However, it is critically important to remember that desulfurization costs will increase sharply when approaching content that is less than 0.005%. The preceding discussion used a value of approximately 0.007% sulfur, which is also highly desirable from a quality control point of view. This minimum value influences the total cost of sulfide shape control and the desulfurization process.

5.6.5 Summary of Solutions to Problems Associated with REM Addition to Liquid Steel

This section briefly summarizes solutions to problems associated with REM addition to liquid steel. Firstly, it is essential to try to operate at much lower sulfur content than what is normally practiced in the United States, where sulfur contents vary between 0.005 and 0.007% prior to rare earth metal addition. For ingot casting, delayed mold addition, particularly when coupled with low sulfur concentrations, is more than sufficient and economical for most commercial and industrial applications. For exceptionally stringent property requirements and continuous casting, rare earth metals should be introduced in the ladle, using improved refractoriness and slogs, as well as an efficient stirring procedure after its treatment. The requirements stated above are essentially the same, which are imposed by the suppliers of the competing calcium practice. Wire feeding in the slab casting mold is neither required nor essential.

5.7 Techniques for Manganese Sulfide Control

Because of serious nozzle blockage problems, comprehensive development efforts have focused on the desulfurization of steel to a point at which shape control would no longer necessary. Among the several practical techniques existing today, those combining the injection of calcium alloys, basic slag mixtures, and improved ladle linings have eliminated the need for REM additions in 80 to 90% of the steel deployed by European and Japanese steel corporations.

Originally, it was thought that when sulfur content is close to 0.005%, no more shape control is required. However, it is now thought that calcium also provides total shape control that is as good or better than REMs. The Kawasaki research team mentioned earlier concluded that there is also a calcium-to-sulfide ratio requirement, as with the rare earth metals for total manganese sulfide substitution. They proposed reliable estimates of retained calcium requirements for homogeneous sulfide shape control, which are summarized in Table 5.6. The manganese levels shown in this table are typical of English-resistance welded pipe at 1% and utility pipe at 1.5%.

This is a newly developed unique shape control technique for the manufacture of very large pipe made of heavy-plate steel and the manganese-molybdenum-cobalt (Mn-Mo-Co) steel developed mainly in the United States by AMAX. Due to the limited solubility of calcium in steel, the only way these ratios can be achieved in high-manganese steels for the most critical property levels is through extreme desulfurization, down to 0.004% sulfur. In the easiest case, it could be down to 0.001% maximum in the most favorable situation. Most steel is currently produced with low sulfur content, typically close to 0.005%, and does not show complete substitution of manganese sulfide inclusions. Fortunately, the improvement in properties is more than sufficient to satisfy almost all specifications for commercial and industrial applications. There is a remote chance of a partial return to a final rare earth metal addition at low sulfur levels achieved by calcium injection, which

Table 5.6 Minimum Calcium-to-Sulfide Ratio Requirements for Shape Control

Critical Level of Steel Properties	Manganese Content		
	1%	*1.5%*	*2%*
Shape control level			
Improvement in toughness			
Partial	0.65	1.1	1.5
Total	1.3	2.5	3.2
Prevention of hydrogen cracking			
Partial	1.3	2.5	3.2
Total	1.7	3.1	4.5

Source: Luyckx, L.A. 1981. *The Rare Earth Metals in Steel.* New Castle, PA: American Chemical Society.

will make it much easier to obtain shape control at much lower costs and without any adverse side effects, when critical performance levels are essential.

5.8 Hydrogen-Induced Cracking during REM Application

There could be several failure mechanisms under hydrogen-induced cracking during the REM application practice. Catastrophic failures of large-diameter gas and oil pipelines in seawater and in hydrogen-sulfide contaminated water surfaces have been traced to disintegration of the steel structure by hydrogen-induced cracking. At the surface of the pipe, hydrogen atoms are liberated by an electrochemical reaction and penetrate quickly by diffusion inside the steel structure to recombine in the molecular hydrogen as soon as they meet with a discontinuity in the steel, such as nonmetallic inclusions. This molecular hydrogen can reach local hydrogen pressures of 10,000 atm, thereby developing tremendous localized stresses. Elongated inclusions, frequently caused by manganese sulfide, will tend to open up under applied stress and initiate cracking at their tips, ultimately joining other cracks in a stepwise fashion until microcracks develop, thus initiating the catastrophic failure. When steel is under constant stress due to natural gas pressure, hydrogen-induced cracking is greatly accelerated. The same phenomenon occurs in deep sour-gas wells, only at even greater speeds on account of hydrogen sulfide concentrations.

5.8.1 Critical Effects due to Sulfide Shape Control

Japanese and Italian researchers have demonstrated that sulfide shape control is relatively more important than sulfur reduction. Research studies performed by these scientists show a progressive improvement in crack appearance after the standard exposure time using progressively improved steelmaking practices, in which the REM treatment contributes the most significant progress. Other research activities showed similar results. However, in this case the results observed were on commercial seamless steel as opposed to line pipe steel, and the results indicate that REM addition can bring very significant improvements in resistance in hydrogen-induced stress corrosion cracking, even at 0.003% sulfur. Scientific conclusions based on these research studies can be briefly summarized as follows:

- The replacement of elongated manganese sulfide inclusions by various small REM oxysulfide and sulfide inclusions multiplies precipitation points for the transformation of hydrogen from an atomic to molecular form and eliminates the severe weakening effect at the tips of elongated inclusions.
- The above scientific conclusion is correct as long as the REM-to-sulfide ratio is maintained in the order of 3:6.

5.8.2 Rare Earth Metal Hydride Effect

Kortovich [3] broke new ground when he added very large excessive rare earth materials, such as lanthanum and cerium, to commercial steel designated as 4340 melt containing 0.004% sulfide. He retained the REM-to-sulfur ratios in five laboratory heats, which were treated with up to 0.2% lanthanum and cerium, with the REM-to-sulfur ratios varying from 8:1 to 46:1. This large ratio variation resulted in the precipitation of the low-melting lanthanum-sulfide or cerium-sulfide eutectic at the primary grain boundaries (as described previously) and in hot shortness and loss of impact properties. However, Kortovich did not properly identify the eutectic as a grain boundary oxide—the same error was made by several steel mill metallurgists. Nevertheless, a dramatic improvement in resistance to hydrogen embitterment was obtained by moving from the low cerium and lanthanum levels deployed by other research scientists focusing only on sulfide shape control (from three to five times that level), with the principle objective of REM hydride formation. It was a major scientific breakthrough.

In addition, there were many inconsistent observations that the properties of highly doped steel improved after hydrogen charging, which can be recognized as startling scientific observations. Further laboratory tests confirmed the formation of hydrides, as suspected by Kortovich [3]. It is also highly possible that these hydrides formed at the site of grain boundary eutectics, which typically contain 92.5% cerium in a cerium ferrous system and 95% lanthanum in a lanthanum ferrous system. This area of research and development infuenced several new applications of REMs in steelmaking during the 1980s and beyond.

5.8.3 Summary of the Metallurgical Effects of Rare Earth Metals in Steel

Significant improvements in the physical properties of steel have been observed from the addition of REMs to steel. The qualitative effects of these additions can be briefly summarized as follows:

- Qualitative plotting of the effects of REM addition to steel reveals cold formality, impact resistance, hot workability, and enhanced weld integrity.
- The weld integrity quickly reaches the maximum with fairly small, economical additions of rare earth metals.
- All these improvements are directly related to sulfide shape control in conventional steels.
- All these improvements are directly related to tramp element control in stainless steels.
- Most technical literature published in the 1950s specifically described the improvements in stainless steels, with major emphasis on mechanical and physical-chemical properties of the material.

- Two other properties—high-temperature oxidation resistance and hydrogen cracking noise—follow an explicitly separate course because the mechanism of action has little or nothing to do with manganese sulfide inclusions.
- The latter two effects are thought to severely conflict with each other concerning the basic requirements of hot workability and plain strength of most steel grades.
- It is further believed that ingenious manipulation of the as-cast structure may be one way to minimize the deleterious effects of the grain boundary eutectic while capitalizing on higher additions to the higher impact resistance of future steels to higher temperature oxidation, hydrogen embrittlement, or both.
- It is extremely important to state that the industrial deployment of mischmetal in steel has significantly improved the quality of steels and their physical properties at elevated temperatures.

5.9 Role of Rare Earth Materials as Noncracking Catalysts

Since the 1960s, material scientists have focused research efforts on the role of rare earth materials to stabilize zeolite cracking catalysts for their benefits, particularly in the petroleum industry. Today, a number of new and unique applications with potential commercial significance are receiving great attention. One of the most important of these applications is the deployment of cerium in catalyst devices for automobile exhaust emission control. The use of these devices has significantly improved air quality in major cities, which has demonstrated health benefits for their residents. A number of oxides have distinguished properties that are well suited for catalyst applications. The oxides are generally basic compared to alumina. Lanthanum oxide (La_2O_3) is the most basic oxide for this particular application; it is widely used as a catalyst due to its moderate cost and abundant availability.

Other rare earth oxides have demonstrated excellent thermal stability, which is a suitable characteristic for most industrial applications where reliability and cost-effective performance are the principal requirements. Other popular rare earth elements and their oxides include praseodymium and terbium, which, along with cerium, are considered to be nonstoichiometric oxides. Nonstoichiometry is an important property of many good oxidation catalyst sensors. These mixed-valance rare earth compounds are typically polymorphic compounds.

Note that cost and abundance are the most critical properties to be considered for any commercial application. Typical cost and abundance data on rare earth oxides are summarized in Table 5.7. The rare earth oxides shown in the table were chosen based on their affordability and abundance. Cerium oxide and lanthanum oxides are available at the lowest costs and in abundant quantities, which is why these two rare earth oxides are widely used in catalyst devices.

Table 5.7   Typical Cost and Abundance Data on Popular Rare Earth Oxides from Various Ore Sources

Pure Oxide	Cost ($/lb.)	Abundance in Ore Source (%)	
		Monazite	Bastnäsite
Cerium oxide (CeO_2)	7.50	45	49
Lanthanum oxide (La_2O_3)	7.25	20	32
Neodymium oxide (Nd_2O_3)	18.00	18	13
Yttrium oxide (Y_2O_3)	30.00	2.1	0.1
Praseodymium oxide (Pr_6O_{11})	32.00	5	4
Samarium cobalt oxide (Sm_2O_3)	32.00	5	0.5
Gadolinium oxide (Gd_2O_3)	55.00	2	0.3

Source: U.S. Bureau of Mines. 1977. *Mineral Yearbook.* Washington, DC: U.S. Bureau of Mines.

Cerium oxide is widely selected for monitoring smog performance because of its low cost and abundance. The typical performance parameters of cerium-based catalyst sensors are summarized in a table from reference [5]. Note that recent cost and abundance data on individual rare earth elements and their oxides derived from major ores are shown in a table from reference [5]. Cerium is by far the most abundant and least expensive among the catalytically rare earth–forming nonstoichiometric oxides.

5.9.1 Potential Catalytic Commercial and Industrial Applications of Nonstoichiometric Oxides

Potential catalytic commercial and industrial applications of the nonstoichiometric oxides can be identified as follows:

- Automobile exhaust emission control
- Polymerization
- Isomerization
- Ammonia synthesis
- Oxidation
- Hydrogenation
- Dehydrogenation
- Applications of perovskite

Studies have also summarized some experimental results in the fields of isomerization, hydrogenation, and oxidation, which might be useful in various commercial and industrial applications.

Concentrates of cerium and lanthanum oxides are available at low costs: The price of cerium concentrate is $0.85 per pound, whereas lanthanum concentrate costs $1.05 per pound.

5.9.2 Performance of Cerium-Promoted Lummus Catalysts

Performance parameters, such as the average reactor temperature and percent of ammonia (NH_3) for equilibrium, conventional, and cerium-promoted catalysts are summarized in Table 5.8.

5.9.3 Importance of Ammonia Synthesis

Ammonia synthesis plays a critical role in the formation of catalytic reactions. The synthesis also determines the removal of ammonia (NH_3) from the catalyst surface, which is considered very important for the catalyst reaction. A typical ammonia synthesis catalyst contains iron oxide in addition to 1% potassium oxide, 1 to 2% alumina, and approximately 1% calcium oxide on the catalyst surface. After the fusion and reduction phases, the surface is largely metallic iron and reduced promoters are concentrated on the surface. A catalyst washed with $Ce(NO_3)_3$ and subsequently reduced is much more active than a conventional catalyst, as shown in Table 5.9. Note that mischmetal salts can be substituted for the cerium salt.

Catalyst applications of the materials to hydrogenation, methanation, and ammonia synthesis have been described in the literature, along with some critical information concerning the structures of these materials. Note that rare earth compounds are less active than the conventional cobalt, nickel, molybdenum, and tungsten combinations; Raney nickel alloys; and the noble metal catalysts.

Table 5.8 Performance Parameters of Various Catalysts

Reactor Temperature	Percentage of Ammonia in Product		
	Equilibrium	Conventional	Cerium Promoted
910°F	16.8	12.7	15.6
840°F	22.4	12.8	17.8
710°F	39.0	10.4	13.5

These performance parameters were recorded under the following experimental conditions: pressure was 150 atm (14.7 × 150 = 2,205 psi); gas space velocity was 16,000 ft³/h; and hydrogen-to-nitrogen ratio was approximately 2:8.

Table 5.9   Substitution of Mischmetal Salt for Cerium Salt

Rare Earth Compound	Hydride
Cerium nickel (Ce_3Ni)	Ce_3NiH_8
Lanthanum nickel (LaNi)	$LaNiH_3$
Lanthanum nickel ($LaNi_5$)	$LaNi_5H_6$
Yttrium iron (YFe_3)	YFe_3H_4
Yttrium iron (YFe)	YFe_2H_4

5.9.4 Polymerization Process

Compounds known as "super slurpers" were developed by the U.S. Department of Agriculture. Based on starch-polyacrylonitrile copolymers, they are able to absorb as much as 500 to 1,000 times their weight of water, depending on the purity of the water. The formation of these copolymers is catalyzed by a cerium ion (Ce^{4+} ion). These polymers have potential applications in the agricultural industry, including water storage additions, as well as consumer and industrial areas.

5.9.5 Oxidation Process

The oxidation process is a redox (oxidation-reduction) mechanism that involves lattice oxygen (originally proposed in 1954) for hydrocarbon oxidation over vanadium oxide (V_2O_5). This oxidation process can be applied to a variety of catalytic oxidation reactions. A lattice redox mechanism for carbon monoxide (CO) oxidation is used to demonstrate CO adsorption. CO oxidation by lattice oxide can be used for catalyst reduction.

Rare earth oxides possess a higher degree of oxygen mobility compatible with a lattice redox (oxidation-reduction) mechanism. The oxides of praseodymium, neodymium, and terbium exchange oxygen readily with gas-phase oxygen molecules; these same compounds are active oxidation catalysts for hydrogen. Similar results were obtained for the oxidation of other gases, except that the cerium is active in this case. However, the activity of cerium has not been proven conclusively.

5.9.6 Potential Application of Catalytic Agents to Determine Automobile Exhaust Emissions

A catalytic agent is a substance that, by its mere presence, alters the velocity of a chemical reaction. It may be recovered unaltered in nature or at the end of the chemical reaction. Rare earth oxides have been widely used for decades to determine automobile exhaust emissions. Monitoring automobile and truck exhaust emissions by catalytic converters is essential for public health, as well as to identify the presence of concentrations of harmful hydrocarbons and other toxic gases in the lower atmosphere. Details on various catalytic converters designed to meet the exhaust emissions standards are published elsewhere.

Automobile exhaust emissions standards are summarized in Table 5.10. Smog tests performed at approved California gas stations indicate the hydrocarbon and nitrogen oxide parameters in parts per million, whereas the CO content is a percentage. These data are collected at a specified speed, depending on the weight of the vehicle. However, the data presented in Table 5.10 are expressed in grams per mile.

Due to greenhouse gas effects, catalytic converters have been deployed as a part of emission control systems since 1975. One approach has been to use a dual-bed catalytic converter device, in which the reduction of nitrogen oxide to nitrogen gas occurs over the first bed; the excess oxygen is provided to the second bed to oxidize the carbon monoxide and hydrocarbons more completely. The exhaust emissions shown in Table 5.10 may contain some poisonous or unsafe chemical elements, such as lead, phosphorus, and sulfur. The principal aim of the catalytic converter is to reduce concentrations of hydrocarbon, carbon monoxide, and nitrogen oxides (NO_x) to legally acceptable and safe levels as determined by monitoring agencies. The catalytic converter is designed in such a way that an oscillation from rich (oxygen deficiency) to lean (oxygen excess) occurs roughly once per second, or 1 Hz/s [6]. The following typical net exhaust gas reactions illustrate some of the chemical reactions that are believed to occur on the rich and lean side over an active three-way exhaust catalytic converter:

1. Hydrocarbon + oxygen → [CO_2 + H_2O] = lean side
2. [CO + H_2O] = rich side
3. Carbon monoxide + oxygen → carbon dioxide = lean side

Table 5.10 State and Federal Exhaust Emission Standards for Passenger Vehicles

Vehicle Model Year	Exhaust Emission Standard (g/mile)		
	Hydrocarbon	Carbon Monoxide	Nitrogen Oxide
1968	10	80	5
1975	1.5	1.5	3.1
1977	1.5	1.5	2.0
1980	0.41	7.0	2.0
1981	0.41	3.4	1.0
1983	0.41	3.4	1.0
1985	0.41	3.4	1.0

Note: Data for the years 1990, 1995, 2000, 2005, and 2010 were unavailable.

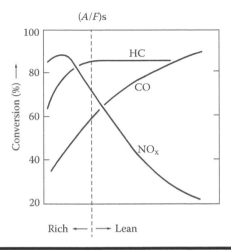

Figure 5.6 **Demonstration of a three-way catalyst's performance in a cycling test at 1 Hz. An air-to-fuel weight ratio of 14.6 is assumed at the stoichiometric point (dotted line).**

4. Carbon monoxide + water → carbon dioxide + hydrogen = water-gas shift
5. Hydrocarbon + water → carbon monoxide + hydrogen = steam reforming
6. Carbon monoxide + nitrogen monoxide → carbon dioxide + 1/2 nitrogen
7. Nitrogen monoxide + hydrogen → ammonia + water
8. Nitrogen monoxide + 2/3 ammonia → 5/6 nitrogen + water

It is evident from these chemical reactions that oxidation will be inhibited on the rich side, whereas nitrogen oxide reduction to nitrogen will be limited on the lean side. Conversion efficiencies for a three-way catalyst over a rich to lean cycle can be seen in Figure 5.6. This figure demonstrates a three-way catalyst's overall performance in a cycling test at 1 Hz/s. Note that the air-to-fuel weight ratio shown by the dotted line is approximately 14.6 at the stoichiometric point. The hydrocarbon detection efficiency is greater than 82% at the air/fuel ratio, equal to 14.6, whereas the nitrogen oxide detection efficiency is approximately 63% at the same air-to-fuel ratio. In summary, the air-to-fuel weight ratio determines the catalytic device efficiency.

5.9.7 Description of Potential Catalyst Shapes

Three distinct catalyst shapes are available for monitoring exhaust emissions. Typical automobile exhaust catalysts, along with their associated critical components and shapes, are shown in Figure 5.7. Note the most active and durable catalysts deploy noble metals such as rubidium, palladium, and platinum on an aluminum support structure, which is very inexpensive with moderate thermal and mechanical properties. The active catalytic material is thinly coated on the walls of the monolithic

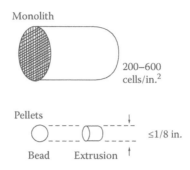

Figure 5.7 **Typical configuration for an automobile exhaust catalyst as used in automobile exhaust tests.**

support. Alternatively, the catalytic metals can be placed at or near the surface of either beads or excursions for maximum activity. Typical cell surface area and pellet dimensions are shown in Figure 5.7

A role for cerium oxide or another suitable rare earth oxide catalyst occurs as a result of the oscillation of the exhaust emission gases converting from lean to rich. It is important to mention that there is always a lack of oxygen on a rich side, as shown in reaction 2 above. However, if cerium oxide can provide lattice oxygen for the chemical reaction, improved levels of carbon monoxide conversion to carbon dioxide can be materialized [5]. At a certain carbon monoxide conversion level, the oxygen source can be shut off. However, after a certain delay, the carbon monoxide conversion drops. A catalyst with a cerium oxide or another oxygen storage component could exhibit a much longer delay time. Note that the time will determine the measure of oxygen storage. There may be some effects due to cerium oxide content on the carbon monoxide conversion delay time. The cerium oxide content could also promote the water-gas shift reaction as a function of cerium oxide. The water-gas shift reaction can operate through a lattice redox (oxidation-reduction) mechanism similar to the oxidation mechanism.

The effect of ceria on the noble metal catalyst may underestimate the performance of the emission monitoring sensor. In addition, cerium oxide can significantly degrade the activity of a platinum-supported catalyst after a thermal aging cycle, but it does not degrade the performance of palladium-supported catalysts.

5.9.8 Effects on Catalytic Activity from Materials with Perovskite Structures

Calcium titanate is a material with a perovskite structure, but it cannot be used as a catalytic agent because it does not belong to the noble metal category, such as platinum or palladium. Only materials with perovskite structure can be prepared in a form containing the rare earth metal lanthanum. There are some perovskite materials containing cobalt, strontium, lanthanum carbonate, and strontium lanthanum carbonate ($Sr_{0.3}La_{0.7}CoO_3$), which have been considered as potential oxidative automobile exhaust catalytic sensors. Other compounds containing lead and lanthanum lead manganese oxide ($La_xPb_{1-x}MnO_3$) have demonstrated NO_x reduction activity and can provide active and stable support to provide needed mechanical integrity for the platinum, appearing to impart resistance to sulfur dioxide poisoning. Compounds such as rare earth–copper double oxides have been investigated for determining the catalytic activity. However, any rare earth–based compound that shows high catalytic activity for automobile catalytic sensors must be investigated in great detail in terms of cost, detection efficiency, sensors performance, and catalytic activity.

More than thirty patents have been published describing the use of rare earth compounds in automobile exhaust gas purification and monitoring. Assuming that all automobile catalysts in the United States were to deploy rare earth oxides, the market volume for these oxides would be close to 100 tons per year. If cerium oxide was used for this application, its consumption level could be between 1% and 10% just for catalytic applications.

5.9.9 Critical Role of Delayed Conversion of Carbon Oxide Conversion

The delay time indicates a measure of oxygen storage. Catalysts using different rare earth oxides will have longer delay times for the carbon oxide conversion. Cerium oxide can significantly degrade the chemical activity of a platinum-supported catalyst sensor compared to a palladium-supported catalyst after thermal aging adverse effects. Therefore, catalytic device designers must take into account the thermal aging effects when selecting a rare earth element for the support structure.

5.10 Summary

This chapter summarizes the critical roles of rare earth oxides and alloys needed in the manufacture of conventional steel, stainless steel, HSLA steel, carbon steel, and various types of irons for industrial applications. Potential properties and applications of different irons, such gray iron, malleable iron, and nodular iron, were briefly discussed, with an emphasis on cost and mechanical strengths. The interface properties of iron-graphite were identified in great detail. Gray iron is a useful engineering material, with exceptionally high casting tonnage shipped to various parts of the industrial world for commercial and industrial products requiring strength and longevity. There is a relationship between a nodule's shape, size, physical properties, and structural distribution of the iron produced as a result of incorporating an REM as a nucleating agent. Elimination of iron carbides could cause deterioration of the physical properties of the iron, which could be eliminated by the addition of cerium as mischmetal in right amount.

White iron, in general, has poor physical properties, but carbon diffusion through the matrix results in a breakdown leading to the generation of some form of spheroids rather than flakes. White iron produced in this way offers high tensile strength, ranging from 52,925 to 101,500 psi. This illustrates that improvements in mechanical properties are strictly due to the change in graphite morphology from flakes to spheroids. The addition of cerium may result in spheroidal graphite structures in nickel carbon steel alloys, which are best suited for applications where smooth surface and high tensile strength are the principal requirements.

Properties and applications of malleable iron are discussed in detail, with an emphasis on tensile strength and ductility. Tests performed by various companies have confirmed that tensile strength and ductility of malleable iron are best suited for specific industrial applications where reliability and mechanical performance are of critical importance. Magnesium is added as a nickel-manganese alloy to iron to produce nodular iron. Both hypo- and hypereutectic base irons have been treated successfully, particularly when using the nickel-manganese alloy.

Mischmetal played a critical role in the production of various kinds of steel. The use of mischmetal led to great demand for the bastnäsite ore, which contains various alloys with some content of rare earth elements such as cerium, lanthanum, neodymium, and praseodymium. It is important to mention that 50% of rare earth alloys contain cerium-based materials. The most deleterious effects contributed by lead, bismuth, antimony, and titanium can be significantly reduced by the addition of specific REMs in the alloys, which were summarized with an emphasis on cost and material quality. Estimates of mischmetal annual output capacity and cost were provided, and density and melting temperatures of the alloying elements were summarized. Five distinct physical properties of mischmetal were identified, with emphasis on vapor pressure, total solubility, and alloying density, which are introduced in the liquid steel. A low-melting-point eutectic develops when excess metallic REM is present in the steel towards the end of the solidification process; it concentrates at the as-cast grain boundaries.

The cost of mischmetal is much higher than the cost of rare earth silicide alloy. This particular alloy is composed of one-third mischmetal, one-third silicon, and one-third iron. This alloy is widely used in the production of HSLA steel. Physical chemistry and metallurgical aspects of rare earth metals were discussed. Manganese can be assigned for strengthening and phase control of HSLA steels; such steels are best suited for various industrial and commercial applications with minimum cost.

Surface defects on the finished rolled steel products can lead to quality control problems in precision steel products, such as bearings for high-speed superheated steam turbines. Deoxidation with zirconium, titanium, aluminum, and calcium must be avoided because this could develop accumulation in the throat sections of the nozzles, leading to serious nozzle blockage. Significant improvements in steel properties have occurred due to the growing addition of REMs in the steelmaking process. The exact amount of REM additions in steel will lead to significant improvement in shape control in conventional steels and tramp element control in stainless steels. The role of REMs as noncracking catalysts was described in great detail. A number of rare earth oxides were recommended for catalyst applications in which reliability, consistent parameter recording, and sensor performance are of critical importance. Cost and abundance data on the most widely used oxides, such as cerium and lanthanum, were summarized for automobile emission control devices. Cerium and lanthanum oxides are widely used by state and federal emission control agencies because of their minimum cost.

Ammonia synthesis plays a critical role in the formation of catalytic reactions. An ammonia synthesis catalyst contains calcium oxide, iron oxide, and potassium oxide and has been observed on the catalyst surface. Catalysts are vital in the determination and measurement of automobile exhaust emissions, which are considered to be extremely harmful to human beings and animals. The adverse effects of catalytic activity from catalytic agents with perovskite structures, which have unique benefits in terms of precision recording of exhaust emissions, were also described.

In summary, the role of rare earth elements is not limited to iron, steel, or alloy industries; electro-optical sensors, such as lasers; or automobile emission control devices. Rare earth materials have also played important roles in nuclear power plants, nuclear-powered submarines, nuclear aircraft carriers, and medical research and cancer treatments involving isotopes of some specific rare earth elements.

References

1. Lineberger, H.F. and T.K. McCluhan. 1981. *The Role of Rare Earth Elements in the Production of Nodular Iron.* Niagara Falls, NY: Union Carbide Corporation.
2. Howard W. Sams & Co. Engineers. 1970. *Reference Data For Radio Engineers,* 5th ed. New York: Howard W. Sams & Co. Engineers, 4–12.
3. Kortovich, C.S. 1977. *Inhibition of Hydrogen Embrittlement in High-Strength Steel.* Cleveland OH: TRW, Inc.
4. U.S. Bureau of Mines. 1977. *Mineral Yearbook.* Washington, DC: U.S. Bureau of Mines.
5. Peters, A.W., and G. Kim. 1982. *Rare Earths in Non-cracking Catalysts.* Columbia, MD: Chemical Division, 118.
6. Luyckx, L.A. 1981. *The Rare Earth Metals in Steel.* New Castle, PA: American Chemical Society, 43–65.

Chapter 6

Applications of Rare Earth Intermetallic Compounds, Hydrides, and Ceramics

6.1 Introduction

This chapter investigates the possible commercial and industrial applications of rare earth intermetallic compounds, hydrides, and high-temperature ceramics. In addition, the physical properties and unique characteristics of intermetallic compounds are summarized, with an emphasis on reliability and affordability. The magnetic and electrical properties of rare earth hydrides are of significant importance. Structural aspects of intermetallic compounds are discussed, with particular emphasis on energy conservation, national defense, and safety. Rare earth intermetallic compounds offer the promise of improving human conditions through medical and dental procedures.

Current applications of rare earth intermetallic compounds include high-power, high-temperature samarium-cobalt permanent magnets, microwave devices with improved performance, high-efficiency and lightweight electric motors, biomedical materials for unique medical sensors, energy conservation, new and improved energy sources, fuel for heterogeneous catalysts, and coal gasification and liquefaction. Near-term future and projected applications may include transducers with improved gain performance, automobile ignition systems with improved fuel economy, hydrogen getters, and safety devices for nuclear reactors.

6.2 Rare Earth Intermetallic Compounds and Rare Earth Hydrides

This section discusses the critical properties and applications of rare earth intermetallic compounds and rare earth hydrides. Later, three distinct topics—the magnetic and electrical properties of hydrides, the deployment of rare earth intermetallic materials as hydrocarbons from hydrogen gas, and the magnetism of cobalt intermetallic compounds—will be discussed in great detail, along with their commercial and industrial applications.

Some major benefits and applications of rare earth intermetallic compounds include the following:

■ Samarium-cobalt (Sm_2Co_5 and Sm_2Co_{17}) magnets offer the optimum energy product (HB_{max}) at operating temperatures as high as 300°C. These magnets are widely used in high-power traveling-wave tube amplifiers in military radars and electronic warfare systems. These rare earth permanent magnets provide high efficiency and reliability under harsh mechanical and thermal environments.

■ Rare earth intermetallic compounds are widely used in the design and development of microwave components for communication and national defense applications, where electrical performance, reliability, and longevity are of critical importance.

■ Rare earth permanent magnets using intermetallic compounds are well suited for electric motors and generators, where light weight and high conversion efficiency are the principal design requirements. Rare earth intermetallic compounds may also be used for energy conservation applications.

■ Some intermetallic compounds have potential medical and dental applications for which reliability and safety are of critical importance. Structural types and their typical examples are shown in Table 6.1.

Table 6.1 Structural Types of Intermetallic Compounds and Typical Examples

Structural Intermetallic Compounds	Typical Example
Thorium caesium chloride (ThCaCl)	Gadolinium silver (GdAg)
Magnesium copperate ($MgCu_2$)	Praseodymium aluminate ($PrAl_2$)
Chromium boron (CrB)	Cerium nickel (CeNi)
Ferrous carbide (Fe_3C)	Erbium nickel (Er_3Ni)
Calcium copperate ($CaCu_5$)	Lanthanum nickel ($LaNi_5$)
Nickel ($ThNi_{17}$)	Holmium cobalt ($HoCo_{17}$)

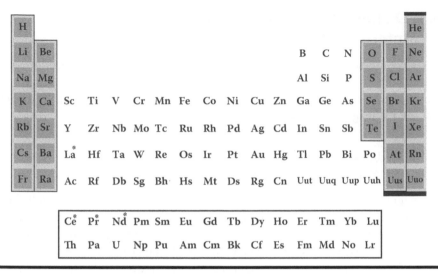

Figure 6.1   Periodic Table of Elements showing the heavy elements and light lanthanide elements (as indicated by asterisks).

Heavy elements and light lanthanide elements shown in Figure 6.1 play critical roles in the development of intermetallic compounds. Lanthanum-nickel ($LaNi_5$) intermetallic compounds are able to store as much hydrogen as an equal volume of liquid hydrogen at low pressure and room temperature, and they will release gaseous hydrogen safely on demand with minimum cost and complexity. In addition, the storage of gaseous hydrogen in conventional forged steel cylinders at 2,300 psi requires three times as much storage space as the same quantity of hydrogen in an $LaNi_5$ intermetallic compound. Because hydrogen holds tremendous promise as the prime gaseous fuel of the future, rare earth intermetallic compounds related to lanthanum-nickel, in which mischmetal replaces lanthanum, will likely be used as a means of achieving cost-effectiveness. Hydrogen gas associated with lanthanum-nickel as a fuel has potential as an energy source for heterogeneous catalysis, coal gasification, and liquefaction. Projected or near-term applications of rare earth intermetallic compounds are shown in Table 6.2.

Table 6.2   Applications of Rare Earth Intermetallic Compounds

Intermetallic Compound	Near-Term Application
Transducers	Reliable automotive components
Ignition systems	Improve fuel economy
Hydrogen getters	Safety improvement in nuclear reactors

6.3 Distinct Properties of Rare Earth Intermetallic Compounds

This section describes the three important aspects of rare earth intermetallic compounds and hydrides—the magnetic and electrical properties of hydrides, the deployment of rare earth intermetallic materials as hydrocarbons from hydrogen gas, and the magnetism of cobalt intermetallic compound [1].

6.3.1 Magnetic and Electrical Properties of Rare Earth Hydrides

The magnetic and electrical properties of rare earth hydrides have been studied for a number of years, particularly the fundamental exchange mechanism operating in the elemental rare earth materials. In the materials investigated, the distance of separation of adjacent ions is larger compared to the range of the f-shell. Therefore, the overlap of the f-orbitals centered on the adjacent atoms and the direct exchange is negligible. Coupling arises from the polarization of the conductive electrons, which is known as the Ruderman-Kittel-Kasuya-Yosida (RKKY).

The RKKY interaction was initially observed by several researchers who concluded that hydrogenation of a rare earth element results in electrical conductivity. The conductivity falls by many orders of magnitude during the hydrogenation process; in a fully hydrogenated material, it approaches the conductivity of the insulator. During the hydrogenation process, electrons are absorbed from the conduction band; in fully hydride materials, the conduction band is completely depopulated— that is, the electrical conductivity disappears. In other words, the rare earth hydride material is seen as a nonconducting analog of the elemental rare earth intermetallic hydride material. The hydride lacks conduction electrons, which is essentially known as the RKKY mechanism. In this situation, a sharp fall in the magnetic ordering temperature of the hydride is expected, compared to the element from which the hydride is formed (Table 6.3).

Richard Heckman first determined the Hall coefficient for hydrogenated cerium. The increasingly positive Hall coefficient is the most reliable indication that the hydrogenation application is close to completion of the process. A positive Hall coefficient is characteristic of the full band structure, so the experimental data seem to indicate the band-filling products rather than band-depopulation effects. Proper resolution of the electrical and magnetic properties of the rare earth hydrides can be achieved using the band structure information provided by the calculation. Conduction band depletion and new band formation due to antibonding should be given appropriate consideration.

It appears that a new process is developed during the hydrogenation process, which lies below the conduction band of the rare earth elements. During the hydrogenation process, the hydrides are transferred from the conduction band to the hydrogenation process. Due to this, there is indeed a band-filling effect, which

Table 6.3 Magnetic Ordering Temperatures of Rare Earth Elements

Parental Material	x^a Coefficient	Magnetic Ordering Temperature (K)
Neodymium	2.75	19
Samarium	2.91	106
Terbium	2.97	235
Dysprosium	2.97	184
Holmium	3.06	135
Erbium	3.02	88
Thulium	2.95	56

Source: Wallace, W.E., and F.H. Spedding. 1972. *Rare Earth Intermetallic Compounds and Their Applications.* Pittsburgh, PA: Department of Chemistry, University of Pittsburgh.

ultimately accounts for a positive Hall coefficient x^a. The hydrogenation band is formed from the antisymmetrical hydrogen atom. Accordingly, the electron charge must be localized around the nucleus. Hence, the hydrogen in the rare earth hydrides seems to be in accordance with the conclusions drawn on the electrical and magnetic properties of rare earth materials.

6.3.2 Heterogeneous Catalysis with Rare Earth Intermetallic Compounds

Heterogeneous catalysis is widely used in formations of hydrocarbons and hydrogen gas. The process is recognized as the most important aspect of the heterogeneous catalysis phenomenon. Heterogeneous catalysis also involves rare earth intermetallic compounds in the catalysis of hydrocarbons from carbon monoxide and hydrogen gas. This is one of the most practical solutions for coal conversion technology involving gaseous hydrocarbons from coal. Coal produces high-pressure, high-temperature steam, which leads to the production of a mixture of gaseous carbon monoxide and hydrogen, among other things. Carbon monoxide and hydrogen are separated out of this as a combined synthesis gas. However, synthesis gas alone will not react with hydrocarbons; to accomplish this process, a catalyst is needed.

Lanthanum nickel ($LaNi_5$), a rare earth intermetallic compound, is an excellent solvent for hydrogen. When $LaNi_5$ is exclusively chosen, the hydrogen bond is cleaved and monoatomic hydrogen is the choice of the intermetallic compound in large concentrations. It seems that the atomic hydrogen exists at least briefly on the surface of this material. From these observations, it appears that $LaNi_5$ might be responsible for the synthesis of hydrocarbons from the synthesis gas.

Table 6.4   Turnover Numbers for Catalysts and Intermetallic Compounds

Catalyst Type	Turnover Number
Conventional supported catalysts	
Nickel-supported silica (Ni/SiO_2)	0.5–1.0
Transformed intermetallic compounds	
Thorium nickel ($ThNi_5$) $\rightarrow$ Ni/ThO_2	4.7
Lanthanum nickel ($LaNi_5$) $\rightarrow$ Ni/La_2O_3	2.7
Nickel silica (Ni_5Si_2) $\rightarrow$ Ni/SiO_2	11
Oxidized intermetallic compounds	
Nickel silica (Ni_5Si_2) $\rightarrow$ Ni/SiO_2	18
Thorium nickel ($ThNi_5$) $\rightarrow$ Ni/ThO_2	10.6

Source: Wallace, W.E., and F.H. Spedding. 1972. *Rare Earth Intermetallic Compounds and Their Applications.* Pittsburgh, PA: Department of Chemistry, University of Pittsburgh.

Material scientists found that, in the presence of magnetic ordering temperatures ranging from 300 to 350°C, hydrocarbons are available from the synthesis of gas, likely due to the presence of catalytically active material. During the chemical reaction, the active material $LaNi_5$ is transformed into a mixture of nickel and lanthanum oxide (La_2O_3). Some researchers believe that the catalytically active material is not the original material but instead is the elemental nickel resting on a substrate; thus, there may be a new way of making supported catalysts by the virtue of the synthesis gas or the reaction products of carbon monoxide and water.

One can also achieve the oxidation of an intermetallic compound into a transitional metal oxide mixture by preassociation of an intermetallic compound with oxygen. This oxidized material is a highly active catalyst. The turnover numbers shown in Table 6.4 are for conventional supported catalysts; they represent the number of reactions that take place per site per unit of activity. The transformed intermetallic and oxidized intermetallic turnover numbers vary from approximately 5 to 40 times the turnover number for the conventional supported catalyst.

6.3.3 Magnetism of Rare Earth–Cobalt Intermetallic Compounds

This section discusses the magnetic characteristics of rare earth–cobalt compounds, such as $SmCo_5$ and $SmCo_{17}$. These rare earth–cobalt compounds are of critical importance because of their potential applications in various commercial and defense disciplines. As mentioned in Chapter 2, samarium-cobalt permanent magnets offer the maximum magnetic energy product and highest operating temperature (close to 300°C). No other conventional or rare earth material can meet or beat such magnetic characteristics under harsh mechanical and thermal environments.

To demonstrate the importance of cobalt-based rare earth magnets, the stoichiometries of the rare earth–cobalt compounds are shown in Table 6.5. Technological significance is attached to the intermetallic compounds $SmCo_5$ and Sm_2Co_{17} because these compounds are members of the underscored RCo_5 and R_2Co_{17} groups. The magnetization in these rare earth–cobalt intermetallic compounds is largely confined internally, and these compounds are devoid or short of technological advancements. The other intermetallic materials shown in Table 6.5 are of potential technological interest, but so far the potential has been achieved only for the $SmCo_5$ compound.

$SmCo_5$ and Sm_2Co_{17} are examples of the 1:5 and 2:17 classes of materials, respectively, and are of significant technological interest. The coupling mode in the 1:5 rare earth–cobalt systems is useful in certain applications. It should be noted that lanthanum cobalt ($LaCo_5$) is a ferromagnetic material with all its magnetic properties originating with cobalt. Again, the compounds praseodymium cobalt ($PrCo_5$) and neodymium cobalt ($NdCo_5$) are classified as ferromagnetic materials, with magnetization originating from both the cobalt and the rare earth sublattices. In the case of $SmCo_5$, it is not absolutely clear what the nature of the magnetic coupling is. Material scientists suspect that samarium is likely coupled parallel to the cobalt sublattice. However, this material is characterized as a ferromagnetic material; the samarium moment is quite small, and the nature of coupling has not been established explicitly. The heavy rare earth–cobalt compound materials starting with gadolinium, such as $GdCo_5$, are coupled antiferromagnetically with the rare earth cobalt.

Table 6.5 Structure Types for Cobalt-Based Rare Earth Intermetallic Compounds

Compound	Structure Type
R_3Co	Fe_3C
R_4Co_3	Hexane
RCo_2	$MgCu_2$
RCo_3	$PuNi_3$
RCo_5	$CaCu_5$
RCo_7	Hexane
R_2Co_{17}	Th_2Ni_{17} and Th_2Zn_{17}

The Curie temperature of the rare earth intermetallic compound RCo_5 is quite high (1,020 K), which is necessary if this is to be incorporated into permanent magnets to be operated at elevated temperatures. The rare earth permanent magnet $SmCo_5$ has a powerful anisotropy field, close to 600,000 Oe at 0 K. The following magnetic characteristics of $SmCo_5$ make it a leading high-energy magnetic material:

■ Curie temperature: 1,020°C
■ Saturation magnetization: 8.5×10^6 G
■ Anisotropy field (near 0 K): 600,000 Oe

However, the importance of the samarium-cobalt rare earth compound does not lie with its saturation magnetization characteristic. On a per-atom basis, the saturation magnetization is only approximately 60% of that of elemental iron and 80% of that of elemental cobalt metal. The deployment of $SmCo_5$ magnetic material is strictly based on its high anisotropy characteristics. For a material to be ideal as a permanent magnet, there must be something to clamp the magnetic moments in place after the aligning field is withdrawn. In the case of $SmCo_5$ magnetic material, the crystal field interaction plays that critical role.

$SmCo_5$ has a hexagonal structure, and its atomic moments lie along the hexagonal axis. The anisotropy—the magnetic field required to twist the moment to 90 degrees—is approximately 600,000 Oe near absolute zero and approximately 400,000 Oe near room temperature. Work is needed to twist the moment away from the hexagonal axis. This work was measured some years ago by U.S. scientists at the Wright-Patterson Air Force Base. Later, additional work was undertaken on the quantum mechanics of the system to estimate the efforts required to calculate the energy involved when the moments lie parallel and perpendicular to the unique hexagonal axis. The difference between the two energies is the work required to twist moments through 90 degrees, which is often called the stabilization energy. Research undertaken in the mid 1980s by Oxford University scientists investigated the quantum electronics of rare earth systems. Some scientists believe that this calculation procedure cannot be employed for Sm^{3+} material because the J multiples (Racah tensorial coefficient) are not well separated. Therefore, it is necessary to employ Racah tensorial algebra.

Crystal field interactions for Sm^{3+} in an $SmCo_5$ compound can be calculated by the Racah tensor-operator technique involving the Hamiltonian function. The Hamiltonian equation can be written as follows:

$$H = [\lambda L \times S] + [H_{CF}] + [H_{ex}] \tag{6.1}$$

H_{CF} and H_{ex} can be defined as follows:

$$H_{CF} = N_2 A_2 < r^2 > [U^2 + N_4 A_4] < r^4 > [U^4 + N_6 A_6] < r^6 > [U^6 + N^6 A^6] < r^6 > U^6 \tag{6.2}$$

$$H_{ex} = [2\mu_B \, S \cdot H_{ex}] \qquad (6.3)$$

A^k is a lattice sum that defines the crystal field q intensities; N^k is the normalization factors; and U^k is the tensorial operator, which is related to the spherical harmonics in operator form. These calculations can be performed using the Hamiltonian tensor. The crystal field interaction for Sm^{3+} in the rare earth compound $SmCo_5$ when treated with the Racah tensor-operator technique offers the following:

■ The lattice sum, which defines the crystal field intensities
■ The normalization factors
■ The tensorial operators (which are related to spherical harmonics in operator form)

This analysis proves an important fact—the powerful magnetism of $SmCo_5$ is a function of field effect.

Laboratory work by some scientists indicates that the energy products achievable from $SmCo_5$ and $R_{1-x}Fe_x$ are 25 and 30 MGOe, respectively. The samarium-cobalt material Sm_2Co_{17} provides an energy product so small that it is not worth mentioning. This is due to the fact that the anisotropy of Sm_2Co_{17} is roughly 20% of the anisotropy field of $SmCo_5$. However, this anisotropy field can be improved by the replacement of cobalt with iron.

Studies of 2:17 systems (based on the Sm_2Co_{17} compound) have been undertaken by researchers in the United States, Japan, and Switzerland. In brief, the five magnetic species involving samarium plus four types of cobalt were studied. The overall magnetic behavior is a composite of the interactions between the five sublattices, which include both the exchange and anisotropy field interactions. It appears that three of the cobalt sublattices containing some traces of manganese indicate improvement in stabilization energy. Substitutions of cobalt by iron, manganese, or chromium suppress harming effects on a selective basis that arises from a particular sublattice. It is unclear whether this suppression is due to a specific consequence of a single ion or band-structure effects. Further research is required to determine the technological significance for stabilization of energy fields.

6.4 Low-Temperature Magnetic Properties of Hybrids and Deuterides of $Er(Fe_{1-x}Mn_x)_2$ Compounds

$Er(Fe_{1-x}Mn_x)_2$ compounds absorb large amounts of hydrogen and deuterium, leading to the formation of stable hydrides and deuterides. The addition of manganese to $ErFe_2$ results in a significant increase in the amount of hydrogen or deuterium absorbed [2]. X-ray diffraction measurement data of hydrides and deuterides demonstrate a considerable expansion of the lattice. Magnetization measurements of the hydrides and deuterides can be made at cryogenic temperatures ranging from

4.2 to 300 K and in external field intensities up to 20,000 Oe. These measurements confirm that the magnetic properties of hydrides and deuterides are similar. The ferromagnetic ordering of the unhydrated material does not alter by the addition of hydrogen or deuterium. However, the strength of the magnetic interaction is considerably weakened by a reduction of the Curie temperature. The compensation temperatures of the hydrides and deuterides are lower than the corresponding parent materials. The measurements further indicate that the anisotropy effects observed in some unhydrided ternary materials appear to be destroyed by the mere presence of hydrogen or deuterium [2].

A number of rare earth–transition metal intermetallic compounds seem to absorb large amounts of hydrogen to form stale hydrides. Research on the magnetic properties of hydride compounds, such as RFe_2H_x, RFe_3H_x, RCo_5H_x, and RMn_2H_x (where R stands for a rare earth element), revealed that remarkable charging the compounds with hydrogen. A rare earth scientist demonstrated that the compound erbium-iron ($ErFe_2$) has only four atoms of hydrogen, and that the Curie temperature compensation decreases significantly while the magnetic moment expands by more than 25%. The studies further report that the magnetization varies at 4.2 K as a function of external field. The hydride was unusual in the sense that saturation was observed at fields up to 120,000 Oe.

Studies on the structural and magnetic properties of ternaries of the rare earth compound $Er(Fe_{1-x} Mn_x)_2$ reported very useful information on its hydrogen absorption properties. Investigations demonstrated that the substitution of manganese for iron significantly indicates the amount of hydrogen absorbed; the hydride composition of $Er_{0.8}Mn_{1.2}H_{4.6}$ resulted in hydration in the rare earth compound exceeding 7×10^{22}/L. Furthermore, these atoms were found to absorb deuterium. Magnetization measurements were performed in laboratory environments on the hydrides and deuterides of the compound $Er(Fe_{1-x} Mn_x)_2$ in the Curie temperature range of 4.2 to 300 K and in external anisotropic fields as high as 20,000 Oe. The lattice constants, Curie temperatures, and magnetic moments of two ternary compounds are summarized in Table 6.6.

In Table 6.6, it can be seen that the addition of ferrous cerium significantly increases the amount of hydrogen absorption; these increases result in increased lattice constants, ranging from 7.826 to 8.004 for the hydride samples and from 7.780 to 7.935 for the deuteride samples. It is evident from the data presented in Table 6.6 that increasing the manganese concentration in the ternary rare earth compound $Er(Fe_{1-x} Mn_x)_2$ results in an expansion of the cell size or lattice constant when parameter x is varied from 0 to 0.5. It is important to mention that the values of the magnetic moments do not represent the saturation values; rather, the measured values of the magnetic moments [2] were at a cryogenic temperature of 4.2 K with an external anisotropy field equal to 20,000 Oe.

In summary, the measured results shown in Table 6.6 were obtained in magnetization experiments [2] as a function of external anisotropy fields at 4.2 K for the ternary compositions in the absence of saturation up to external fields of 20,000

Table 6.6 Lattice Constants, Curie Temperatures, and Magnetic Moments of Two Ternary Compounds at a Curie Temperature of 4.2 K

Ternary Compound	Lattice Constant (A°)	Curie Temperature (K)	Magnetic Moment (μ)
Hydride compounds			
$ErFe_2H_{3.94}$	7.826	275	1.55
$ErFe_{1.6}Mn_{0.4}H_{4.10}$	7.840	180	2.27
$ErFe_{1.4}Mn_{0.6}H_{4.25}$	7.941	160	1.84
$ErFe_{1.2}Mn_{0.8}H_{4.53}$	7.974	20	2.68
$ErFe_{1.0}Mn_{1.0}H_{4.6}$	8.004	10	1.80
Deuteride compounds			
$ErFe_2D_{3.75}$	7.780	295	1.97
$ErFe_{1.6}Mn_{0.4}D_{3.90}$	7.869	190	2.00
$ErFe_{1.2}Mn_{0.8}D_{4.35}$	7.885	35	1.65
$ErFe_{1.0}Mn_{1.0}D_{4.50}$	7.935	—	1.57

Oe for both the hydrides and deuterides. For the ternary composition $ErFe_2$-H, the applications of an external field of 120,000 Oe at a cryogenic temperature of 4.2 K does not even provide the saturation condition. These results, as well as the neutron diffraction data obtained by other scientists, suggest that part of the rare earth ions are paramagnetic in the hydride and deuteride.

The researchers also observed the behavior of magnetization as a function of temperature for both the hydrides $Er(Fe_{1-x}Mn_x)_2$-H and the deuterides $Er(Fe_{1-x}Mn_x)_2$-D. It can be observed from the behavior data and the data summarized in Table 6.6 that the Curie temperatures of the hydrides and deuterides are much lower than the unhydrided materials. It is reasonable to assume that the variations in Curie temperatures between the hydrides and deuterides could be attributed to slight differences in their material compositions. Both the hydrides and deuterides behave identically in terms of structure and magnetism. Based on the tabulated data, the hydrides and deuterides of the material ErFeMn do not order even at 4.2 K, whereas the unhydrided ErFeMn orders ferrimagnetically at a Curie temperature of 175 K. Researchers believe that the reductions in the Curie temperatures are partly due to a possible weakening of exchange interactions brought about by the sizable expansion of the lattice constants. Similar conclusions may be made for the behavior of the hydrides and deuterides of other rare earth compounds.

6.4.1 Applications of Hydrides

Before discussing the applications of hydrides, this section presents some background knowledge on the hydrides and deuterides of rare earth metals and intermetallics. Importantly, large amounts of hydrogen and deuterium are absorbed by the ternary compound or composition $Er(Fe_{1-x}Mn_x)_2$, where parameter x varies from 0 to 0.5. Additional conclusions can be summarized briefly as follows:

- The substitution of manganese for iron increases the amounts of hydrogen and deuterium to be absorbed.
- Magnetization measurements of the rare earth compositions are critical.
- The hydrides and deuterides essentially exhibit identical magnetic behavior in terms of lattice constants, Curie temperatures, and magnetic moments.
- The Curie temperatures and compensation temperatures are considerably lower for hydrides and deuterides compared with unhydrided samples.
- Some rare earth ions are paramagnetic in nature.
- Nuclear material radiation and neutron diffraction experiments must be undertaken to obtain additional scientific information on the hydrides and deuterides of other rare earth–based compositions or compounds.

6.4.2 Storage of Hydrogen as a Hydride

The storage of hydrogen is of critical importance because it is considered to be efficient and economical for multiple applications. Hydrogen has an optimum energy-to-weight ratio. Furthermore, hydrogen is the lightest of all fuels and yields the best calorific ratio, exceeding 51,500 BTU/lb. of fuel compared with 18,500 BTU/lb. for jet fuel. Most hydrogen is produced using electrolysis technology with an efficiency of approximately 70%. Significant improvement in the electricity-to-hydrogen efficiency may be possible without increasing the capital cost for electrolyzer plants.

6.4.2.1 Production Approach for Hydrogen from Hydrides

The storage of hydrogen as a chemical hydride is a cost-effective approach that is receiving great research attention. Because of the alarming declining rate of the availability of fossil fuel sources, particularly gas and oil, it is important to look for alternate energy sources that are economical solutions without greenhouse gas effects. Other sources, such as nuclear and hydroturbine technology, require enormous capital investment. When various alternatives for future energy sources are considered, the hydrogen energy option emerges as a very favorable choice.

Tradeoff studies undertaken on hydrogen storage technology indicate that rare earth hydrides tend to offer the most convenient and economical solution. Hydrides of magnesium, iron-titanium alloys, and other rare earth ternary compounds can all be formed economically and spontaneously by reacting the finely divided metal with hydrogen gas. The hydrogen gas can then be recovered by heating the hydrides at moderate temperatures. However, because waste is released in the hydrogen-formation process, the storage process is not 100% efficient.

Basic research and development programs to provide a clear understanding of the alloy hydride chemistry must be undertaken in the hope that improved formulations can be developed, with an emphasis on cost and efficiency. Engineering studies on relatively large-scale stationary storage systems using an iron-titanium alloy hydride are in progress according to scientific reports. The conclusions about this energy delivery technology with relatively minimum cost and complexity can be summarized as follows:

- Hydrogen can be produced from water using various technologies such as electrolysis, thermochemistry, and chemical reactions energized by direct solar or nuclear radiation.
- Hydrogen may be transported and stored either in underground rock formations or by a liquefaction technique, which involves higher costs.
- Hydrogen can be delivered to the customers through existing gas distribution pipes.
- Hydrogen can be produced and stored using rare earth hydride technology with the utmost safety and minimum cost and complexity.
- Electrochemical, thermochemical, and radiochemical technologies are feasible for the production of hydrogen, but they require comprehensive tradeoff studies to justify the cost and complexities involved in each case.
- The production and storage of hydrogen as a chemical hydride is receiving the greatest attention by researchers and economists.
- Basic efficiency characteristics and economic aspects must be carefully evaluated for hydride storage technology.
- For portability and storage feasibility, major emphasis must be placed on the low storage cost, low weight, and low reaction enthalpy.
- The timely availability of hydride materials is also of paramount importance.
- Selection of the most appropriate rare earth hydrides for both hydrogen production and storage is critically important.

6.4.2.2 Critical Properties of Hydrides

The following critical properties of hydrides are identified for the benefit of the reader:

- Abundance
- Rate of formation and disassociation
- Physical stability

Table 6.7 Critical Parameters of Hydrides for Energy Storage Applications

Hydride Material	Equilibrium Pressure at 25°C (atm)	Heat of Formation (cal/mol)	Hydrogen-to-Metal Weight Ratio (g; weight % H)
Samarium cobalt ($SmCo_5$)	2–3	6,500	0.67
Lanthanum nickel ($LaNi_5$)	2–3	7,200	1.52
Niobium (Nm)	1–2	7,000	2.1
Iron titanium (FeTi)	5.0	5,500	1.8
Manganese nickel ($MnNi_5$)	10–11	6,500	1.5
Mischmetal	9–10	6,350	1.4

- Tolerance to impurities
- Enthalpy or heat formation speed
- Safety and reliability
- Useful temperature range
- Thermal conductivity
- Equilibrium pressure that is vital for reliability and safety
- Cost, weight, and volume

Weight and reaction enthalpy are considered to be drawbacks for certain hydrides. Hydrides consisting of rare earth materials and iron-titanium oxides offer minimum cost and low heat formation; thus, they are best suited for hydrogen storage systems. The ideal equilibrium pressure, heat of formation, and hydrogen-to-metal weight ratio of potential hydrides for energy storage applications are summarized in Table 6.7.

In Table 6.7, note that mischmetal—a mixture of rare earth metals in their naturally occurring composition—is cheaper than most rare earth metals. The user must select a hydride material that will meet most of the necessary applications. For automobile applications, it is recommended to compare the stored energy densities of the hydrides with those of battery systems. The data summarized in Table 6.7 are a valuable comparison of vehicle storage weights, verifying that the hydride systems are significantly better than batteries [3].

6.4.2.3 Disassociation Pressures of Various Rare Earth–Based Hydrides

The values of disassociated pressures as a function of temperature are the estimated values and may have errors ranging from 5 to 10%.

Table 6.8 Dielectric Properties of Rare Earth–Based Ceramic Magnetic Materials

Material	Dielectric Constant at 1 MHz	Dielectric Strength (V/mil)	Volume Resistivity at 25°C (ohm-cm)	Loss Tangent
Zircon	7.1–10.5	250–400	10^{13}–10^{15}	0.0002/0.004
Titanates	15–12,000	50–300	10^8–10^{15}	0.0001/0.02
Titanium dioxide	14–110	100–210	10^{13}–10^{18}	0.0002/0.005

6.5 Rare Earth Ceramics, Their Properties, and Applications

Ceramic materials fall under various categories, including piezoelectric ceramics, ferroelectric ceramics, magnetic ceramics, and spinel ceramics, which each have distinct characteristics. The rare earth–based ceramics include zircon, titanates of barium, strontium, calcium, magnesium, lead, and titanium dioxide. Among the titanates, barium titanate ($BaTiO_2$) possesses a structure known as perovskite, which has unique dielectric properties. The maximum energy product of the magnet (BH_{max}) is higher for strontium titanate compared to barium titanate [3].

Spinel ceramics, such as zinc and cadmium, are known as the "normal" spinel ceramics. Iron, cobalt, calcium, and nickel are the inverse spinel ceramics. Among these ceramics, cobalt iron ferrite ($CoFeO_4$) spinel exhibits a large positive anisotropy constant. Its anisotropy energy places its magnetic properties between soft ferrite and hard ferrite. It is interesting to mention that the Curie temperature of the magnetic materials changes the magnetic status of the material, such as from a ferrimagnetic material to a paramagnetic material. The Curie temperature for both barium iron oxide ($BaFe_{12}O_{19}$) and yttrium iron oxide ($Y_2Fe_5O_{12}$) garnet materials is approximately 300 K, for example.

The dielectric properties of rare earth–based ceramic magnetic materials are shown in Table 6.8. The dielectric constant, dielectric strength, volume resistivity, and loss tangent of any titanate fall within the ranges indicated in Table 6.8. The dielectric constant for barium titanate is 1,700, which falls in the range of 15 to 12,000.

6.5.1 Most Popular Rare Earth Ceramics and Their Applications

Zirconium oxide (ZrO_2) and cerium oxide (CeO_2) have been recognized as the most popular rare earth ceramics. Other rare earth ceramics such as $18CeO_2$-$82ZrO_2$,

$50CeO_2$-$50ZrO_2$, and $75CeO_2$-$25ZrO_2$ are widely used in commercial and industrial applications. Potential applications of rare earth ceramics include the design and development of magnetohydrodynamic (MHD) power generators. For many years, zirconium oxide stabilized with 10 to 20% calcium oxide, magnesium oxide, and yttrium oxide and was considered to be a prime contender as an MHD electrode material due to its excellent refractory and electrical properties at high operating temperatures (up to 2,700°C). Note that the higher electrical conductivity at elevated temperatures is ionic in nature. Therefore, zirconium oxide will have a tendency to decompose at the cathode wall under the influence of large electric field environments in an MHD channel wall. However, this difficulty can be overcome by using solid solutions of rare earth ceria and zirconia. Experimental results seem to demonstrate that the 100% ionic conductivity changes to 100% electrical conductivity when the amount of added ceria is varied between 0 and 80%. However, selection of the optimum ceria-to-zirconia ratio requires consideration of a number of physical and chemical properties of both oxides. Zicornia is best suited for applications where service temperatures are as high as 2,300°C.

Ferrites can be classified into two categories—soft ferrites and hard ferrites. The hysteresis curves distinguish between soft ferrites and hard ferrites. In general, soft ferrites are magnetized with low magnetic fields. The magnetically hard ferrites retain their magnetization strength, even after the applied field is removed. A rounded S-shape hysteresis loop is found in most soft ferrite materials, such as Mn-Mg and Li-Ni spinel ferrite. Doped yttrium-iron garnet ferrites have rectangle-shaped hysteresis loops. Note that abrupt changes in magnetic flux as a function of the applied field are critically important for the ferrite memory cores suited for high-speed digital computers. The high coercive force arises from the very high strength of the anisotropy field found in hexagonal ferrites and the small grain size of ceramic magnets.

Ferrites are widely used in various commercial and industrial applications. Some of the ferrite materials include rare earth elements such as nickel, magnesium, or yttrium. Microwave ferrites are best suited for the design and development of non-reciprocal microwave devices that are designed around the gyromagnetic behavior of elementary magnetic dipoles or uncompensated electron spins of ferromagnetic materials. Most of these ferrites are polycrystalline ferrite materials. When a ferrite material is placed in a magnetic field, it will have a resonance frequency given by $f_{res} = [R_g \ H_i]$, where R_g stands for the gyromagnetic ratio, which has a value of approximately 2.8 MHz/Oe, and H_i indicates the magnetic field within the ferrite. Faraday rotation devices, such as phase shifters, isolators, and gyrators, operate on this concept [4].

Room temperature critical properties of widely used ferrite materials are summarized in Table 6.9. The gyromagnetic ratio for these ferrites varies from 2.0 to 2.4 at room temperature (25°C). A microwave engineer can select a ferrite for a specific device application, such as a high anisotropy field or narrow line width (field bandwidth) or high or low Curie temperature.

Table 6.9   Room-Temperature Critical Properties of Widely Used Microwave Ferrites

Ferrite	Curie Temperature (K)	Saturation Magnetization (G)	Field Bandwidth (Oe)	Dielectric Constant
$MgFe_2O_4$	352	1,450	900	8.5
$NiFe_2O_4$	587	3,200	500	8.9
$BaFe_{12}O_{19}$	450	4,600	50	—
$Y_3Fe_5O_{12}$	277	1,740	50	12

Source: Harper, C.A., ed. 1984. *Handbook of Materials and Processes for Electronics.* New York: McGraw-Hill.

6.5.2 Advantages of Ceramic Permanent Magnets

Ceramic permanent magnets have several advantages over metallic ones, which are summarized as follows:

- Much higher ohmic resistance (10^8 ohm-cm or higher)
- Higher coercive force, ranging from 1,500 to 2,000 Oe compared with 50 to 1,500 Oe for metallic magnets
- Lower cost and minimum weight
- Minimum chemical reactivity

6.6 Rare Earth–Based Ceramics for Magnetohydrodynamic Power Generators

Intermetallic rare earth compounds and alloys have been used in various commercial and industrial applications in the field of stored energy and steelmaking technology [5]. The following sections identify the potential benefits of rare earth ceramics in the field of MHD power generation. The hot-wall type of open-cycle MHD power generators needs electrode materials that are capable of operating at temperatures as high as 1,800°C in a highly corrosive atmosphere and under severe electrical and thermal environments. For satisfactory and reliable operation of MHD generators, the use of rare earth–based ceramics is highly desirable.

6.6.1 Requirements of Rare Earth Compounds for MHD Power Generators

When combined with conventional steam power generation, an MHD generator has an overall efficiency exceeding 50%, compared with the typical 15% for a steam-power generator. Rare earth compounds containing lanthanum and cerium elements along with suitable oxides are well suited for MHD power generation. The physical, chemical, and structural properties of the materials are of critical importance for the safe and reliable operation of these power generators. Lanthanum chromium oxide ($LaCrO_3$)- and cerium oxide (CeO_2)-based electrodes meet these performance requirements in severe corrosive and electrical operating environments with high safety and reliability.

The characteristics of an MHD power generator can be summarized as follows:

■ A cheap fossil fuel, such as coal, is burned in a combustion chamber.
■ The expanding hot gases (at temperatures between 2,300–2,700°C) are made electrically conductive by seeding with a potassium salt. They are propelled through a channel between the poles of a large magnet, with a magnetic field ranging from 2 to 5 T (1 T = 10^4 G).
■ As a result, an electron current will be produced perpendicular to both the hot gas stream and the magnetic field.
■ By placing rare earth–based electrodes on either side of the channel wall (similar to the carbon brushes of a conventional steam turbo-alternator), this electron current can be collected and finally transferred to an external load, which could be an electrical power distribution system.
■ By using rare earth–based ($LaCrO_2$) electrodes, large-joule heating can be avoided, thereby providing safety and reliability under the harsh operating environments.

The lower temperature limit for $18CeO_2$-$82ZrO_2$ electrodes is approximately 1,200°C (activation energy = 2.2 eV), whereas the $75CeO_2$-$25ZrO_2$ electrodes can be used down to 1,000°C (activation energy = 1.15 eV). Experimental tests indicate that even higher electrical conductivity can be achieved by doping the binary materials with small amounts of tantalum oxide (Ta_2O_5), which forces a certain amount of cerium (Ce) cations into the trivalent state. Improvements in electrical conductivity of a $75CeO_2$-$25ZrO_2$ sample and tantalum-doped are shown in Table 6.10. It is evident that doping with tantalum oxide significantly improves electrical conductivity.

The data summarized in Table 6.10 show an approximate 32% improvement in the electrical conductivity for the $LaCrO_3$-20SrO ceramic sample over the temperature range of 1,727 to 1,496°C, which represents a change of 71%. However, in the case of $78CeO_2 + 20ZrO_2 + 2Ta_2O_5$, an improvement in electrical conductivity of 99.9% is possible over the same temperature range due to the addition of tantalum oxide in the ceramic material. The improvement of 99.99% can be realized in the electrical conductivity of the ceramic compound $75CeO_2 + 25ZrO_2$ over the same temperature range without the addition of tantalum to this particular ceramic compound.

Because the operating temperature of an MHD generator typically ranges from 1,200 to 1,000°C, the percentage improvement in electrical conductivity

Table 6.10 Improvements in the Electrical Conductivity of Various Rare Earth Ceramics as a Function of Temperature

Electrical Conductivity (ohm-cm)			
Temperature (°C)	*LaCrO$_3$-20SrO*	*78CeO$_2$ + 20ZrO$_2$ + 2Ta$_2$O$_5$*	*75CeO$_2$ + 25ZrO$_2$*
1,727	28	7.2	5.1
1,156	27	1.2	0.4
977	26	0.30	0.10
838	25	0.12	0.025
636	21	0.02	0.005
496	19	0.003	0.0003

Source: Hosler, W.R., and H.P.R. Frederikse. 1976. MHD Electrical Power Generation. *Proceedings of the 6th International Conference, Vol. 2.* Washington, DC, 67.

can be even higher than the values stated above. Electrical conductivity calculations over the temperature range from 1,156 to 977°C, which nearly represents the MHD generator operating temperature limits from 1,200 to 1,000°C, showed improvement in electrical conductivity that varies from approximately 3.7% for LaCrO$_3$-20SrO to 75% for 78CeO$_2$ + 20ZrO$_2$ + 2Ta$_2$O$_5$. Reliability and cost-effectiveness studies are not available to determine which ceramic compound yields the optimum electrical performance reliability with minimum cost and complexity. Nevertheless, the fact remains that the addition of tantalum oxide (Ta$_2$O$_5$) to the ceramic compounds significantly improves the electrical conductivity of the ceramic-coated electrodes.

Thermal conductivity of the ceramic coated electrodes is also of critical importance. Studies performed by the author indicate that higher thermal conductivity of the electrodes will provide higher thermal efficiency, improved reliability, and longevity of the electrodes.

6.6.2 Material Property Requirements for the Design of MHD Electrodes

Specific properties of the electrode materials will ensure operational reliability, electrode longevity, and improved efficiency of MHD generators. The thermal conductivity of the electrode material, magnitude of mechanical stress, thermal shock resistance, vaporization rate, and resistance to corrosion must be given consideration to ensure longevity and efficient electrode performance. These properties are strictly dependent on the method of preparation, such as sintering, arc plasma spraying, or hot-pressing. In most cases, tradeoff studies are essential, with a particular emphasis on cost, reliability, and overall electrical efficiency of the MHD generator. For example, good thermal shock resistance requires a moderate density. High porosity in the 15 to 20% range gives rise to corrosion by penetrating the gases and liquids. A good compromise can be achieved by fabricating samples of 92% density, preferably with many closely located pores.

6.6.3 Description of Potential Rare Earth–Based Electrode Materials

Ceramic composite materials such as $LaCrO_3$-$20SrO_2$, $78CeO_2 + 20ZrO_2 + 2Ta_2O_5$, and $75CeO_2 + 25ZrO_2$ are ideal materials for MHD electrodes. Zirconium oxide (ZrO_2), when stabilized with calcium oxide (CaO), magnesium oxide (MgO), or yttrium oxide (Y_2O_3), has been recognized as a potential material for MHD electrodes due to its ideal refractory properties, high melting point close to 2,700°C, and reasonably good electrical conductivity at high temperatures. This conductivity is ionic in nature; therefore, zirconium oxide has a tendency to decompose at the cathode wall under the presence of large electric fields in the MHD channel wall. This difficulty can be overcome by using solid solutions of ceria and zirconia.

Preliminary laboratory experiments indicate that the ceramic compound conductivity changes from nearly 100% ionic conductivity to essentially 100% electronic conductivity when the amount of ceria added is varied between approximately 0 and 80%, as illustrated by the data presented in Table 6.11. Selection of the optimum CeO_2/ZrO_2 ratio requires careful evaluation of the physical and chemical properties of these oxides. In addition, predicting the stability of the rare-earth mixed oxides or binary compounds requires complete knowledge of the phase diagram. The electrical conductivity of these rare earth compositions is sufficient

Table 6.11 Ionic Transport Number Indicating the Conductivity Change from Ionic to Electrical When the Amount of Added Ceria (CeO_2) Is Varied between 0 and 80%

Ceria Added to the System (%)	System Temperature		
	1,070 K (797°C)	1,270 K (997°C)	1,470 K (1,197°C)
0 (ZrO_2)	0.893	0.954	0.835
20	0.775	0.875	0.782
40	0.625	0.293	0.205
60	0.135	0.065	0.054
80	0.068	0.026	0.012
100 (CeO_2)	0.187	0.154	0.078

at the high temperature end of the electrode surface, which could be as high as 1,700°C (designated as $T_{surface}$). The electrical conductivity of the electrode surface decreases rapidly towards lower temperature values; therefore, the electrode material must be physically connected to a high-conductivity lead-out at relatively excessive joule heating.

If one plots the curves indicating the ionic transport number versus the percentage of zirconium oxide (ZrO_2) and cerium oxide (CeO_2), all curves will bottom out at 80% of the values of both oxides, regardless of temperature. This 80% ratio of the two rare earth oxides is determined by phase diagram investigation of three distinct ceramic compositions—$18CeO_2$-$82ZrO_2$, $50CeO_2$-$50ZrO_2$, and $75CeO_2$-$25ZrO_2$. The phase diagrams for these three ceramic compositions have been measured up to 1,600°C by the renowned German rare earth scientists. These scientific investigations indicate that the electrical conductivity of these rare earth ceramic compositions is sufficient at elevated electrode surface temperatures ($T_{surface}$) close to approximately 1,700°C. As mentioned, the electrical conductivity decreases rapidly towards the lower end of the temperature range; thus, the rare earth ceramic material must be connected to a highly conductive lead-out at a relatively high temperature to avoid excessive joule heating.

High conductivity can be achieved by adding small amounts of tantalum oxide (Ta_2O_7), which has a melting temperature close to 2,995°C, to the binary ceramic compositions [5]. These rare earth ceramic compositions are widely used for MHD electrodes. Note that the requirements for MHD electrodes are very stringent because these electrodes must withstand high temperatures (1,500–2,300°C), thermal shock, abrasion, and alkali-based chemical attack. The most critical requirement for rare-earth ceramic electrodes is high electrical conductivity at very high temperatures. Electrode surface conditions must be carefully monitored and verified by electron spectrometry for chemical analysis to ensure the suitability of the electrodes for the MHD generator.

6.6.4 Most Promising Materials for the MHD Generator Applications

Hot gases from the MHD generator are used to produce the additional heat energy needed for operating a steam turbo-alternator to generate electrical energy. Therefore, the materials selected for MHD operation could have dual functions without interruption in either of the power-generating systems. The construction materials used in the erection of an MHD generator and steam turbo-alternator must meet stringent structural, thermal, and chemical properties because the MHD, combustor, and steam turbine all operate under severe operating environments, including high temperature, corrosion attack, destructive ions and salt elements, severe mechanical stress caused by electrical field gradients, and electrochemical effects resulting from various chemicals. Note that gas temperatures in the core generator are extremely high, ranging from 2,300 to 2,700°C. The gas stream temperatures may be as high as 1,900°C, whereas the velocity of the gas stream is on the order of Mach 1. These operating conditions give rise to heat fluxes ranging from 25 to 300 W/cm². Under high heat flux conditions, there are temperature gradients of several hundred degrees per millimeter. Because of high temperature gradients, electrode materials must be capable of passing ultra-high current densities.

The specifications for electrode materials constitute a major reliability problem for MHD power development. Based on this reliability problem, MHD generators can be classified into three distinct categories, each of which may require a different type of electrode. These generators can be described as follows:

■ A generator using clean fuel and a cold-wall structure with strongly cooled electrodes with surface temperatures ranging from approximately 50 to 500°C
■ A generator using clean fuel with a hot-wall structure that is equipped with conducting ceramic electrodes with surface temperatures ranging from approximately 1,500 to 1,900°C
■ An MHD system designed for coal fuel in which the generator will be covered with a thin layer of coal slag most of the time; the temperature of the coal slag could be in the range of 1,400 to 1,700°C

MHD generators using a hot-wall channel fueled with clean fuel or benzene are often recommended. Several rare earth compounds may be promising candidates in MHD generator electrode design, particularly, cerium oxide/zirconium oxide solid solutions and lanthanum chromate alkaline earth oxides. MHD systems employing solid solutions and $LaCr_3$ alkaline earth oxides have been demonstrated as efficient and safe power systems.

6.6.5 Cermet and Chromate Materials and Their Benefits

Another material has demonstrated considerable potential for use in MHD electrodes: lanthanum chromate ($LaCrO_3$) doped with 5 to 18% alkaline rare earth oxides. Research studies indicate that calcium (Ca^{2+}) and strontium (Sr^{2+}) will contribute to the lanthanum (La^{3+}) ion. The magnesium (Mg^{2+}) ion is much smaller and has a strong preference for the octahedral site. These materials have excellent refractory properties, with the melting point of the chromate with these additions being in the 1,000 to 1,500°C range. Substituted compounds, such as $LaMg_{0.05}Cr_{0.95}O_3$, show an electrical conductivity of 4 to 10 ohm-cm at a temperature of 2,000 K (1,723°C), which increases only three- or fourfold at room temperature, as shown in Table 6.12. This implies that the material can be used over the temperature interval from the plasma interface to the water-metal lead-out.

The thermal conductivity of magnesium-doped $LaCrO_3$ will be twofold greater than that of the CeO_2-ZrO_2 materials. Unfortunately, the material also has a few drawbacks. For example, the operating pressure of chromium oxide at a temperature

Table 6.12   Thermal Conductivity as a Function of Temperature in Electrode Materials

Electrode Temperature (°C)		Thermal Conductivity (ohm-cm)		
	$LaMg_xCr_{1-x}O_3$	$78CeO_2 + 20ZrO_2$ $+ 2Ta_2O_5$	$75CeO_2 + 25ZrO_2$
250	0.032	0.021	0.018
500	0.027	0.019	0.016
750	0.024	0.017	0.015
1,000	0.022	0.016	0.014
1,250	0.021	0.014	0.012
1,500	0.019	0.013	0.009
1,750	0.018	0.011	0.008
2,000	0.017	0.010	0.010

Conductivity values from 750 to 2,000°C are accurate within 5%. The values for temperatures from 250 to 750°C might have errors exceeding ±5% because they are projected values.

of 1,700°C or more is much higher than that of La_2O_3. Therefore, excess lanthanum oxide, which is ionic at room temperature and behaves like insulation, is left behind. The other important property with respect to MHD application is the strong interaction of the chromium with potassium to form undesirable materials in the temperature range from 800 to 1,200°C.

Soviet rare earth scientists showed that lanthanum chromate is useful for the fabrication of cermet by mixing the chromate with chromium oxide. The optimum mixing ratio appears to be close to 60% $LaCrO_3$ to 40% Cr. The ionic conductivity of the cermet (CeO_2) is much greater than that of the chromate (CrO_3), while its electrical conductivity is several orders of magnitudes higher. In addition, the thermal shock resistance of the cermet improves significantly. The major advantage of the higher heat conductivity is that cermet electrodes require considerably larger thickness than ceramic types, which will tend to decrease the thermal and mechanical stresses. It should be noted that the oxidation of chromium starts at a temperature range of 1,000 to 1,200°C, which prevents the use of this particular material at temperatures much above the temperature range specified herein.

There has been considerable interest in the physical and thermal properties of a closely related perovskite or lanthanum ferrite ($LaFeO_3$). This rare earth compound has demonstrated excellent electrical conductivity from room temperature to 1,800°C. This ferrite material has poor refractory properties compared to lanthanum chromate ($LaCrO_3$). However, this shortcoming can be overcome by the addition of strontium zirconate ($SrZrO_3$), with which it forms solid solutions at temperatures exceeding approximately 800°C.

6.6.6 Electrode Test and Evaluation Data

Many electrodes and insulator assemblies have been tested and evaluated in small test rigs and medium-sized MHD facilities over test durations ranging from a few hours to more than 100 hours. Electrodes with various configurations and thicknesses were subjected to different electric fields corresponding to current densities ranging from 0 to 1.25 A°/cm². Larger MHD facilities were equipped with strong magnets capable of producing magnetic field intensities on the order of 1.5 to 3.0 T [6]. During the tests, the electrode performance level was accessed from the measurements of cross-channel voltages and heat fluxes. Temperatures at the electrode surface and in the interior of the electrodes were recorded by precision pyrometers and thermocouples, respectively. Temperature values are subject to change under unpredictable thermal and chemical environments within the MHD because of chemical and structural changes in the electrode materials under MHD operating environments. Posttest analyses could include optical and scanning electron microscopy (SEM), x-ray diffraction, and radiographic technology.

An electrode designed at the Massachusetts Institute of Technology (MIT) in a Reynolds test rig consisted of a grooved stainless base, on top of which three ceramic layers were deposited using arc plasma spraying technology [6]. The material composition of three layers starting from the steel base can be summarized as follows:

- The hercynite + magnetite ($FeAl_2O_4$-Fe_3O_4) layer sprayed in batches
- ZrO_2-CeO_2 layer
- ZrO_2-Y_2O_3) layer

The principal objective of the zirconium oxide layers is to prevent vaporization of the ferrous oxide (Fe_3O_4) from the underlying iron-aluminum spinel. The spinel is a hard crystalline mineral represented by the chemical formula $Mg_2Al_2O_4$, which varies from colorless to ruby-red to black. The spinel is widely used as a gem. In a broad sense, this compound mineral could contain any group of minerals that are essentially the oxides of magnesium, ferrous iron, zinc, or manganese. The total thickness of the ceramic, which includes zirconia layers, is close to 2 mm (0.08 in). The purpose of the ceramic layer is to pass a heat flux ranging from 80 to 100 W/cm².

The Reynolds test rig lasted over a test duration of approximately 6.5 hours, including interruptions by two shutdowns. It is important to mention that this rig was not equipped with a high-power magnet. Because no seed was applied, no current could be drawn or monitored. During the MIT-designed electrode tests, an SEM micrograph of a cross-section of the anode structure was obtained. After completion of the tests, cracks in the ceramic were noticed at the spinel-zirconia interface in the middle of the ceramic portion consisting of zirconia as a result of differential linear expansion of the materials involved. Examination of the SEM micrograph indicates that the interface of the zirconia layers gives an indication of a good solid joint. Dispersive x-ray maps indicate that there was slight evidence for the presence of zirconium, yttrium, and cerium during the tests.

6.6.7 Most Ideal Ceramic and Refractory Materials for MHD System Applications

A thorough understanding of the important characteristics of potential conventional ceramics, rare earth ceramics, and refractory materials is absolutely necessary if one is interested in the design and development of magnetohydrodynamic power-generating sources. Studies performed by the author in early 1985 on material properties indicate that realistic values of thermal conductivity, Curie temperature, and electrical conductivity are of critical importance. The room-temperature properties of polycrystalline ferrites are shown in Table 6.13. Critical properties of rare earth–based refractory materials are summarized in Table 6.14.

Comprehensive studies undertaken on MHD electrode materials around 2008 indicate that low electrical conductivity, high thermal conductivity, good refractive properties of the electrode material, and low ionic transport numbers are essential for high efficiency and reliability of the MHD generator. The studies further indicate that reasonably good electrical conductivity at higher temperatures is critically important if both reliability and efficiency are the principal goals for an MHD electrical power generator.

6.6.8 Electrical Conductivity as a Function of Electrode Temperature

The electrical conductivity of various materials are as follows:

- Zirconium oxide (ZrO_2): 0.0001/ohm-cm at 385°C, 0.0002/ohm-cm at 700°C, and 0.0400/ohm-cm at 1,200°C

Table 6.13   Room-Temperature Properties of Polycrystalline Ferrites

	Polycrystalline Ferrite Material		
Properties	$MgFeO_2$	$MnFe_2O_4$	$Y_2Fe_5O_{12}$
Saturation magnetization (G)	1,450	4,800	1,740
Curie temperature (K)	352	302	277
Electrical conductivity(ohm-cm)	0.01	100	0.0001

Table 6.14 Properties of Rare Earth–Based Refractory Materials

Rare Earth Oxide	Property		
	Melting Point (°C)	Density (g/cm³)	Crystalline Form
Cerium oxide (CeO₂)	2,600	7.13	Face-centered cube
Magnesium oxide (MgO)	2,800	3.58	Face-centered cube
Zirconium oxide (ZrO₂)	2,677	5.56	Monoclinic

- Cerium oxide (CeO_2): 0.0006/ohm-cm at 800°C and 0.0320/ohm-cm at 1,200°C
- Magnesium oxide (MgO): 0.0002/ohm-cm at 850°C and 0.0003/ohm-cm at 1,000°C
- Chromium oxide (CrO_2): 0.0010/ohm-cm at 350°C and 0.2320/ohm-cm at 1,200°C

6.6.9 Thermal Conductivity of Various Oxides Used in MHD Electrodes

The thermal conductivity of cerium oxide (CeO_2) is 0.108 W/cm·°C, whereas the thermal conductivity of zirconium oxide (ZrO_2) is 0.041 W/cm·°C.

6.6.10 Performance of Advanced Rare Earth Ceramic and Refractive Materials for MHD System Components

Research on thermal conductivity has focused on the function of operating temperatures of lanthanum magnesium chromium oxide ($LaMg_{0.05}Cr_{0.95}O_3$) and cerium oxide/zirconium oxide (CeO_2/ZrO_2) materials. The magnesium ion is much smaller and indicates a strong preference for the octahedral site. Therefore, the magnesium-substituted rare earth compound is most likely to be $LaMg_xCr_{1-x}O_3$. This compound and other materials mentioned in this chapter have excellent refractory properties, particularly at high temperatures. The substituted compounds have electronic conductivities of 4 to 10 ohm-cm at temperatures as high as 1,727°C (2,000 K), increasing by a factor of three or four at room temperature (300 K). Therefore, this material can be used over the temperature interval from the plasma interface to the water-connected lead-out. The thermal conductivity of magnesium-doped lanthanum chromate ($LaCrO_3$) is higher by a factor of two over the cerium oxide–zirconium oxide ratio. This material has a few drawbacks, however. The pressure of chromate oxide at a temperature of 1,700°C and above is much higher than of lanthanum oxide, leaving excess lanthanum oxide (La_2O_3) behind, which is ionic at room temperature and also insulating. In addition, the strong interaction of chromium with potassium forms an undesirable material, especially in the temperature range of 800 to 1,200°C, which could impact the MHD generator efficiency by 2 to 3%.

Lanthanum chromate lends itself to fabrication of cermets by mixing the chromate with chromium. The optimum ratio appears to be 60% $LaCrO_3$ to 40% Cr. Note that the electrical conductivity of the cermet is much greater than that of the chromate, and its electrical conductivity is several orders of magnitude higher. In addition, the thermal shock resistance of the cermet is significantly improved.

6.6.11 Comments on the MHD Electrode Materials

Cermet is a complex compound consisting of ceramic and metal. This material is considered a strong alloy compound with high heat resistance capability similar to that of titanium carbide and a strong metal such as nickel. This material is widely used in the manufacturing of high-speed steam turbine blades. Such blades offer high strength and reliability under harsh thermal and vibration environments. Chromite is a mineral compound represented by a chemical symbol such as $FeCr_2O_4$. This compound consists of iron oxide and chromium oxide. This mineral compound is also known as an oxide of bivalent chromium. Chromate is a type of salt made from chromic acid. It is widely used in dealing with rare earth oxides and compounds.

The major advantage of the higher thermal conductivity of the cermet electrodes is that the MHD generator electrodes made with cermet material require significantly larger thickness compared to the ceramic types, thereby reducing the thermal and mechanical stresses considerably. Furthermore, the oxidation of cermet starts at temperatures over the temperature range from 1,000 to 1,200°C or even higher.

Based on its thermal, electrical, and refractory properties, stabilized zirconium oxide (ZrO_2) with 10 to 20% calcium can be considered as a prime contender for MHD applications. Its melting point of 2,700°C and excellent electrical conductivity at high temperatures are responsible for the reliable and safe operation of magnetohydrodynamic electrical power generators. The conductivity of the electrode materials changes from nearly 100% ionic conductivity to essentially 100% electrical conductivity when the amount of ceria (cerium oxide) is varied between 0 to 80%, as shown in Table 6.11.

There is significant interest in the properties of a closely related perovskite: lanthanum ferrite. This particular compound ($LaFeO_3$) has demonstrated excellent electrical conductivity as a function of temperature ranging from room temperature (27°C) to 1,800°C. Note that the lanthanum ferrite ($LaFeO_3$) is less refractory compared to $LaCrO_3$. However, this difficulty can be overcome by the addition of strontium zirconium oxide ($SrZrO_3$), with which it forms solid solutions as long as the temperature is above 800°C.

Tests performed over a period exceeding 100 hours yielded useful information on the MHD performance. The tests were performed on wall sections involving the cathode and anode elements of the MHD generator, constructed in the United States, under an applied magnetic field close to 1.7 T (17,000 G). During these tests, six different electrode materials, including a set of five fabricated from magnesium-doped lanthanum chromate ($LaCrO_3$), were evaluated. The temperatures of the top and bottom surfaces of the chromate were measured around 800°C and 500°C, respectively. Excessive thermal and mechanical stresses were monitored, which caused partial detachment of the ceramic from the leading edges of the electrodes. The temperature of the $LaCrO_3$ increased well above the estimated values, causing massive overheating and frequent cracking. Monitoring equipment, such as a scanning electron microscope and x-ray, indicated Cr-oxide losses near the surface and in the cracks and chemical reactions, producing potassium chromate (K_2CrO_4) and other harmful oxides.

As mentioned, in ionic conductivity environments, zirconium oxide (ZrO_2) has a greater tendency to decompose at the cathode wall due to the presence of large electrical fields in an MHD channel wall. Decomposition of the cathode wall under the influence of large electric fields, with the problem of overheating and cracking under excessive thermal and mechanical stresses, will seriously undermine the safety and the reliability of MHD electrical power generators. Further research must be undertaken to root out the most serious design problems associated with MHD walls and electrode problems. The author does not see solid evidence for safe and reliable performance of MHD electrical power generators in the tests performed to date.

6.7 Summary

This chapter strictly focuses on the properties and applications of rare earth intermetallic compounds, hydrides, and ceramics in various commercial and industrial fields. Applications and benefits of rare earth–based intermetallic compounds, rare earth hydrides, and ceramic materials were identified, with major emphasis on the structural integrity of the MHD electrodes under harsh thermal and mechanical stress environments. Structural types and descriptions of intermetallic compounds and their potential applications were summarized, with emphasis on cost and complexity. Magnetic and electrical characteristics of rare earth–based intermetallic compounds were summarized for high thermal and mechanical stress environments. Magnetic order temperatures of rare earth elements and Hall coefficients for the hydrogenated rare earth metal ceramic materials were specified. Note that a positive Hall coefficient is a performance indicator of full-band structures.

Rare earth intermetallic compounds are well suited for industrial heterogeneous catalysis applications to produce hydrogen with minimum cost and complexity. The magnetic characteristics of cobalt play a critical role in the production of rare earth intermetallic compounds, such as $SmCo_5$ and $SmCo_{17}$, which are widely used as permanent magnets with high energy products. Low-temperature magnetic properties of hydrides and deuterides were provided to identify their suitability for specific commercial or industrial applications. Lattice constants, Curie temperatures, and magnetic wards of binary compounds were provided for the benefit of the reader.

Potential commercial and industrial applications of hydrides were discussed, with particular emphasis on the economic and technical aspects. Cost-effective methods for the storage of hydrogen produced from hydrides were summarized, with emphasis on reliability and safety. The production of hydrogen gas from hydrides is considered to be a reliable method. Disassociation pressures as a function of temperature for various rare earth hydrides were provided. Dielectric properties of rare earth ceramic-based magnetic materials and their commercial and industrial applications were summarized. The principal advantages of ceramic permanent magnets were identified, with particular emphasis on maximum energy product.

Rare earth ceramic magnets may be cost-effective for MHD power generators with small or moderate capacities. Material requirements for the MHD electrodes were briefly mentioned, with major emphasis on reliability and electrical conductivity at higher temperatures ranging from 1,200 to 1,725°C. Electrical conductivity data for the most promising rare earth ceramic compounds, such as $LaCrO_3$-$20SrO_2$, $78CeO_2 + 20ZrO_2 + 2Ta_2O_5$, and $75CeO_2 + 25ZrO_2$, were summarized as a function of electrode surface temperatures, with emphasis on reliability and survivability under excessive thermal and mechanical stress environments.

The benefits of the stabilization of zirconium oxide (ZrO_2) with calcium oxide (CaO), magnesium oxide (MgO), or yttrium oxide (Y_2O_3) were identified. Ionic transport numbers were identified for the conductivity change from ionic conductivity to electrical conductivity, when the amount of CeO_2 addition is varied between 0 and 80% [6]. The most promising materials for an MHD power generator were identified, with major emphasis on operational reliability and structural integrity under excessive thermal and mechanical stress environments. The major advantages of cermets (CeO_2) and chromate (Cr_2O_2) oxides were identified, particularly under harsh thermal and mechanical stress environments.

Thermal conductivity experimental data obtained by various scientists as a function of temperatures were summarized for electrode surfaces using various rare earth oxides. Hard ceramic and refractory materials are sometimes recommended for MHD electrodes and hot walls, which are constantly subjected to higher temperatures and higher thermal flux levels under excessive thermal and mechanical stress conditions.

Room-temperature values of saturation magnetization, electrical conductivity, and Curie temperature were summarized for polycrystalline ceramic materials. Melting temperature, electrical conductivity, and thermal conductivity of specific rare earth–based refractory materials were summarized, with particular emphasis on the structural integrity and reliability of the MHD system components in high-temperature environments. Values of electrical conductivity as a function of electrode surface temperature were provided. Performance data on some specific rare earth ceramics and refractory materials were summarized for the benefit of the reader. General comments on the materials best suited for MHD applications were summarized, with major emphasis on the structural integrity and reliability of MHD electrodes and hot-walls.

References

1. Wallace, W.E., and F.H. Spedding. 1972. *Rare Earth Intermetallic Compounds and Their Applications*. Pittsburgh, PA: Department of Chemistry, University of Pittsburgh.
2. Sankar, S.G., W.E. Wallace, and D.M. Gualteri. 1972. *Low-Temperature Magnetic Properties of the Hydride and Deuteride of Ternary Compound*. Pittsburgh, PA: Department of Chemistry, University of Pittsburgh.
3. Gregery, D.P. 1976. *Prospects for the Use of Hydrogen as an Energy Carrier*. Chicago, IL: Institute of Gas Technology.
4. Harper, C.A., ed. 1984. *Handbook of Materials and Processes for Electronics*. New York: McGraw-Hill.
5. Fredeukse, H.P.R., W.R. Hosler, and T. Negas. 1975. *Rare Earth Ceramics for MHD-Power Generators*. Washington, DC: National Bureau of Standards.
6. Hosler, W.R., and H.P.R. Frederikse. 1976. MHD Electrical Power Generation. *Proceedings of the 6th International Conference, Vol. 2*. Washington DC, 67.

Chapter 7

Contributions of Rare Earth Materials in the Development of the Glass Industry, Crystal Technology, Glass Polishing, Electro-Optical Devices, and the Chemical Industry

7.1 Introduction

This chapter discusses the applications of rare earth compounds and alloys that were not included in previous chapters. Approximately 98% of rare earth elements are consumed in commercial and industrial applications, such as in metallurgy, chemicals and catalysts, glass-based products, and polishing materials. The contribution of rare earth materials to the glass industry will be discussed in great detail. The critical roles played by rare earth materials as various types of crystals for commercial, industrial, and scientific disciplines will be summarized. The role of rare earth metals in the production and deployment of mischmetal in various industrial applications will be identified, with particular emphasis on the unique mechanical properties that are best suited for maintaining the structural integrity of commercial highrise buildings. Mischmetal and its alloys have unique burning characteristics that make them very useful in ordinance applications. Furthermore, when the ignition of a mischmetal alloy particle is achieved by a high-energy explosion or impact, the metal will continue to burn until it is totally consumed. Because of this, mischmetal can be developed in the form of shell linings, fighter aircraft penetrating devices, tracer bullets, bomblet markers, and other defense-related products such as tanks and fighters.

Rare earth materials best suited for the growth of the optical and glass industries are identified, with emphasis on their suitability for commercial, industrial, scientific, and medical products. The optical properties of the heavier rare earth metals are reviewed, including the most practical and accurate methods for the measurement of optical properties and a discussion of some errors arising from the samples. A survey of optical data is presented, including the data from electron-energy loss measurements, magneto-optic effects, and modulation spectroscopy. The optical properties of rare earth metals are strictly those of the transition metals; the 4f electrons and magnetic order play minor but observable roles. The optical properties of rare earth metals yield vital information on the electronic structure of the rare earths that are several electron volts or more away from the Fermi level; therefore, they complement the Fermi surface measurements by providing a test for calculated or intuitive band structures.

In addition to standard optical measurements, this chapter briefly discusses magneto-optical effects and modulation spectroscopy. Photoemission measurements can yield additional information on the electronic states located far from the Fermi level. A review on band structure, optical effects, and photoemission properties is available in a technical paper published in 1963 [1]. When co-doped with trivalent rare earth elements, such as thulium and holmium, fluoride can be used for efficient diode-pumped lasers operating at room temperature with minimum cost. This chapter identifies such solid-state lasers and briefly describes their unique applications in commercial fields.

This chapter also identifies the critical roles of some rare earth elements, such as lanthanum and niobium, in the development of high-temperature, high-voltage, high-permittivity ceramic capacitors, which are desirable in electronic devices. Multilayer and laminated high-temperature ceramics, such as $BaNd_2Ti_4O_{12}$, offer optimum sintering temperatures close to 1,350°C. Neodymium plays a critical role in achieving high sintering temperatures.

7.2 Critical Role of Rare Earth Materials in the Development of Various Glass Categories

The use of rare earth oxides in glass is relatively new when one considers that glass has been produced for more than 4,000 years for various domestic, commercial, and industrial applications. Modern glass technology started around 1880 when Otto Scott of Germany studied the effects of various oxides as constituents in glass. German science pioneers Winkleman and Straubel investigated rare earth fluorescence during the same period. German scientist Drossbach was granted a patent on a mixture of rare earth for decolorizing glass, which was the first commercial use of cerium oxide. However, it was used in a crude form with other rare earth oxides, including neodymium oxide. English scientist Crookes undertook systematic studies on eye protection glasses, which confirmed that cerium oxide is effective in absorbing ultraviolet rays without adding any color to the glass. Continued research activities on rare earth oxides led to the first use of lanthanum oxide in optical glasses in 1935. The rapid growth of rare earth oxides began right after World War II. Since 1946, new glass technologies have been developed using more purified rare earth oxides. Refined materials have been successfully obtained using more advanced separation techniques with purity levels close to 99.009%.

The rare earth oxides can be used in various glassmaking applications, including [1]:

- Eye protection glass
- Cathode ray tube (CRT) faceplate glass
- Radiation-shielding glass
- Optical glass
- Light bulb glass
- Window glass for residential, commercial, and industrial use
- Decolorization glass

This section identifies rare earth oxides capable of meeting glass requirements for various residential, commercial, and industrial applications. Each glass category is briefly described, and the rare earth oxide best suited for that category is also identified.

7.2.1 Cerium Oxide

Approximately 350 tons of rare earth oxides are used per year by the glassmaking industry. Among these oxides, cerium makes up the biggest share of the glassmaking market in the form of a mixed rare earth material containing approximately 88% cerium oxide (CeO_2), with the balance made of erbium-oxide (Er_2O_2), lanthanum oxide (La_2O_3), praseodymium oxide (Pr_6O_{11}), and neodymium oxide (Nd_2O_3) [1]. Lanthanum oxide represents the next largest market share, followed by cerium oxide with a purity level of 95 to 99%. Other oxides are less commonly used. The typical transmission efficiency of a glass containing erbium oxide is shown in Figure 7.1.

Cerium has been increasingly used as a stabilizer against the browning of glass by cathode and gamma rays. Most CRT faceplates are made from cerium-stabilized

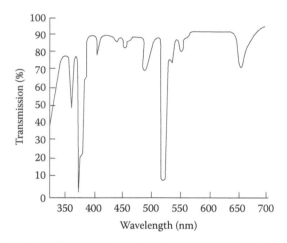

Figure 7.1 Transmittance characteristics of a glass containing erbium oxide (Er$_2$O$_3$).

glass. The nuclear industry has specific requirements for their glasses, including large quantities of radiation-shielding windows that provide very high light transmission without darkening as a result of the formation of color centers. Most of the research and development work to understand the technology of browning by gamma radiations was carried out during the 1950s and early 1960s in conjunction with the work on the use of nuclear reactor energy [1].

The glass container industry uses special types of glasses that deploy cerium oxide. Cerium oxide is best suited for glass containers. It is widely used to decolorize glass and to stabilize against solarization caused by ultraviolet rays. The transmission efficiency of cerium-based glass and cerium-oxide/titanium-oxide glass can be seen in Figure 7.2.

7.2.2 Lanthanum Oxide

Photographic and optical equipment may use very sophisticated lenses to provide high resolutions. Lanthanum-based optical glasses with a high index of refraction and low dispersion are best suited for such photographic and optical equipment.

7.2.3 Neodymium Oxide

Neodymium, erbium, and praseodymium are often used as glass coloring agents, particularly in handcrafted glasses. Neodymium oxide is also used to dope laser crystals, particularly highly efficient solid-state lasers that operate at room temperature and emit at specific infrared wavelengths.

For coloring glass, neodymium oxide yields a delicate pink tint with violet reflections. The hue of the color varies with glass thickness, the concentration of

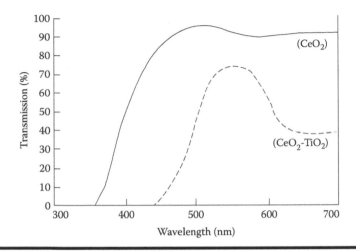

Figure 7.2 **Transmittance curve of a glass containing cerium oxide (CeO_2) and titanium oxide (TiO_2).**

neodymium, and the source of illumination. The color intensity can range from a light pink in thin sections of the glass sample to a beautiful blue-violet in thicker sections of the glass sample—a characteristic known as *dichroism*. It is used in art glass and special filters, with concentrations ranging from 1 to 5%.

Neodymium welding glasses are used by welders and lamp workers to protect their eyes from the yellow flare emitted by sodium vapors originating from hot glass section or fluxes. This is strictly due to the narrow absorption peak of neodymium, which ranges from 500 to 590 nm, because the sodium atoms emit yellow light. A typical transmittance curve, which indicates the transmittance as a function of wavelength, can be found in any physics book for graduate students.

7.2.4 Praseodymium Oxide

It is important to mention that praseodymium oxide (Pr_6O_{11}) is the next strongest rare earth oxide, which gives off a green color very similar to the eye as chromium containing glasses elements. Because of the high cost of praseodymium oxide compared to chromium oxide, this particular oxide is not often used except in special filter glasses. Furthermore, this oxide is not used in combination with neodymium in welding glasses because of the high cost.

7.2.5 Coloring Effects of Rare Earth Oxides

When transparent glasses absorb portions of the visible light spectrum from 400 to 700 nm, they will appear colored. The color is determined by the transmitted portion of the visible spectrum after subtracting certain wavelengths by absorption from the illumination source. Rare earth oxides have absorption spectra consisting of a large number of narrow bands involving both the visible and invisible portions of the spectrum. There is a slight change in the characteristic absorption spectra of rare earth oxides in various rare earth compounds, solutions, and glass media.

The colors in rare earth oxide glasses are strictly caused by the ion being dissolved. They behave uniquely because the 4f electrons are deeply buried. The colors in the glass medium depend on the transitions that are taking place in an inner electronic shell; in the other elements, such as the transition metals, the chemical forces are restricted to deformation and exchanges of electrons within the outer shell. Because the sharp absorption spectra of the rare earth materials are insensitive to glass composition and oxidation-reduction conditions, it is relatively easy to produce and maintain specific colors in the glassmaking process.

7.3 Classification of Glass Materials Free from Rare Earth Oxides

Glass is a unique material—appearing as a solid although often referred to as a supercooled liquid surface. Many definitions of glass can be found. However, from a technical point of view, it can be considered an inorganic product of fusion that has been cooled to a rigid condition without any crystallization.

7.3.1 Types of Glass That Do Not Contain Rare Earth Oxides

The most common commercial glasses are soda-lime glasses, which are made from silica. The melting temperature of silica is reduced by the addition of soda and potash, which are the fluxing agents. Lime, magnesia, and alumina (Al_2O_3) are added to promote chemical durability and stabilization. After the addition of agents, the glass must be refined to remove gaseous inclusions. The refining process can be chemically done by the addition of sodium nitrate and sulfate, arsenic oxide, and antimony oxide. Soda-lime glasses are free from rare earth oxides and are generally used for plate and window glass, containers, light bulbs, and ophthalmic lenses. Other types of glasses include lead-alkali-silicate, borosilicate, barium-borosilicate, and aluminosilicate. In addition, there are a broad range of optical and other technical glasses using a variety of raw materials.

Color is one of the most important properties of glass, which can be easily achieved by the addition of various metal oxides. The strongest of these oxides are titanium, vanadium, chromium, manganese, selenium, iron, cobalt, nickel, and copper. Silver and uranium can also add weak color. Some of the rare earth metals can also be used as colorants with sharp absorption bands, in contrast to broad spectral bands normally given by the previously mentioned colorants.

7.3.1.1 Interaction between Colorant Ions

If two or more metal oxides are involved, the color will be dependent on the interaction between the ions of two coloring oxides. For example, the ferric oxide iron (Fe_2O_3) changes to ferrous oxide iron (Fe_2O_2), with the oxidized manganese forming a purple color. This is the characteristic of bottles that have been subjected to sunlight for a specific period of time. The manganese is added initially to the glass as a decolorizer to offset the yellow color of the iron.

In the case of other ions, especially arsenic in combination with iron, the color is enhanced by solarization. For example, clear window glass solarizes to a yellow-brown color. More than 0.005% cerium oxide in combination with arsenic will cause this yellowing effect, with the strongest color occurring at 2.5% cerium content. Higher cerium contents will tend to reduce the solarization effect by filtering out the actinic radiation. By removing the arsenic completely, small amounts of cerium can be added as an oxidizing agent for decolorizing the glass and will further stabilize against the solarization effect.

Browning is another type of discoloration, which is especially caused by x-rays, gamma rays, and cathode rays. Cerium oxide is an important ingredient in specialty glasses just to reduce the browning characteristic. The cerium ions act as electron traps in the glass medium and absorb electrons liberated by this high-energy radiation, which keeps the color centers from forming. Cerium ions, which have little visible color, are formed and protect the glass from discoloration from energy and nuclear radiation. Some of these nonbrowning glasses are used in radiation-shield windows, televisions, and other cathode ray faceplates.

Nonbrowning radiation-shield windows offer good protection for nuclear scientists who are conducting nuclear and radiochemical tests within a reactor or laboratory. These windows are carefully designed with particular emphasis on the radiation leakage through the structural joints and glass cover or plate on the window. Radiation-shielding windows are made from high-density lead glass because lead offers optimum protection against a radiation leak from the reactor. When used as viewing windows, they are placed in walls made from thick concrete and lead. Cerium-stabilized borosilicate cover plates are used on the hot side for thermal protection. Unstabilized glass will darken from a radiation leak containing cobalt-60, which will decrease the transmission through the window, thereby making it difficult to view the interior region of the reactor.

The intensity of the darkening effect is strictly a function of the energy of ionization radiation, the intensity of radiation, and the radiation dose level. Nevertheless, the darkening effect is not stable and will fade slightly after radiation exposure. These viewing windows can be as thick as 3 feet or more to provide the required maximum transmission. Figure 7.3 illustrates the results of transmission through lead glass stabilized with 1.5% cerium oxide that was subjected to various radiation dose levels compared to a similar glass that was not stabilized. Cerium oxide may be used in amounts up to 2.5% in order to produce the required nonbrowning stability.

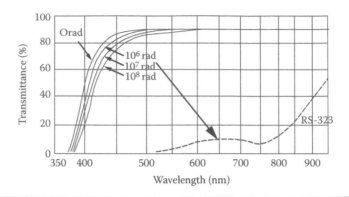

Figure 7.3 **Effects of radiation levels on cerium-stabilized lead glass (solid line) and nonstabilized lead glass (labeled RS-323).**

In high-lead glass containing more than 60% lead oxide (PbO), cerium additions could make the glass amber prior to irradiation and reduce the overall transmission efficiency. These glasses are naturally resistant to coloration by irradiation, and any color that develops thereafter fades rapidly. Therefore, it is fair to assume that cerium stabilization is necessary only for lower-density glasses. Stabilized glasses are frequently deployed towards the hot side, with the denser glasses on the cold side of the windows. A typical glass window cross-section using a cerium-stabilized borosilicate window with a density of 2.53 g/cm^3 located on the hot side is shown in Figure 7.2. A 3.23 g/cm^3 density stabilized window glass is on the hot side, with 5.2 g/cm^3 and 2.53 g/cm^3 density unstabilized windows towards the cold side.

The effects of various irradiation levels on cerium-stabilized and nonstabilized lead glass transmission efficiencies are shown in Figure 7.3 as a function of wavelength. The transmission efficiency varies from 80 to 90% at all irradiation levels ranging from 0 rad to 10^8 rad at a wavelength of 500 nm. The efficiency remains constant for cerium-stabilized glass over the spectrum from 500 to 1,000 nm. The transmission efficiency for nonstabilized lead glass remains well below 60% regardless of wavelength.

7.3.1.2 Decolorizing Effects due to Glass Impurities

Iron oxide (Fe$_2$O$_3$) is always present as an impurity in glass. It is generally introduced through natural raw materials, such as sand and limestone. Both of these contain iron oxide as an impurity. Another source of impurity is from trapped iron mixed in the cullet and abraded metal as a result of mishandling the batch. All these sources of impurities add up to several hundred parts per million (ppm), which causes light absorption at the ends of the light spectrum rather than the middle, as well as a yellow-green color in the glass sample. This adverse effect can be overcome using a process known as decolorization. There are two types of decolorization techniques—chemical decolorization and physical decolorization—as described in the following sections.

7.3.1.2.1 Chemical Decolorization

In chemical decolorization, oxidizing materials are added to change the iron from the ferrous to the ferric state. Essentially, this technique shifts the maximum light transmission towards yellow by absorbing more blue. Several popular materials have been considered, such as arsenic and manganese. However, cerium oxide has been used as a substitute for these materials. The principal advantage of cerium oxide is that the solarization does not occur as long as arsenic is not present. Furthermore, when the cerium is used in conjunction with selenium and cobalt, even the addition of 2 to 3 ounces per ton of the batch can yield a reduction in the use of arsenic and manganese.

7.3.1.2.2 Physical Decolorization

Physical decolorization is accomplished by masking the color of ferric oxide in the glass with complementary colors. This makes the transmittance across the spectrum fairly constant, giving a neutral color. Physical decolorization reduces part of the transmittance of the glass to get rid of the green color. A true colorless glass, such as an optical glass, must be made with very low iron content because the decolorization agents would tend to reduce the transmission efficiency. The principal physical decolorizers are manganese, selenium, cobalt, and neodymium oxides. Note that manganese with minimal cobalt will be more effective in complementing the iron in the ferric state.

Selenium seems to be one of the best decolorizers in tank melting. The iron is neutralized by the pink tint of the selenium oxide. Because the yellow shade is still present, the decolorizing process can be completed with the addition of a small amount of cobalt oxide. Arsenic oxide helps to stabilize the decolorization process with selenium oxide. However, as arsenic oxide increases, more selenium oxide is required, thereby making the process costlier.

The third (and last) rare earth oxide used for achieving physical decolorization is neodymium oxide. Its absorption curve, which indicates the transmission efficiency as a function of wavelength, is very close to that of an average mixture of ferrous oxide and ferric oxide, particularly with the strong absorption band at a wavelength of 590 nm. Neodymium oxide is equally stable against any state of oxidation change in the furnace. Neodymium oxide is most effective as a decolorizing agent for glass containing potassium silicate and lead. However, if the redox balance is not absolutely correct for the iron, appropriate corrections could be made with small amounts of oxides of manganese, cobalt, nickel, selenium, and erbium.

Very small quantities of rare earth oxides are used for decolorizing. For example, 25 g of neodymium oxide is required per 100 kg of sand. Note that the cost of cerium oxide is higher than that of neodymium oxide. Therefore, to reduce the cost of cerium oxide, there must be a limit of 60% cerium oxide content.

7.3.2 Effects of Rare Earth Materials on the Properties of Optical Glass

In the previous section, the technical discussion was limited to the influence of rare earth oxides on the color and transmission characteristics of glass. It is important to point out that the use of rare earth materials in optical glass is different because they must not give any absorption indication in the visible spectrum. The rare earth elements are often used as major ingredients, although not exceeding 40% of the batch, compared with the fractions of a percent required for use as decolorizers and coloring oxides (not exceeding 5%).

7.3.2.1 Major Characteristics of Optical Glasses

The properties of optical glasses differ significantly from other types of glasses. Optical glasses are used for optical lenses and elements to provide specific functions. They are classified by their index of refraction and Abbe number, which is reciprocal of its dispersive power. In addition, these glasses must meet special quality requirements, such as a high degree of transparency or high light transmission efficiency, being reasonably free of bubbles, and resisting abrasion. Such glasses must be homogeneous, free from striations, and free from alternating dark and light cross-bands of a myofibril of striated muscle. In addition, these glasses must be annealed to keep the internal stresses to a minimum for eye comfort.

Optical glasses are classified by their specific chemical compositions. Originally, there were two types of optical glasses—crown glass, which is essentially a soda-lime-silicate, and flint glass, which is a lead-alkali-silicate. The additions of other oxides, including a couple of rare earth oxides such as barium oxide, boron oxide, zinc oxide, and lanthanum oxide, have created new families of glasses—borosilicate crowns, zinc crowns, barium crowns, barium flints, and lanthanum crowns and flints. Each of these glass families possesses its own optical characteristics, which are shown in Figure 7.4.

The raw materials used in optical glass differ considerably in terms of purity levels compared to other commercial glasses, which actually drive the cost of optical glasses. Because of the high degree of light transmission required, it is absolutely necessary for the chemical purity of the materials to be extremely high. For example, a normal silica sand used in glass may contain iron within the prescribed purity levels, which could be approximately 300 ppm or slightly more; in optical glass, it is in the range of 10 to 50 ppm or less. Furthermore, material impurities from any of the transition metal oxides, such as chrome oxide, cobalt oxide, and nickel oxide, which could add color to the optical glass, must be limited to less than 1 ppm to preserve the optical performance and quality of the optical glass.

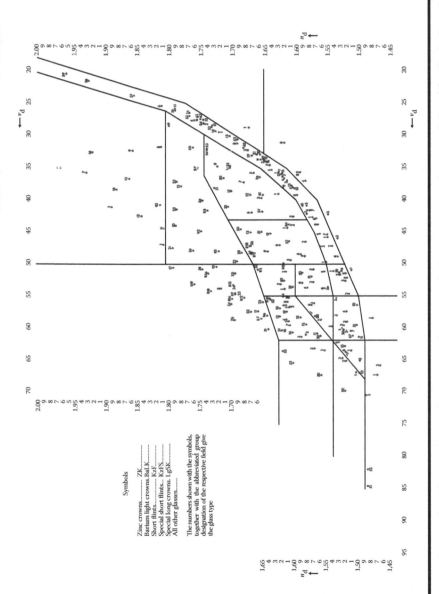

Figure 7.4 Optical map identifying various glass groups (letters) and glass types (numbers) as defined by the index of refraction (n_d) and Abbe number (v_d).

The purity requirements for rare earth materials are very stringent. Rare earth oxide purity level must be between 99.9% and 99.995% [2]. It is absolutely essential that no other color occurs in the optical glass from the absorption bands of neodymium and praseodymium. Furthermore, traces of cerium oxide must be eliminated to minimize the absorption of ultraviolet rays.

Because of these stringent requirements, optical glasses are generally more expensive than other commercial glasses. The principal rare earth material generally used is lanthanum oxide due to its lower cost. Another potential material is thorium oxide, which is not characterized as a rare earth oxide but rather as an element extracted from monazite sand; it has been in use along with lanthanum oxide as a major ingredient. However, because thorium is a radioactive material, it is no longer being used. Newer glasses have been developed using other materials to replace the thorium-based glasses. Other rare earth oxides can be used in very small amounts, such as gadolinium oxide (Ga_2O_3), ytterbium oxide (Yb_2O_3), and yttrium (Y_2O_3). However, most of these oxides are very expensive compared to other oxides.

Lanthanum glasses have low silica content and consist significantly of zirconia, boric acid, barium oxide, and lanthanum oxide, with other additives to provide stability in the viscosity and crystallization temperatures. Lanthanum oxide may be added due to its lower cost, high index of refraction, and high Abbe number, which are considered most ideal for special optical glasses. These properties yield low dispersion in glass, which is best suited for specific applications. The optical glass map in Figure 7.4 demonstrates that the highest index refraction glasses are extradense flints due to their high lead oxide content. These glasses generally have low Abbe numbers. On the other hand, the borosilicate glasses have a low index of refraction but high Abbe number. Barium crowns have high Abbe numbers but an intermediate index of refraction. From these characteristics, one can select the glass that is best suited for a particular application.

7.3.2.2 Fluorescence of Glass

Fluorescence in glass is the result of atoms being excited by the absorption of light resulting from light emission. This is a phenomenon strictly associated with certain rare earth minerals and other crystalline minerals when they are exposed to ultraviolet light. Most rare earth ions exhibit fluorescence when added to glass.

The practical application of the fluorescence phenomenon in glass is rather limited. However, with the development of optical lasers in the 1960s, neodymium-doped glass crystals became very important in the operation of high-power lasers for selected commercial and military applications. Neodymium-based lasers emitting at 1,060 nm received the greatest attention because these lasers can operate at room temperature with relatively high conversion efficiencies, thereby avoiding the cryogenic cooling that is required for most lasers. These rare earth–doped crystals are made from alkali-alkaline-earth silicates. However, fluorophosphates significantly contribute to the high efficiency of these lasers. Note that the purity of the material used plays a critical role in laser development. The neodymium-oxide-doped crystals must be 99.9% pure to achieve such high efficiency; these crystals have a purple-colored glass appearance. Recent technical articles published on lasers and glass technology provide a comprehensive review of this subject.

7.4 Use of Rare Earth Elements in the Development of Polishing Compounds

Early applications of cerium oxide and cerium-rich rare earth oxide mixtures in glass polishing began in the European glass industry around 1933. Some glass polishing efforts were made in 1940 by the Canadian optical industry. German-American scientists working for W.F. and John Barnes Company of Rockford, Illinois in 1943 deployed a rare earth cerium-oxide with 45% purity as a polishing compound known as Barnesite. This particular polishing compound demonstrated immediate success in the polishing of precision optical components, such as bombsights, range finders, periscopes, and other fire control systems. A chemical company of west Chicago introduced a high cerium oxide substance with 90% purity called Cerox for ophthalmic applications. A few more companies focused on glass polishing applications between 1950 and 1960. Commercial manufacturing published reports indicate that by 1960, more than 340 metric tons of cerium-based compounds per year were used for polishing optical mirrors, plate glasses, television CRTs, ophthalmic lenses, and precision optical components. The use of rare earth cerium oxide polishing compounds have significantly improved the performance levels of optics deployed in various sophisticated military and space sensors, where high resolution, minimum tracking error, and precision ranging are of critical importance.

The Pilkington process, invented in the early 1970s for large-scale plate glass manufacturing, severely reduced the demand for cerium oxide polishing agents for large-scale plates. Currently, more than 100 metric tons of cerium oxide are sold per year in the United States. The principal source of polishing substances from 1940 to 1965 was the mineral known as monazite, which contains 46% thorium. Note that there is essentially no market for thorium, except for limited use in conducting scientific research using thorium-based research reactors. The huge cost of separating out the thorium-free rare earth products from the monazite mineral and the mild radioactivity of the thorium are the principal reasons for not using this particular mineral as a polishing substance.

Glass polishing compounds of high quality have been produced since 1965 from bastnäsite concentrates, with appropriate mechanical, chemical, and heat treatments. The United States has the largest supply of bastnäsite, a rare earth fluorocarbonate, at a mine located at Mountain Pass, California that is owned and operated by Molycorp. At the end of December 1978, the ore reserve was estimated at close to 365,000 metric tons, which amounts to a reserve exceeding 3 million metric tons of rare earth oxide. Current mine production capacity is estimated around 27,000 metric tons per year for the bastnäsite-mineral concentrate, which is produced in three distinct grades: a 60% unleached concentrate; a 70% rare earth–oxide leached concentrate less strontium oxide and calcium oxide, calcined concentrate with carbon-dioxide removed; a 90% rare earth oxide calcined concentrate less carbon dioxide. The company indicated that the total bastnäsite shipment in 1977 contained 13,521 metric tons of rare earth oxide. However, the polishing compounds consumed about 10% of this production capacity.

7.4.1 Three Distinct Theories of the Polishing Mechanism

This section discusses the principal theories of the polishing mechanism with reference to different applications, with emphasis on quality of the surface with minimum invisible defects. In an ophthalmic laboratory dealing with prescription lenses, both the quality and accuracy of the optical lens surface are of critical importance because these parameters must satisfy the optical lens prescription requirements that were defined by the eye examination. Therefore, in the case of a prescription laboratory, an optical glass blank is firmly fastened to a lens chuck, which is then pressed down on a curved grinding precision tool and rotated at high speed. The grinding tool must be completely free from vibrations in all three dimensions if the prescription numbers obtained during the eye examination are to be satisfied for the patient's comfort and ability to see clearly, both near and far.

One or two stages of the diamond grinding will achieve the lens curvature. Alternately, one diamond toll-generating step followed by grinding or refining with a loose powdered abrasive, such as corundum, emery, garnet, or silicon carbide suspended in water slurry, may be used to speed up the smooth grinding process. This is considered to be a quasi-rough treatment; one might expect to find some slight subsurface irregularity or damage, which hopefully can be repaired or corrected during the polishing phase. The refined lens, which is still attached to the lens chuck, is rinsed free of any adhering abrasive particles and ultimately is placed in polishing. Here, the tool contacted by the lens is known as the "lap"; it may consist of a variety of materials depending on the surface quality and lens readings to be satisfied. In an ophthalmic factory or machine shop, for example, thick, hard thermoplastic pads have excellent surface quality, curve control, long life, and the ability to operate well under high speeds and pressures without undesired vibrations.

A prescription laboratory prefers to use a paper-thin plastic or smooth cloth pad that is used only once. The polishing compound consists of ceric oxide, zirconium oxide, ferric oxide, or silica, which is slurred in water with a concentration ranging from 5 to 25% by weight; it is recirculated constantly over the lap and lens surfaces. In a factory operation, large central systems collect the polishing slurry, and the pumps furnish the slurry to the polishing bowls constantly [3]. The lens weight must be measured once in a while because the glass removal rate is of principal consideration in measuring the efficiency of a polishing compound. A higher measuring efficiency decreases the polishing compound requirement. The polishing slurry will have an indefinite life because the glass products gradually dilute and contaminate the slurry. Furthermore, the buildup of alkali ions is so great that daily pH level adjustments are essential in large central system slurry tanks. A buildup of alkali ions is of critical importance to maintain the high efficiency of the polishing compound with minimum polishing compound requirements.

7.4.2 Benefits of Chemical Etchants during Polishing Procedures

Centuries ago, Sir Isaac Newton concluded, based on experimental polishing procedures, that polishing is just a fine-scale abrasion operation. Furthermore, in the early twentieth century, Lord Rayleigh found that a polished surface was entirely different from and abraded surface, suggesting that the polished surface must be considered smooth on a molecular scale [3]. Later, the British chemist Sir George found that original grinding scratches disappeared during the application of chemical etchants to a polished surface. He further noticed that a molecular flow of the material from high to low spots took place, which essentially covered the scratches. Two professors from the University of Cambridge stated that if abrasion was the fundamental mechanism, then the hardness of the polishing agent should correlate with the ability to polish the surface. However, they found a remarkable correlation between the melting point of the polishing agent or material and the rate of polishing. Because of this, they concluded that the polishing operation was a melting phenomenon rather than an abrasion.

Testing their hypotheses against the melting points of known polishing oxides, which included rare earth oxides, such as zirconium oxide (ZrO_2) with a melting point of 3,000°C, cerium oxide (CeO_2) with a melting point of 1,950°C, silicon oxide (SiO_2) with a melting point of 1,700°C, ferrous oxide (Fe_2O_3) with a melting point of 1,565°C, and stannic oxide (SnO_2) with a melting point of 1,898°C, their hypotheses were correct except for the stannic oxide. These high-temperature melting points support the hypothesis of Bowden and Hughes.

Comprehensive discussions took place on abrasion technique versus melting technique before 1930. A few years later, a third hypothesis was proposed, which gradually took precedence over the first two techniques mentioned above. In 1931, Grebenschikov found that the presence or absence of water influenced the polishing of glass. He suggested that a layer of silicic acid would build up on the glass surface being polished. This buildup on the glass surface would protect the glass from further erosion were it not for the fact that the polishing agent was at work to sweep away this layer, exposing a fresh surface.

Cornish and Watt [4] established the active role played by the presence of water. They also concluded that a chemical-mechanical hypothesis would fit the observed findings by the above mentioned authors. These authors further concluded that in the case of ceric oxide for polishing the glass, they suggest the formation of a cerium-oxide-silicon (CeO_2-Si) activated complex, which permits the rupture of the oxygen-silicon-oxygen (O-Si-O) bonds through hydrolysis. The cerium-oxide-silicon complex then breaks apart, the hydrated silica is swept away along with the alkalis released from the glass surface, and the whole process repeats again.

The rupture of O-Si-O bonds through hydrolysis presents a good example of cerium oxide acting like a catalyst. Cornish and Watt [4] suggested another possibility that might involve the chemical theory. The most efficient polishing compounds mentioned previously not only have exceptionally high melting points, but they also have large coordinate valences. These two chemical parameters lead to a high concentration of hydroxyl radicals, which are readily available at the glass surface to accelerate the hydrolysis chemical process.

7.4.2.1 Worldwide Market for Cerium Oxide

The annual consumption of the polishing agent cerium oxide is approximately 4,400 metric tons. The suppliers of the cerium oxide polishing compound do not expect significant growth potential in future sales, despite an 11% growth per year in the sales of optical lenses. There are two major reasons for its poor growth. First, the most efficient polishing compounds are available at lower costs and can be easily used in slurry concentrations, using only half of the materials used a few years ago. Second, half the market for ophthalmic glass lenses has been captured by plastic lenses made from CR-39 polymer, which are extremely light and provide safety and reliability under accidental mishandling. Cerium oxide is not suitable for polishing such plastic lenses. However, other materials, such as alumina or stannic oxide polishing compounds, can be used for lenses made from moderately hard plastic materials, provided they meet other polishing requirements. Note that plastic lenses can be easily scratched from mishandling, thus disqualifying them from use as ophthalmic lenses.

7.4.2.2 Consumption of Cerium Oxide as a Polishing Compound

Estimates on the use of cerium oxide as a polishing compound by various countries are shown in Table 7.1 and summarized for various applications in the United States in Table 7.2. The estimates were prepared from data available in 1979. Cerium oxide content in polishing compounds varies from 45 to 90%.

7.4.2.3 Economic Benefits of Rare Earth Materials

It is not the cost per pound of cerium oxide that matters; what matters is your cost per thousand polished surfaces. High-quality cerium oxide polish may be priced at $3.60 per pound; a good-quality zirconium oxide–based polish may be $1.60

Table 7.1 Estimated Worldwide Consumption of Cerium Oxide as a Polishing Agent

Country	Cerium Oxide Consumption (Metric Tons per Year)	Annual Percentage (%)
United States	1,600	36.5
Western Europe	1,400	31.8
Far East	850	19.3
South America	350	7.9
Canada	200	4.5
Worldwide consumption	4,400	100

Source: Cornish, D.C., and J.M. Watt. 1966. The mechanism of glass polishing. Presented at the American Ceramic Society Meeting, Washington, DC, May 11, 1966, 4–7.

Note: Values are estimates with possible errors within approximately ±8%.

per pound; red rouge or ferrite oxide may be priced around $0.50 per pound; and white rouge or precipitated silica may be $0.30 per pound. Cerium oxide may have a competitive advantage as a polishing compound. Based on the cost and quality of the polished surface, one of the largest lens manufacturers in the United States was persuaded to switch from zirconium oxide to a high-quality cerium oxide. This particular switch realized a savings of $400,000 in 9 months using only 50%

Table 7.2 Estimated Cerium Oxide Consumption in the United States by Application

Application	Consumption in Metric Tons per Year (%)
Glass ophthalmic lenses	720 (45)
Optical mirrors	320 (20)
Television faceplates	240 (15)
Precision glass lenses	192 (12)
Miscellaneous products	128 (8)

Source: Cornish, D.C., and J.M. Watt. 1966. The mechanism of glass polishing. Presented at the American Ceramic Society Meeting, Washington, DC, May 11, 1966, 4–7.

Note: Values are estimates with possible errors within approximately ±8%.

of the polishing compound. In addition, this switch shortened the time required for polishing each ophthalmic lens and increased the yield of finished lenses with minimum cost.

7.4.2.4 Other Factors Influencing Glass Polishing

This section summarizes some critical factors that influence glass-polishing efforts and improve the surface quality. The polishing agent zirconia has a tendency to settle out rock-hard, particularly in tanks and pipes; in addition to material loss, cleanup costs are very high. Ceria or cerium oxide polishing compound will settle eventually, but it will always be soft and easy to resuspend. Ferric oxide is an excellent polishing agent, but its slow polishing speed and effects as a pollutant due to its irreversible staining quality are serious drawbacks. White rouge is a very slow polishing agent; therefore, it is rarely used today.

High polishing speeds are essential, and the latest equipment employs much higher spindle speeds and pressures than those used just a few years back. Cerium oxide is considered to be the most ideal polishing agent under current market conditions. For example, a spherical lens that required 8 minutes to polish 15 years ago now can be polished in less than 1 minute, indicating an operation time reduction ratio of 8:1. In another example, a cylindrical lens that previously took 15 minutes to polish now requires slightly less than 4.5 minutes, indicating a time reduction of 10.5 minutes. These reductions in polishing time indicate the savings in polishing labor requirements. The parameters presented in Table 7.3 illustrate the progress made in polishing machines over a 50-year span.

7.5 Role of Rare Earth Elements in the Development of Crystals for Lasers and Minilasers

This section is dedicated to the development of rare earth–based crystals for possible applications in lasers and minilasers, with an emphasis on ion interactions [5]. Research efforts are focused on how to decrease the number of fixed ions while achieving a higher gain per ion. Ion interactions are generally self-quenching, which reduces the yield. Research studies indicate that a higher gain per ion is possible for a given concentration level, as demonstrated by lanthanum oxysulfide (La_2O_2S). Published technical literature identifies a new concept of laser materials with rare earth ions that is suited for minilasers using small-size crystals containing high active ion concentration and pumping with a monochromatic source.

7.5.1 Theory and Design Concept for Minilasers

The general design concept for a minilaser was developed in the mid-1970s. The minilaser design uses small crystals of praseodymium chloride ($PrCl_3$), which are

Table 7.3 Progress Made by Spherical and Cylindrical Polishing Machines from 1930 to 1980

Manufacturer	Model	Year of Origin	Spindle Speed (rpm)	Surface Friction (lb.)	Time Required to Complete Job (min)
Spherical lenses:					
Bausch & Lomb	Hand Pan	1930	300	Variable	15
Robinson-Houchin	Greyhound	1950	450	30	8
CMV	ICM-10	1970	1,200	90	1
Coburn	608	1970	2,400	100	0.75
Cylindrical lenses:					
Bausch & Lomb	Hand Pan	1930	300	Variable	30
American Optical	427	1940	400	20	15
Optek	400	1960	400	30	5
American Optical	Super-Twin	1970	550	40	4.5

excited by a laser dye pumped by a nitrogen laser. The preferred material for the laser crystal is neodymium-doped ultraphosphate, for which the continuous-wave laser threshold is on the order of a few milliwatts.

Two types of laser materials are available in which a high gain per unit can be achieved. The first type has a high number of ions per unit length, and the material is rich in rare earth ions that enter as a constituent at a given concentration. This material is used in the design of Pr_6Cl_2 lasers. The other type of material has high gains with strong transition probabilities. This type is best suited for a neodymium-based lanthanum oxysulfide laser (LOS).

7.5.2 LOS Crystal Growth

This section describes the critical procedures for the crystal growth of lanthanum oxysulfides. Single crystals of these compounds can be synthesized at a temperature of 2,000°C, providing vital information on internal stresses and deviations due to sulfur loses or tungsten inclusions from the methods described before. The temperature of the melting point is considerably high and the cooling rate is very slow, which allows the crystals to be free from imperfections. The equipment used to grow single crystals is simple. However, the presence of sulfur enhances the dissolution of the oxysulfide, particularly at low temperatures.

7.5.2.1 Procedure to Form Minicrystals from Polycrystalline LOS Powder

The procedure for minicrystal formation consists of placing the binary mixture of La_2O_3-S (50% by weight of each constituent) in a pure alumina crucible inside a furnace under 15 bars of argon pressure at a temperature of 1,200°C, which must be maintained over a 4- to 5-hour duration. Chemical analysis and x-ray fluorescence indicate that the mixture is only phase constant. The application of argon pressure to minimize the loss by evaporation excludes the use of classic fluxes such as Bo, PbF_2, and Bi_2O_3. Note that the classic fluxes provoke the reduction of the metal and reaction with sulfur.

Good results in the formation of the crystal can be obtained by using the ternary mixtures K_2S-S-La_2O_3S. The combination of two factors—namely, the presence of sulfur and high temperature—will cause La_2O_3S to be replaced by La_2O_3; the synthesis (La_2O_3 + S = La_2OS) and crystal growth takes place in the same run. The ternary diagram and the phase diagram are very important to determine the quality of crystal structure. Finally, a partial melting of the ternary mixture or strong flux losses due to evaporation can provide nontransparent crystals of poor quality. The best results in the formation of minicrystals are possible using the following compositions [5]:

- Potassium sulfide (K_2S), 40% by weight
- Sulfur, 30% by weight
- Lanthanum oxide (La_2O_3), 30% by weight

7.5.2.2 Formation of Good-Quality Ultraphosphate Single Crystals (NdP_5O_{14}) for Laser Applications

The formation of single, small crystals of good quality requires knowledge of the phase equilibrium, which is considered to be the pseudoternary mixture between the apatite and $PbO/PbCl_2$ by the exploratory technique of single-crystal growth. $PbO/PbCl_2$ is a simple eutectic composition that does not introduce impure ions. Volumetric composition of approximately 200 g is placed in a platinum crucible and heated to 350°C for approximately 2 hours to eliminate water. After that, the batch is heated to 1,000°C for a period of 3 hours, then is cooled slowly to 600°C at a cooling rate of 2°C per hour. The first single crystals obtained were small and not of good quality. Phase equilibrium must be considered as pseudoternary between apatite and $PbO/PbCl_2$ using the exploratory technique of single-crystal growth.

One can obtain good-quality ultraphosphate single crystals of NdP_5O_{14} when the concentration of CLAP is 30% in weight and the PbO-to-$PbCl_2$ ratio is 3:1. Under these conditions, small, nontransparent CLAP crystals are formed at the top of the crucible and the transparent needle is shaped at the bottom of the crucible. The best zone for the quality and size of the crystal lies at the top of the phase diagram of the system; the zone containing the NdP_5O_{14} crystals is located at the extreme left corner of the diagram. As far as growth is concerned, the flux melt technique permits rare earth–based crystals of optical quality, which could be used for laser applications where enhanced efficiency and high stability are principal design objectives.

7.5.2.3 Fluorescence Results and Laser Effects of the Crystals

CLAP lasers are designated as $Na_2Nd_2Pb_6 (PO_4)_6Cl_2$, whereas LOS lasers are designated as $Nd^{+3}La_2O_2S$. As mentioned, these crystals are best suited for the design and development of minilasers with minimum threshold current levels. LOS crystals are ideal for lasers requiring longer wavelengths, higher conversion efficiencies, improved optical stability, and stronger transition effects. LOS has the advantage of a compact spectrum with a stronger line, which indicates a larger laser cross-section for self-quenching. This characteristic of the laser indicates fluorescence intensities for the transitions. Fluorescence intensities for both the lasers are impressive. In the case of CLAP lasers, the fluorescence intensity rises up to a maximum after quenching. This is in accordance with previous lifetime measurements demonstrated in studies as a function of oscillator strength. The lifetime measurements of these lasers are dependent on measuring the oscillator strength.

The performance of these lasers is generally discussed in terms of line width, laser cross-section ratio, threshold level, and the optimum Nd^{3+} concentration. The Nd^{3+} concentration level can be 10 times more in CLAP crystal laser than in the LOS laser, which makes the CLAP crystal the most ideal material for the threshold level. The threshold level was estimated to be approximately 25 mW in the presence of incident power from an ionized argon laser with a beam width of 0.5 A° and pumping longitudinally for a sample 200 μm in length. With the same measurement conditions, the threshold for the LOS laser was observed to be approximately 18 mW. The measurements concluded that the presence of scattering centers in the first CLAP and LOS laser versions demonstrated significantly improved results in the crystal characteristics.

7.6 Trivalent Praseodymium-Doped:Yttrium Lithium Fluoride Laser Operating at Room Temperature

The operation of lasers at room temperature (300 K) yields maximum economy with minimum size and weight. In addition, room-temperature operation offers optimum reliability. Room temperature laser operation of praseodymium:lithium yttrium fluoride ($Pr:LiYF_4$) in the blue-spectrum region offers the most efficient performance at a wavelength of 640 nm. The diode-pumped rare earth–based lasers were designed and developed in the early 1990s to operate at 2 μm, which offers high efficiencies even at room temperature for potential applications in altimetry, low-altitude wind shear detection, atmospheric remote sensing, water vapor profiling, and a host of other applications. High-power Q-switched holmium 2-μm lasers are best suited to provide a pumping source for optical parametric oscillators. In addition, these 2-μm lasers provide several medical applications, such as laser angioplasty to clear coronary arteries, ophthalmic procedures, arthroscopy, laparoscopic cholecystectomy, and precision refractive surgeries. The following sections describe two such lasers, their critical performance parameters, and their principal applications.

7.6.1 Trivalent Praseodymium-Doped Lithium-Yttrium-Fluoride Laser Operating at 0.64 μm

The room-temperature laser action in a triply ionized praseodymium-doped lithium-yttrium-fluoride (YLF) laser at 0.64 μm (640 nm) corresponds to the $^3P_0$–$^3F_2$ transition illustrated in Figure 7.5. The stipulated emission corresponds to this transition of a $PrCL_3$ laser at room temperature (300 K), according to German laser scientists. The room-temperature laser operation of a Pr:YLF laser in the blue

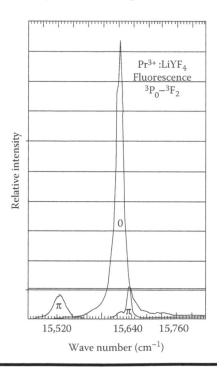

Figure 7.5 The $^3P_0$–$^3F_2$ fluorescence spectra of a praseodymium-doped lithium-yttrium-fluoride (Pr:YLF) laser.

spectrum has been observed previously by other laser scientists. Other laser transitions of a Pr^{3+} ion in the orange and red spectral regions have been observed by other scientists in lanthanum chloride and lanthanum ferrite host crystals at cryogenic temperatures below 77 K. According to laser scientists, the hygroscopic nature of the chloride crystals has very little use for many applications. On the other hand, the YLF crystal has generated significant interest as a laser host crystal because of its superior optical properties.

The pump source and laser transitions associated with the 640-nm laser operation in Pr^{3+}:YLF are illustrated in Figure 7.6. The pump radiation from a tunable flashlamp-pumped dye laser at 479 nm excites the Pr^{3+} ion directly from the $^3H_4$

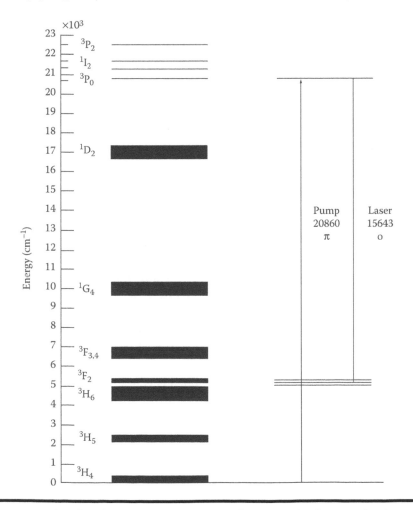

Figure 7.6 **Simplified energy-level diagram of triply ionized praseodymium element in a YLF laser.** *H,* **ground state as appears in the multiplet** $^3H_4$; *P,* **power state as appears in the singlet** $^3P_0$; *F,* **crystal field state as appears in multiplet** $^3F_2$.

ground state into the $^3P_0$ state, as illustrated in Figure 7.6. Stipulated emission is observed from the $^3P_0$ singlet level to the $^3P_2$ singlet level. The point group of the symmetry of the Y^{3+} site into which the Pr^{3+} ion goes substitutionally is S_4. Note that the $^3F_2$ multiplet is split by the crystal field into five levels, with group theoretical notations of various levels. The group theory allows the electric transition from $^3P_0$ into three of these levels.

All three transitions between these two levels can be observed in the fluorescence spectra, as shown in Figure 7.5. The high-resolution spectra in this laser was obtained using a double monochromator with a spectral resolution better than 0.01 cm. One transition has the electric field vector **E** polarized perpendicular to the crystal-axis or c-axis (σ); the other two transitions have **E** parallel to c (π). The transition $^3F_2$ levels can be determined experimentally and are shown in Table 7.4, along with their group theoretical notations and polarizations.

The transition $^3P_0$ is used for resonant pumping of the laser transition because its absorption coefficient ($\alpha = 3.8$/cm) is larger than those of other potential excitation spectral lines. This absorption band at 479 nm is π-polarized and its measured spectral width is approximately 0.8 nm. $^3P_0$ can be populated by pumping the $^3P_2$ multiplet using π polarized 444 radiation. Pumping at this wavelength is less efficient because $\alpha = 1/1.6$ cm. Other absorption lines can be used for laser excitation; they are relatively weaker, but they can be used for flashlamp pumping applications.

Experimental equipment to measure performance parameters includes a coaxial flashlamp-pumped dye laser, which longitudinally pumps the laser crystal Pr:YLF (mounted at Brewster's angle for optimum performance). The dye laser contains a polarizer inside the laser cavity and a differential grating for fine-spectral tuning. A cooled methanol solution of Coumarin dye is used.

Laser measurements indicate that the spectral width of the dye laser output is about two times narrower than the 0.8-nm line width. The pump beam is focused at the center of the laser crystal in order to measure the full width at half maximum (FWHM) beam width of approximately 200 nm. Approximately 70% of the pump radiation, polarized parallel to the crystal axis, is absorbed. The input optical mirror through which the pump beam passes has a transmission efficiency

Table 7.4 Experimentally Determined Levels and Polarizations of $^3F_2$ Levels

Experimental Energy and Polarization (cm^{-1})	Calculated Energy (cm^{-1})	Group Notation
5,167, π	5,198	G1
5,201, π	5,250	G2
5,221, π	5,267	G3,4
5,342, π	5,349	G2

of better than 90% at a wavelength of 489 nm and a reflectivity exceeding 99.7% at a wavelength of 639 nm. Both input and output mirrors must have low reflectivities at a wavelength of 607 nm, which corresponds to another laser transition between the two levels of the Pr:YLF laser, as shown in Figure 7.5. Laser action at the orange wavelength can be observed for other crystal orientations, but it has a higher threshold than the $^3P_0$ and $^3H_6$ transitions.

The laser measurements indicate the temporal dependence of the lasing emission from this laser at 639 nm with that of the pump radiation. The pump pulse width as measured is approximately 200 ns. The Pr:YLF crystal lases near the end of the pump pulse and shows a spike, which is a characteristic of relaxation oscillation. The laser output waveform can be obtained at a pump energy that is approximately 30% above the threshold. Spectral analysis of the lasing output energy reveals narrow lines contained in a total spectral width of 0.08 nm. To visualize the spectral line width, which is called the FWHM, the fluorescence emission is measured to have a lifetime of 50 ms and a spectral line width or FWHM of 0.7 nm. The term FWHM indicates that the line width is measured at half-power points or at 3 dB.

The threshold for a four-level laser is directly proportional to the fluorescence lifetime. The branching ratios can be obtained by integrating the areas of the fluorescence spectra. The branching ratios shown in the spectral decay from the $^3P_0$ level of the Pr:YLF laser at room temperature (300 K) have been obtained. The critical laser design parameters can be summarized as follows:

- Laser pulse emission wavelength: 639 nm
- Design half-power spectral bandwidth: 0.7 nm
- Emission lifetime: 50 ms
- Focused threshold density-based Gaussian profile: 1.5 J/cm²
- Threshold energy requirement for a longitudinal mode of operation: 0.5 mJ or 0.0005 W/s
- Operating temperature: 300 K (27°C)
- Radioactive lifetime: 800 μs (0.800 ms)
- Pump pulse width: 200 ns

As stated previously, the threshold for a four-level laser is directly proportional to the fluorescence lifetime (~50 μs) and is also equal to the fluorescence decay time of the $^3P_0$ transition. The branching ratios for decay from the $^3P_0$ level can be obtained by integrating the areas of the fluorescence spectra shown in Figure 7.6. The calculated values of the branching ratios for various multiplets at room temperature are shown in Table 7.5 [6].

7.6.2 Threshold Energy Density for a Four-Level Laser

For a four-level laser operating in the longitudinal mode, the threshold energy density (J/cm²) can be calculated using the following equation:

Table 7.5 Branching Ratios for Decay Time from the $^3P_0$ Level

Multiplet	Branching Ratio
$^3H_4$	0.452
$^3H_5$	0.042
$^3H_6$	0.351
$^3F_2$	0.058
All others	0.100

$$E_0 = [8\pi f^2 \, n^2/c^2] \, [h f_p \, R_{LT} \, \ln(1/R) \, L_W] \qquad (7.1)$$

where f is the laser frequency (1/639 nm), n is the refractive index, c is the velocity of light (3×10^{10} cm/sec), h is the photon energy of the pump radiation, f_p is the pump frequency (1/639 nm), R_{LT} is the radiation lifetime (800 µs), R is the optical mirror reflectivity (99.7%), and L_W is the line width (18 cm). Neglecting mismatch losses and cavity losses, assuming the values given in parentheses, assuming appropriate values of other parameters, and inserting them into Equation 7.1, the threshold energy density of the Pr:YLF laser comes approximately to 0.615 J/cm^2. If the losses due to mismatch and cavity structure are taken into account, the threshold energy density will be reduced by at least 10%.

7.6.3 Two-Micron Rare Earth Lasers Operating at Room Temperatures

This section is dedicated to 2-µm lasers using rare earth–doped crystals operating at room temperature. Researchers [7] have identified diode-pumped holmium-thulium-lutetium-lithium-ferrite (Ho:Tm:LuLF) and holmium-thulium-yttrium-aluminum-garnet (Ho:Tm:YAG) operating at room temperature. As mentioned, lasers operating at room temperature offer the lowest cost, minimum weight and size, high reliability, and lowest design complexity. Absorption spectra and lifetimes measured as a function of pump energy indicate that these rare earth–doped crystals are best suited for altimeters, ranging sensors, remote sensing systems, and low-altitude wind shear sensors. In addition, Q-switched holmium lasers are very useful as a pump source for optical parametric oscillators. Furthermore, these lasers operating at room temperature have potential medical applications, such as ophthalmic procedures, arthroscopy, and refractive surgeries [7]. These lasers have very similar threshold energies and slope efficiencies—close to 1.3% for LuLF and YLF under the normal modes of operation and at room temperature (300 K/27°C).

7.6.3.1 Performance Parameters of Diode-Pumped Ho:Tm:LuLF Lasers

Experimental performance data [7] indicate that Ho:Tm:LuLF lasers have demonstrated an output energy of close to 80 mJ per pulse and an optical conversion efficiency exceeding 9.4% while operating in normal mode and at room temperature (300 K). Other performance parameters can be briefly summarized as follows:

- Optical-to-optical efficiency: 9.4% at room temperature (300 K)
- Estimated optical-to-optical efficiency: 15.6% at a cryogenic temperature of 125 K
- Tendency to produce more favorable thermal occupation of both upper and lower laser levels
- Demonstrated improved level-to-level branching ratios, which were obtained using both electric and magnetic dipole transition probabilities

With this information, the thresholds of various Tm- and Ho-doped fluorides and oxides can be calculated and compared. Various fluoride laser materials have been modeled and investigated, including LuLF co-doped with trivalent Tm and Ho ions, for possible laser performance improvement and material growth. The modeling has established the correct threshold energy values for the LuLF lasers operating at various temperatures [7].

7.6.3.2 Performance Parameters of Ho:Tm:YLF Lasers

A Ho:Tm:YLF laser is a diode-pumped solid-state device that performs satisfactorily in room-temperature environments. The YLiF$_4$ or YLF laser operates at 2 μm; its optical efficiency is lower than that of the LuLF laser, which was discussed in Section 7.6.3.1. Its slope efficiency is approximately 1.3%, which is similar to that of LuLF. Performance of both YLF and LuLF lasers has been optimized at a center wavelength of 794 nm in room-temperature environments.

Three closely coupled laser diode arrays, placed 120 degrees apart around the perimeter of the laser rod, were deployed for lateral pumping. The laser diode array consists of six laser diode array devices capable delivering a peak power of 300 W with a center wavelength of 794 nm at room temperature and a FWHM spectral bandwidth of 4 nm. The laser diode arrays must be tuned to a temperature such that the center wavelength is 793 nm to match the absorption peak of the Tm ions. It is important to mention that both the threshold energy and slope efficiency vary as a function of mirror reflectivity at a constant room temperature of 300 K. Normal-mode laser threshold level and slope efficiency of LuLF and YLF lasers as a function of coupling mirror reflectivity are summarized in Table 7.6.

Data presented in Table 7.6 seem to indicate that the threshold level goes down with an increase in mirror reflectivity in both lasers. The slope efficiency goes up

Table 7.6 Normal-Mode Laser Threshold Level and Slope Efficiency as a Function of Output Mirror Reflectivity

Mirror Reflectivity (%)	Ho:Tm:LuLF		Ho:Tm:YLF	
	Threshold (J)	Slope Efficiency (%)	Threshold (J)	Slope Efficiency (%)
98	0.4124	0.1776	0.4727	0.1451
94	0.4547	0.1931	0.5162	0.1518
90	0.5037	0.1731	0.5653	0.1476
82	0.6194	0.1485	0.6637	0.1114

Source: Jani, M.G., N.P. Barnes, K.E. Murray, D.W. Hart, G.J. Quarles, and V.K. Castillo. 1997. Diode-pumped Ho:Tm:LuLiF$_4$ laser at room temperature. *IEEE Journal of Quantum Electronics* 33:112–114.

and down in both laser as a function of mirror reflectivity. Both laser performance parameters were recorded in room-temperature environments (300 K). Critical performance parameters such as optical-to-optical efficiency, laser output energy, and other parameters for these diode-pumped solid-state LuLF and YLF lasers can be seen in Table 7.7.

Q-switched laser performance levels for both lasers have been compared at a mirror reflectivity of 82%. Threshold energy and slope efficiency under normal mode and Q-switched mode operations are summarized in Table 7.8.

7.7 Critical Role of Rare Earth Elements in the Chemical Industry

This section discusses the critical role of rare earth materials in the development of the chemical industry. The chemistry industry is a key component in the industrial development of a nation and also in the design and development of energy sources. For example, magnetohydrodynamic (MHD) generation of electrical energy converts thermal energy to electrical energy by passing a high-velocity, alkali-seeded plasma through a magnetic field and picking up electrical current with electrodes projecting into the flow. In this case, the requirements for the electrodes are very severe because they must withstand high temperatures ranging from 1,500 to 2,300°C, abrasion, thermal shock, and alkali attack. Furthermore, electrode conductivity must be very high to minimize the electrical and thermal losses.

Research studies performed in 1976 by the scientists at the General Refractories Company determined that the electrodes made of strontium, calcium, or

Table 7.7 Critical Performance Parameters of Diode-Pumped Solid-State Lasers under Normal Modes of Operation and at Room Temperature

Parameters	Ho:Tm:LuLF	Ho:Tm:YLF
Operating wavelength (μm)	2	2
Optical-to-optical efficiency (%)	9.4	6.1
Dopant concentrations (%)	5 for Tm; 0.5 for Ho	6 for Tm; 0.5 for Ho
Quality of doped crystal	Excellent	Superior
Tm ion concentration (ions/cm³)	7.3×10^{20}	8.4×10^{20}
Absorption efficiency at a wavelength of 793 nm	5	6
Fluorescence lifetime (ms)	12/8	12.5/8
Laser output energy under Q-switched operation (mJ)	14.7	11.0

Source: Jani, M.G., N.P. Barnes, K.E. Murray, D.W. Hart, G.J. Quarles, and V.K. Castillo. 1997. Diode-pumped Ho:Tm:LuLiF₄ laser at room temperature. *IEEE Journal of Quantum Electronics* 33:112–114.

Table 7.8 Performance Parameters for Both Lasers under Normal Mode and Q-Switched Mode

Laser Type	Threshold Energy (J)	Slope Efficiency (%)	
		Normal Mode	Q-Switched Mode
Ho:Tm:LuLF	0.61	2.2	4.1
Ho:Tm:YLF	0.69	11.9	3.9

Source: Modified from Jani, M.G., N.P. Barnes, K.E. Murray, D.W. Hart, G.J. Quarles, and V.K. Castillo. 1997. Diode-pumped Ho:Tm:LuLiF₄ laser at room temperature. *IEEE Journal of Quantum Electronics* 33:112–114.

magnesium-doped lanthanum chromite are best suited for MHD applications. The researchers investigated potential rare earth–based chemical materials for the electrodes capable of meeting stringent performance requirements under severe thermal, mechanical, and chemical environments. This a classic example of rare earth material playing a critical role in the chemical industry.

7.7.1 Lanthanum Chromite Electrodes

In the design and development of electrodes for MHD applications, several chemical materials and processes are involved. A strong aqueous solution of chromic acid is formed. With vigorous stirring, the mixture of lanthanum oxide and strontium, calcium, or magnesium carbonate in appropriate proportions is dissolved. The solution is evaporated and the calcined material is ground to a fine particle size. The finely ground, calcined powder can be formed by conventional ceramic forming techniques, such as extrusion with binder, dry pressing, or isostatic pressing. The formed body or sample is fired in an oxidizing atmosphere to a temperature ranging from 1,500 to 2,300°C.

An alkaline rare earth–doped lanthanum chromite sample prepared in this fashion will have exceptionally high density. Lanthanum chromite composition contains strontium, calcium, or magnesium-doped lanthanum chromite formulations [8]. Lanthanum chromite is prepared with 1.02 to 1.05 moles of chromium oxide per mole of lanthanum oxide. Strontium, calcium, or magnesium doping can vary from 0 to 30%, depending upon the desired electrical properties. The high material density leads to the development of potential properties (both electrical and mechanical) of the material and should provide remarkable performance for use in an MHD power generator.

7.7.2 Role of Rare Earth Elements in the Development of Catalytic Materials

A catalyst comprises one or more materials having a composition within a general formula. It has a perovskite structure. Cations A and B have different valence states and occupy sites with C cations. Cation A may be strontium and Cation B may be lanthanum. Cation A can be barium or calcium. Cation B may have any cation of a rare earth element with an atomic number from 57 to 71. The following sections provide examples of catalysts for various applications.

7.7.2.1 Catalytic Material for Oxygen Electrodes

A material with a lanthanum perovskite structure has catalytic activity, which may be demonstrated at least in relation to the reduction of oxygen in the alkaline solution and its ability to set up a reversible oxygen electrode capable of meeting a wide range of temperatures and oxygen partial pressures. The resistivity of such powders is in the range of 0.1 to 1.0 ohm-cm at room temperature. This kind of material is widely used in making oxygen electrodes in the polarographic method for determining the oxygen partial pressure in a solution or as an oxidation catalyst.

7.7.2.2 Lanthanum Perovskite on a Spinel-Coated Metal

Researchers at Exxon developed a lanthanum perovskite material that contains at least one transition metal composited with a spinel-coated metal oxide. The catalyst was prepared by forming a surface spinel on a metal oxide and subsequently co-impregnating or co-depositing the appropriate perovskite precursor component on the spinel-coated metal oxide followed by calcinations at a temperature of approximately 540°C. The preferred catalyst is lanthanum-cobalt oxide ($LaCoO_3$) supported on a spinel-covered alumina. Note that the term *transition metal* is used here to designate elements having atomic numbers of 21 to 30, 39 to 48, 57 to 80, and 89 to 92. Therefore, a number of rare earth elements could be involved in the development of transition metals.

7.7.2.3 Preparation of a Catalyst Using Lanthanum with Cobalt and Nickel

This particular catalyst was developed in 1978 by scientists working with Siemens AG, Germany. This oxidic catalyst is best suited for the conversion of water gas. It consists of oxides of metals such as aluminum, lanthanum, cobalt, nickel, and uranium, with aluminum oxide predominating as alpha-Al_2O3. The lanthanum and cobalt content varies from 5 to 30% by weight, that of nickel from 0.1 to 4% by weight, and that of uranium from 0.1 to 2% by weight, related to total weight of the catalyst.

Note that the atomic ratio of lanthanum to cobalt is most ideal between 1:4 and 4:1. The finished catalyst contains roughly 12.0 wt% lanthanum, 5.1 wt% cobalt, 0.7 wt% nickel, and 0.4 wt% uranium, as the weight of the metals relative to the total weight of the catalyst.

7.7.2.4 Advantages of Complex Metal Silicates

Some rare earth elements play critical roles in synthesizing complex metal silicates from solutions containing reactive silicates and reactive forms of the desired metal, including the soluble salts or oxides and hydro-oxides. The reactants are combined in an alkaline medium under moderate temperature and pressure. The sources of silica and metal are used in appropriate quantities to provide a reaction mixture having approximately 1.4 to 1.6 moles of a mixture of metals (with at least one rare earth metal, such as cobalt or nickel) per mole of silica. The complex metal silicate is best suited for manufacturing thin-wall tubes, rods, or curls for industrial applications.

7.7.3 Formation of Solid Solutions from Catalysts with Lanthanum

In 1978, scientists at B.F. Goodrich developed a process involving ethylene and vinyl chloride from ethane. The use of a solid-solution catalyst in the process resulted in high yields of ethylene and high combined yields of ethylene and vinyl chloride. The solid-solution catalysts would exhibit their activity for hundreds of hours without any need to interrupt the process for catalyst regeneration.

A solid-solution catalyst containing iron cations indicates a direct substitution of iron (Fe_2O_3) ions for host lattice ions. A solid-solution catalyst containing iron ions can be further stabilized against iron loss by using lanthanum or any lanthanide element. Lanthanum and lanthanide elements do not enter the solid solution with the host lattice as does the iron. The solid solution can be regenerated in situ by the addition of an iron source to the ongoing chemical process. After the addition of an iron source, the activity of the catalyst increases dramatically. This catalyst is best suited for various industrial applications.

7.7.4 A New Catalyst Involving Nickel and Silver

In 1977, Hitachi scientists developed a new catalyst involving nickel and silver with lanthanum, which demonstrated excellent characteristics as a catalyst [8]. Steam reforming of hydrocarbons uses a catalyst of 10 to 30% by weight of nickel, per weight of the catalyst, 2 mg atoms of silver per 100 g of catalyst, and at least one rare earth element such as lanthanum, cerium, yttrium, praseodymium, neodymium, or a mixture of rare earth elements in the atomic ratio of the rare earth elements, and a heat-resistant oxide carrier such as alumina, silica, magnesia, titania, zirconia, or beryllia. Note that the heat-resistant oxides mentioned here provide the highest heat-resistant capabilities.

The raw materials for nickel used as the main active component of the catalyst include nickel oxide, nickel nitrate, nickel carbonate, nickel oxalate, and so on. The raw materials for silver used as the promoter include an oxide, nitrate, and carbonate of silver. Quantitative proportions of the catalyst components are as follows:

■ Nickel in an amount of at least 3% by weight
■ An amount ranging from 10 to 30% by weight if nickel oxide is used

Note that these percentage weights are based on the catalyst weight.

When a rare earth element is further added to the catalyst, the effect of suppressing carbon deposition and the heat resistance can be increased significantly. In the catalyst, an atomic ratio of the rare earth element to silver is 10 or less [8]. The catalyst works well without any rare earth element, but the effect of the rare earth element appears when the catalyst contains 2 mg atoms or more of silver per 100 g of catalyst. This is a very interesting point in developing a catalyst consisting of nickel and silver with lanthanum. This particular catalyst is used in various commercial and industrial applications.

7.7.5 Formation of Mixed Oxide Compounds for Casting Super-Alloy Materials

The advanced super-alloy materials, such as NiTaC-13 and other similar metal eutectic alloys, are cast and directionally solidified at temperatures around 1,700°C and above after 30 hours of exposure [8]. Therefore, cores and molds must have high temperature strength and nonreactivity with the molten metal. In other words, the mold and core material must not dissolve in the cast molten metal nor form an excessively thick interface compound with the molten metal. In addition, the cores must be compatible with the advanced super-alloy to prevent hot tearing or hot cracking during the solidification process.

Studies performed by researchers at General Electric in 1978 showed that there are three distinct materials belonging to the super-alloys, which are summarized as follows [9]:

■ Lanthanum oxide–aluminum oxide ($La_2O_3 \cdot 11Al_2O_3$)
■ Neodymium aluminum oxide ($NdAlO_3$)
■ Magnesium aluminum oxide ($MgAl_2O_4$)

Note that magnesium aluminum oxide belongs to the nonstoichiometric category, which means that an $MgAl_2O_4$ super-alloy is in equilibrium with aluminum oxide (Al_2O_3).

The contents of these super-alloys are as follows:

■ The lanthanum oxide–aluminum oxide super-alloy contains La_2O_3 from 8 to 9 mol%.
■ The neodymium aluminum oxide super-alloy contains NdO_3 at 50 mol%.
■ The magnesium aluminum oxide super-alloy contains Al_2O_3 from approximately 50 to 80%.

According to the General Electric scientists, these super-alloys can be prepared using three distinct methods: 1) mechanical mixing of each of the two oxides in appropriate proportions; 2) mechanical mixing of the oxides strictly in proper amounts, which is then calcined at elevated temperatures ranging from 600 to 1,700°C over 1 to 6 hours; 3) the preparation of the material compositions in proper amounts of the oxides, which are then fuse-cast by heating them to their melting temperatures. The fuse-cast material is refined into the desired particle size (10–150 μm) using suitable milling techniques. The desired core configurations can be prepared from this material. Complicated shapes can be made from these super-alloys by deploying suitable manufacturing techniques, such as injection molding, transfer molding, or other molding techniques. These molded products have potential commercial applications.

7.7.6 Zinc Aluminate for Water Treatment Catalysts

Scientists at Phillips Petroleum Company have discovered a unique catalyst for water treatment. The catalyst lifetime can be extended by including a promoting amount of one cerium group of a rare earth material, such as lanthanum. The activity of the zinc aluminate catalyst promoted with a group-IB metal and a group

VII-B metal can be enhanced, and the catalyst's lifetime can be increased significantly by promoting around at least one cerium group rare earth metal, such as lanthanum. The studies further revealed that the most effective catalyst compositions include copper-zinc aluminate, nickel-zinc aluminate, copper-nickel zinc aluminate, copper-manganese-zinc laminate, cerium-zinc aluminate, platinum-zinc aluminate, copper-platinum-zinc aluminate, and lanthanum-copper-manganese-zinc aluminate. One can select the composition that offers optimum effectiveness at minimum cost. The concentration of the promoting metal on the zinc aluminate after calcinations will generally be in the range of 0.05 to 20 wt% based on the weight of the zinc aluminate support material.

The aqueous wastes applicable as feeds for the process are those that contain minor amounts of dissolved and/or suspended organic materials. The process is particularly applicable to feeds in which the organic materials are hydrocarbons and/or oxygenated hydrocarbons. The organic materials can be present in the predominantly aqueous stream in a broad range of concentrations. However, the presence of organic material will be less than 10% by weight. The aqueous waste streams can be derived from any source, such as chemical or biological sources. However, for best results, the aqueous waste streams should contain very little nonvolatile and/or nonoxidizable inorganic materials.

The catalyst described here is best suited for water treatment applications. The catalysts may be prepared by simple conventional means. For example, compounds or components of the catalyst that are capable of being converted into oxide when heated may be evaporated in a solvent (particularly in water), dried, and calcined in the air at a temperature ranging from 200 to 260°C. Rare earth elements such as molybdenum, tungsten, and vanadium may be used, if desired, in the form of the ammonium salt; the remaining components may be used as nitrates, oxalates, chlorides, or sulfates.

7.7.7 Lanthanum Alloys for Hydrogen Storage with Hot Gas Engines

Engineers at U.S. Philips Corporation developed a system for converting calorific energy into mechanical energy. The system could be a hot-gas turbine, hot-gas engine, or an internal combustion engine. The system elements include at least one combustion chamber for hydrogen and at least one cooler incorporated in a system of ducts for the cooling medium. The combustion chamber or system can be made to communicate with a hydrogen container containing an alloy of *A* and *B*, in which the ratio *A:B* can vary from 1:3 to 2:17 in powder form, where *A* is calcium or one or more of the rare earth elements, combined with thallium, and/or zirconium and/or hafnium, and where *B* is mainly nickel or cobalt rare earth metal.

The hydrogen container and a part of the system of ducts are constructed for the heat exchange between the alloy in the hydrogen container and the medium in the system containing the ducts. The researchers at U.S. Philips Corporation found that alloy *AB* is capable of absorbing very large quantities of hydrogen gas at room temperature and at very low pressures. For example, 0.08 g of hydrogen gas is absorbed by 1 cm^3 lanthanum nickel (LaNi$_5$) powder with a packaging density of 65% at a hydrogen gas pressure of 5 atm (or slightly less than 70 psi) and at an approximate temperature of 40°C. The density of the hydrogen gas in the alloys may be on the order of magnitude of that of liquid hydrogen or more.

The absorption of hydrogen is also possible with lanthanum-nickel alloy, according to research studies performed by U.S. atomic energy scientists in 1974. The research studies revealed that hydrogen will be selectively absorbed from a gaseous product, especially carbon dioxide in substantial amounts, by contacting the gas with a distributed form of an alloy of lanthanum and nickel in an active state. Because of this, a metal hydride is formed in which the rare earth nickel alloy is hydrided. Other rare earth elements, such praseodymium cerium, give indications of being of similar usefulness. In another embodiment where cobalt is present in significant amounts, the inhibiting effect of this gas is overcome by replacing some of the nickel present in the alloy with copper.

In a preferred embodiment using this method, an ingot of LaNi$_5$ at a pressure at least 100 pounds per square inch absolute (psia) at room temperature and a pressure of 200 psia is contacted by hydrogen. The ingot disintegrates when hydrided and then is heated over a temperature ranging from 100 to 200°C to drive off the hydrogen. The LaNi$_5$ is then in an active state. Being in an active state means that LaNi$_5$ is capable of absorbing hydrogen gas quickly.

A mixture of hydrogen and other gases, including carbon dioxide and water vapor, is then passed into the bed of active LaNi$_5$ at a pressure of at least 100 psia at room temperature. After a short period of time, the alloy becomes hydrided, forming a compound LaNi$_5$H$_x$, where x can be any number as high as 7. This hydride readily gives up its hydrogen constituent at room temperature when the pressure is lowered to approximately 25 psia. The hydrogen so produced can be used wherever desired. One of the distinct advantages of this process is that the alloy will function effectively in the presence of large amounts of carbon dioxide, typically approximately 25% by volume. This process is considered to be most effective when hydrogen is mixed with several other gases. Furthermore, the process is considered to be cost effective. It is best suited for industrial applications, where cost and separation speed are of critical importance.

7.8 Role of Rare Earth Materials in Microwave and Electronic Components Best Suited for Military and Medical Applications

This section discusses the applications of rare earth materials in microwave and electronic components that play critical roles in military and medical applications. Applicable rare earth elements are identified under specific categories of applications.

7.8.1 Role of Yttrium-Iron-Garnet in Military System Applications

Yttrium-iron-garnet (YIG) is a principal material used in the design and development of microwave filters and dispersive delay lines. These devices are best suited for radar systems and electronic warfare equipment where minimum size and weight, low insertion loss, and low time sidelobe levels are of importance for critical electronic warfare missions. YIG-based delay lines are best suited for microwave chirp filters, which are widely used in pulse compression filters and are essential components of airborne moving target indicator radars, airborne side-looking radars, and electronic-countermeasure systems (ECMs).

These YIG filters are critically important to airborne radars, where high range and angle resolutions are the principal system performance requirements. In ECM systems, chirp filters provide high clutter rejection and an improved signal-to-jamming ratio, which is considered essential for an active ECM system operating at microwave and millimeter-wave frequencies. Pulse compression filters designed and developed using YIG chirp filters have achieved pulse compression ratios exceeding 50:1, which has the excellent range and angular resolutions needed for military missions.

7.8.2 Role of Samarium Cobalt Permanent Magnets for ECM and Radar Systems

Neodymium-iron-boron (Nd-Fe-B) and samarium-cobalt (Sm_2Co_7) appear to be suitable materials for the permanent magnets needed for critical military missions under harsh mechanical, thermal, and hostile environments. The material characteristics of these two rare earth–based permanent magnets are summarized in Table 7.9.

Periodic permanent magnets (PPMs) made from samarium cobalt play a critical role in the design of high-power traveling wave tube amplifiers (TWTAs),

Table 7.9 Material Characteristics of Rare Earth–Based Permanent Magnets

Characteristics	Neodymium-Iron-Boron	Samarium Cobalt
Operating temperature (°C)	80	300
Curie temperature (°C)	320	830
Compressive strength (psi)	160,000	116,688
Flexural strength (psi)	36,465	21,870
Tensile strength (psi)	10,939	5,105
Material cost per pound ($)	225	250

which are widely used by high-power radars and ECM systems. These PPM focusing magnets are most effective in providing high stable gains, improved TWTA efficiency, focused coherent electronic beams over the entire helix length of the TWTA, and ultralow AM-to-PM conversion ratios needed to retain TWTA performance over an extended duration, which is vital for successful missions. Samarium cobalt offers the highest energy product (BH_{max}), which is a product of flux density and magnetic field under operating temperatures as high as 300°C. Samarium-cobalt periodic-spaced permanent magnets provide significant improvements in electronic efficiency and considerable reductions in the weight and size of high-power TWTAs.

Thermal tests performed in the laboratory on TWTAs using samarium-cobalt PPM-focused magnets indicate that these permanent magnets have demonstrated reliable and safe TWTA performance, even at temperatures slightly greater than 800°C. The deployment of rare-earth PPM-focusing permanent magnets could lead to a significant reduction in weight and size, which is of critical importance in airborne radars and ECM systems. In addition, cathodes in these TWTAs are coated with rare earth oxides, which will further enhance the TWTA's overall radiofrequency performance.

The electrons emitted by the cathode must be focused into the electron beam under a strong magnetic field provided by the samarium-cobalt PPM permanent magnets. Strong interaction is essential between the magnetic field and the electrons with a slow-wave helix structure. Rare-earth PPM magnets provide strong magnetic fields with minimum fluctuations, which is highly desirable to keep the TWTA gain flat and AM-to-PM conversion to a minimum.

7.8.3 Rare Earth–Based Garnets and Mixed Garnets

Garnets are hard ceramics with unusual dielectric and radiofrequency (RF) that are widely used in RF devices. Rare earth ions form a third magnetic sublattice in the garnets. Because the ground configurations of the trivalent rare earth ions generally have large orbital moments (except the gadolinium iron garnet), the magnetoelastic coupling in these materials is generally large. With a few notable exceptions, rare earth garnets seem to have a large bandwidth at room temperature (300 K). For example, the europium iron garnet (EuIG) single crystal has a bandwidth around 290 Oe. Because of its wideband width, it is not used exclusively in RF or electro-optical devices. However, the mixed garnets shown in Table 7.10 are widely used in magnetoelastic devices because of their narrow magnetic bandwidths.

Close examination of the material characteristics indicate that YIG and lithium-ferrous oxide ($Li_{0.5}Fe_{2.5}O_4$) can be used in the design and development of magnetoelastic devices because of their narrow magnetic bandwidth and high figures of merit compared to other materials shown in Table 7.10. The unique characteristics of YIG make it suitable for several applications. It is widely used in the design of microwave band-pass filters, pulse compression filters, YIG-tuned microwave

Table 7.10 Properties of Selected Magnetoelastic Garnets Known or Ferrimagnetisms

Mixed Ferrite	Temperature (K)	$4\pi M$ (G)	Bandwidth (Oe)	Figure of Merit
YIG	300	1,750	0.2	1.0
E_uIG	300	1,100	250	0.052
$Li_{0.5}Fe_{2.5}O_4$	300	3,600	3.5	3.85
$NiFe_2O_4$	300	3,400	20	0.192

oscillators, parametric amplifier for acoustic pulses, and a host of other magneto-acoustic devices. YIG filters are best suited for tunable band-pass filters over a wide tunable range. Such filters have demonstrated low insertion loss and high rejection in stop-band regions, in addition to minimum weight and size. The YIG disc has solved the inherent problem of spurious responses, particularly due to higher-order magnetostatic modes.

7.8.4 Potential Applications of Mixed Garnets

7.8.4.1 Europium Iron Garnet

The ferromagnetic resolution line width of the europium iron garnet (EuIG) has been observed to be very small (less than 1 Oe at a cryogenic temperature of 4.2 K), although the magnetoelastic coupling is not small. Applications of this material are very limited because of the high cost of the cryogenic temperature requirement. In addition, the coupling constants are opposite in sign to those for YIG.

7.8.4.2 Gadolinium Iron Garnet

Trivalent gadolinium is a small-state ion. Like trivalent iron, it might be expected to have small magnetostriction, but its spin orbit coupling in G^{3+} is much larger than for Fe^{3+}; also, its magnetostriction is much larger. Because of these properties, its application is very limited in magnetostrictive and magnetoelastic devices.

7.8.4.3 Terbium Iron Garnet

Terbium with a ground state of $^7F_6$ has a large orbital angular momentum and can be expected to have a large magnetoelastic coupling. Furthermore, the ferromagnetic resonance is extreme in pure TbIG material, which can be lowered with a doping of 0.02% terbium in single-crystal YIG. Then, the doped material has some applications in RF devices.

7.8.4.4 Yttrium Aluminum Garnet

Yttrium aluminum garnet (YAG) is optical crystal and is widely used in the design and development of neodymium-doped YAG lasers operating around 2.0 μm. NASA scientists used neodymium-doped YAG crystal in the successful design and development of Nd:YAG laser transmitters for space optical communications in the 1980s. This laser provides space-to-space communication, space-to-aircraft communication, and space-to-ground communication for the transmission of data. This laser has demonstrated a data transmission capability of 1,000 bits/second, a bit error rate of 10^{-6}, interterminal angle tracking accuracy of less than 1 μrad, station-to-station acquisition in less than 6 s, and a communication range of 40,000 km.

7.8.4.5 Lithium Ferrite Spinel

The ordered state the spinel of lithium ferrite is the only one that has technical merits for further discussion. In the ordered state, trivalent lithium ions alternate with ferric ions in the ratio of 1:3 on the octahedral sites while in the most common inverted spinel, which is designated as nickel ferrous oxide ($NiFe_2O_4$). This material has half the ferric ions that are randomly distributed with all the Ni^{3+} ions on the octahedral sites. This ordering property offers narrow ferromagnetic resonant line-widths in lithium ferrite grown in the ordered state. The magnetoelastic coupling constant of lithium ferrite has been measured as a function of cryogenic temperature ranging from 100 to 300 K. The magnetoelastic constants do not exhibit a rapid rise at low cryogenic temperatures, as observed in the rare earth iron garnets. The figure of merit of this material exceeds the value for YIG at room temperature (300 K). However, it is rather difficult to grow large single crystals, which reduces the technical applications of this material in designing other components.

7.9 Summary

This chapter identified rare earth oxides and compounds that have played critical roles in the development of glass, solid-state crystal technology, glass polishing, development of electro-optical technology, and development of chemical industry. Critical roles of rare earth compounds and oxides in the development of glass industry were summarized. Matured techniques in the production of eye protection glasses, CRT faceplate glasses, light bulb glasses, decolorization glasses, and window glasses for residential buildings, commercial building, and research laboratories were summarized, with particular emphasis on cost and quality. The largest share of the glass market is taken by cerium oxide at 88%; the balance is made of lanthanum oxide, erbium oxide, neodymium oxide, and praseodymium oxide. However, lanthanum oxide represents the next largest market share following cerium oxide.

Because the sharp absorption spectra of rare earth elements are insensitive to glass composition and oxidation reduction conditions, it is relatively easy to produce and maintain certain colors in glassmaking products using specific rare earth oxides and compounds. A type of glass that does not contain rare earth oxides is soda-lime glass, which is widely used in making bulbs, plates, containers, and ophthalmic lenses. Lime, magnesia, and alumina (Al_2O_3) are generally added to rare earth oxides to stabilize the glass structure. Important effects between the interaction of one colorant ion and another colorant ion were described briefly. Arsenic ions in combination with iron ions enhance the color from solarization. Various types of decolorization effects are due to impurities present in the glass materials. These impurities can be due to the presence of sand, lime, or iron microparticles.

The properties of optical glasses could be affected by the presence of certain impurities in the rare earth oxides. Major characteristics of optical glasses were summarized, with an emphasis on glass quality and surface finish. Some specific characteristics of optical glasses can be seen in the map of optical glasses. The purity level in rare earth oxides should be between 99.9% and 99.995%. Furthermore, it is absolutely essential that no other colors occur in optical glass from the absorption bands of the praseodymium and neodymium rare earth oxides. Traces of cerium oxide must be eliminated to minimize the ultraviolet ray absorption and also to satisfy the highest quality control specifications of the optical glass.

Lanthanum oxide is added to optical glass to lower the cost, obtain a high index of refraction, and achieve a high Abbe number, which is considered to be a figure of merit. In brief, optical glasses of high quality must have a high index of refraction as well as a high Abbe number. Fluorescence in glass is strictly due to atoms being excited by the absorption of light resulting from light emissions.

The use of rare earth cerium oxide in polishing compounds has demonstrated significant performance improvement of the optics deployed in military and space sensors, where high resolution and minimum tracking and errors are of critical importance. Three distinct polishing mechanisms have been identified to achieve high-quality glass surfaces. The benefits of chemical etchants during the polishing procedure were discussed.

The estimated worldwide and U.S. consumption of cerium oxide as a polishing agent were identified. The economic benefits of rare earth oxides as polishing compounds were summarized, with emphasis on finished polished surface and cost. The uses and advantages of spherical and cylindrical polishing machines for polishing optical lenses were summarized. The role of rare earth oxides in the development of optical crystals for lasers was discussed, with particular emphasis on polishing speed, cost, and quality control of the surface. Procedures to form minicrystals from polycrystalline LOS were discussed. Techniques to form good-quality ultraphosphate single crystals of neodymium phosphate oxide were summarized, with emphasis on quality control and economic aspects.

The performance capabilities and applications of praseodymium-doped YLF lasers operating at room temperatures were summarized. The potential medical and scientific applications and major benefits of Pr:YLF lasers operating at room temperature and at 640-nm spectral wavelength were briefly summarized. The calculated values of branching ratios for various multiplets of the Pr:YLF laser were summarized for room-temperature environments. A formula to compute the threshold energy density was derived using appropriate values for laser parameters such as the refractive index, operating wavelength of 640 nm, photon energy of the pump radiation, optical mirror reflectivity, and line-width.

The performance capabilities and limitations of some lasers operating at room temperature were summarized, with emphasis on electrical-to-optical efficiency, beam width, and output power. The performance parameters of diode-pumped Ho:Tm:LuLF and Ho:Tm:YLF lasers operating at room temperature (300 K) were summarized, with particular emphasis on optical stability and laser efficiency.

The critical roles of rare earth elements in the development of the chemical industry over the last few decades were summarized, with emphasis on MHD power generators and their commercial and industrial benefits. The advantages of lanthanum chromite electrodes for MHD power generation applications were identified, with emphasis on cost, performance, longevity, and reliability.

The roles of rare earth elements in the development of new and advanced catalysts were discussed, with particular emphasis on spinel-coated metal oxide, highlighting the reliability and cost-effective aspects. The advantages of complex metal silicates were summarized, with emphasis on the manufacturing aspects concerning thin-wall tubes, rods, and curls for various commercial and industrial applications.

Mixed rare-earth compounds for casting super-alloy materials were identified, with particular emphasis on NiTaC-13 and $NdAlO_3$ super-alloy materials. Commercial and industrial applications of various super-alloys were briefly discussed, with emphasis on cost, reliability, and longevity.

The major benefits of zinc aluminate catalysts were summarized for water treatment, with major emphasis on cost-effective operation, safety, and public health aspects. Other potential catalysts for water treatment were briefly discussed, highlighting the principal benefits of the process.

Rare earth–based lanthanum alloys were identified for hydrogen storage with hot-gas engines. The commercial and industrial benefits of lanthanum alloys were summarized. Rare earth materials widely used in the design and development of microwave and millimeter-wave components and systems, which are best suited for military and space applications, were identified with emphasis on reliability, safety, and affordability.

The major benefits of rare earth permanent magnets made from samarium cobalt were summarized. The potential applications of these magnets in high-power radar transmitters and TWTAs, which are critical components of electronic warfare equipment, were identified, with particular emphasis on reliability and safety under harsh thermal and mechanical environments. Applications of YIG material in microwave band-pass and pulse-compression filters, YIG-tuned transistorized microwave oscillators, and a host of other microwave components best suited for military and space systems were discussed with particular emphasis on reliability, RF performance, and reduction in weight and size.

References

1. Stern, F. 1963. Optical properties of heavier rare earths. *Solid State Physics Review* 15:299.
2. Riker, L.W. 1981. *The Use of Earths in Glass Composition.* Washington, DC: American Chemical Society, 81–93.
3. Horrigan, R.V. 1981. *Rare Earth Polishing Compounds.* Washington, DC: American Chemical Society, 95–99.
4. Cornish, D.C. and J.M. Watt. 1966. The mechanism of glass polishing. Presented at the American Ceramic Society Meeting, Washington, DC, May 11, 1966, 4–7.
5. Fadly, M., J. Ostoreno, K.C. Mitchel, et al. 1971. Single crystal growth from flux melt of neodymium-doped materials for mini-laser. *Applied Physics Letters* 18:169.
6. Allen, R., M. Kruer, et al. *Stimulated Emission at 640 nm in Trivalent Praseodymium-Doped LiYF$_4$ Crystal.* Washington, DC: Naval Research Laboratory.
7. Jani, M.G., N.P. Barnes, K.E. Murray, D.W. Hart, G.J. Quarles, and V.K. Castillo. 1997. Diode-pumped Ho:Tm:LuLiF$_4$ laser at room temperature. *IEEE Journal of Quantum Electronics* 33:112–114.
8. Vallani, F.L., ed. 1981. *Rare Earths Technology and Applications.* Park Ridge, NJ: Noyes Data Corporation, 35–41.
9. Peattie, C.G. 1963. A summary of practical fuel cell technology to 1963. *Proceedings of the IEEE* 51(5):795–806.

Chapter 8

Industrial Applications of Pure Rare Earth Metals and Related Alloys

---

8.1 Introduction

This chapter describes the potential industrial applications of pure rare earth metals and related alloys, such as mischmetal or compounds. The worldwide consumption of these metals, alloys, and compounds is presented and their applications in various categories are identified. Light lanthanide elements, which include four light rare earth elements such as cerium, lanthanum, praseodymium, and neodymium, as well as the partitioning in rocks and minerals containing various heavy rare earth elements, are identified. Their potential commercial and industrial applications are discussed, with particular emphasis on the cost and improvements in mechanical properties of metals.

Mischmetal is a well-known rare earth alloy produced by fused electrolysis of four light lanthanide elements, which constitute more than 90% of the rare earth metals (REMs). This particular alloy is widely used for steelmaking due to great demand for various industrial and commercial applications. Approximately 4,000 metric tons of mischmetal, worth around $46 million, are added to liquid steel every year. Typical mischmetal content for the steelmaking industry varies from 95 to 98% REM content, with 0.5 to 5% iron to lower the melting point temperature and to make casting easier for any shape. Residual impurities range from approximately 0.1 to 1% and include magnesium, aluminum, silicon, calcium, nitrogen, hydrogen, and oxygen, among others.

The concentration of the four light lanthanide materials—lanthanum, cerium, praseodymium, and neodymium—and the heavy rare earth elements can vary significantly depending on the origin of the ores. Lanthanum content may vary from approximately 17 to 30%, cerium from 45 to 58%, praseodymium from 4 to 8%, neodymium from 10 to 20%, and the heavy rare earth metals (samarium, gadolinium, and yttrium) from less than 0.1 to 2%. The effects of these individual elements on the physical properties of the steel produced are not fully known.

The percentage content of heavy rare earth metals, such as samarium, gadolinium, and yttrium, in the ores is equally important to the steelmaking industry. The impact of these heavy rare earth metals in the ores is not known. However, the cost of chemical separation prior to electrolysis cannot be fully justified in terms of improved physical properties, particularly for high tonnage carbon steel materials. In the case of superalloys, the percentage content of lanthanum is clearly specified.

The important properties and applications of mixed-valance compounds containing selected lanthanide elements are summarized in this chapter, with an emphasis on chlorides and bromides. Mixed-valance compounds may exist for some of the actinide elements. Experimental results will be reviewed for the benefit of the reader. A systematic experimental investigation of anhydrous halides and oxyhalides was carried out using spectrophotometry and x-ray diffraction techniques. During the course of the experimental investigation, the divalent halides of mixed-valance compounds were evaluated. Experimental results obtained from hydrogen-reduced lancet mixtures and actinide halides are summarized.

In the 1980s, experimental investigations were carried out by the distinguished scientist M.B. Maple at the University of California, San Diego, on the coexistence of superconductivity and long-range magnetic order in ternary rare earth compounds. The investigations were focused on studying the coexistence of superconductivity and long-range ordering of rare earth magnetic moments. The results of investigations of the interaction between the superconductivity and low magnetic order in the rare-earth molybdenum selenides ($RE_x Mo_6 S_8$) and rare-earth rhodium borides ($RERh_4 B_4$) are reviewed.

Finally in this chapter, the industrial applications of mischmetal and its alloys are discussed in great detail. A chronological history of mischmetal applications from the early 1930s onwards is summarized, with an emphasis on the production capacity of mischmetal by the industrialized nations.

8.2 Trends in Industrial Applications Using Mischmetal and Its Alloys

The use of a particular metal or alloy strictly depends on the availability of the material at a reasonable cost, as well as its desirable thermal and mechanical properties for a specific application. This section first discusses the cost of mischmetal and its alloys before exploring their potential commercial and industrial applications.

The price of mischmetal in the 1950s was approximately $6.50 per pound. As time progressed, the market price for mischmetal drifted down, close to $2.50 over the period from 1967 to 1973, then rose again to $5.40 due to inflation. Price fluctuations for industrial metals, including rare earth metals, can be expected depending on the demand for certain strategic metals. Market analysts believe that there is plenty of rare earth ore and large production capacity on a worldwide basis.

8.2.1 Mischmetal Production Capacities of Industrialized Countries

Currently, the total mischmetal production capacity in industrialized countries is estimated to be greater than 7,500 metric tons per year, mainly because of greater consumption in the manufacture of Korean and Japanese automobiles. U.S. manufacturers are now producing more than approximately 55% of the world's total output. The mischmetal production capacities of industrialized countries are summarized in the following sections.

8.2.1.1 U.S. Companies

The leading U.S. companies involved in manufacturing mischmetal are Reactive Metals and Alloys Corporation and Ranson Metals. Their total mischmetal production capacity is close to 2,500 metric tons per year when working close to full capacity. The combined production capacity of these two companies will be able to meet 85 to 90% of the requirements of steel mills in the United States.

8.2.1.2 European Companies

The leading European mischmetal production companies include Treibacher Chemische Werke in Austria, with a production capacity of 1,000 metric tons per year; Th. Goldschmidt AG of Germany, with a production capacity of 1,000 metric tons per year; Ronson–British Flint of Great Britain, with a production capacity of approximately 150 metric tons per year; and Rhone Poulenc–Pechiney of France, with a production capacity of 250 metric tons per year. Most of these mischmetal production companies are working at much less than 50% capacity because of the substitution clause introduced in the 1970s in the steel industry. During this period, the demand for the mischmetal was not great.

8.2.1.3 South American Companies

Corona, Colibri, and Fluminense are three Brazilian companies that are deeply involved in the production of mischmetal. Their combined production capacity is approximately 1,000 metric tons. Furthermore, these companies are working below their capacities. Their mischmetal output is exclusively for export purposes because there is little demand for this material locally.

8.2.1.4 Asian Companies

Santoku Metal Corporation is the only Japanese company manufacturing mischmetal for the steelmaking industry. Their production capacity of mischmetal was slightly more than 200 metric tons in 1979. Since then, most mischmetal has been converted to the oxide-fluoride electrolysis process. The latest published reports on mischmetal reveal that production capacity in Japan has increased to more than 2,000 metric tons per year due to the use of mischmetal for other commercial applications, particularly in the manufacturing of high-quality steel for all-electric vehicles.

8.2.2 Potential Industrial Applications of Mischmetal

Mischmetal and its alloys have unique burning characteristics that make them very useful in ordinance applications. When ignition of a mischmetal alloy particle is achieved by a high-energy explosion or impact, the metal will burn until it is completely consumed. As a result of this phenomenon, mischmetal has been used in shell linings, tracer bullets, aircraft penetrating devices, bomblet markers, and a host of other ordinance components that are vital for national defense [1]. Most of these products were developed, tested, and evaluated in defense laboratories. However, during the Vietnam War, a mischmetal magnesium alloy was used in the manufacturing of the above-mentioned products for military applications. Currently, mischmetal and its alloys are useed by research and development laboratories associated with government-owned ordinance companies.

Industrial applications of mischmetal alloys include refractory linings, high-speed turbine blades, protective shrouding, jet engine nozzles, certain all-electric automobile components, deep-well drilling tools for oil and gas exploration, and a host of other industrial products that require high mechanical strength. Metals or alloys with critical properties such as toughness under shock and vibration environments, improved ductility, and significant resistance to long-term shelf life are suited for defense products frequently used in battlefield applications; mischmetal and its alloys meet these performance requirements.

The demand for mischmetal and its alloys in commercial and industrial applications is dependent on the economic situations of a country. However, economic conditions will not restrict the use of mischmetal and its alloys for military applications because national security and battlefield requirements are not dependent on the price of the mischmetal and its alloys. Sometimes, an energy crisis could affect the use of mischmetal and its rare earth alloys for industrial applications in peacetime periods, as illustrated by the data presented in Table 8.1.

Table 8.1   Industrial-Use Patterns of Mischmetal and Its Alloys during Peacetime

Industrial Application	Estimated Consumption of Mischmetal and Its Alloys per Year (%)			
	1968	1978	1988	1998
Ductile iron	56	13	15	45
Magnesium alloys	20	1	2	35
Mischmetal for steel	18	83	88	90
Miscellaneous alloys	7	4	18	45
Rare earth magnets, electric motors, and generators	—	2	58	84

Source: Hirschhorn, I.S. 1980. Trends in the industrial uses for mischmetal. In *The Rare Earths in Modern Science and Technology,* ed. G.J. McCarthy, J.J. Rhyne, and H.B. Silber. New York: Springer, 527–532.
Note: Data are estimated and may have errors of ±10%.

As mentioned, the demand for mischmetal and its alloys is dependent on the current industrial applications requiring rare earth magnetic materials and heat-resistant steel, which are vital for critical components in electric and hybrid electric vehicles. Rare earth magnets and electric motors and generators require heat-resistant steel to provide high reliability, long-term shelf life, and longevity under extreme thermal and mechanical environments. The demand for heat-resistant steel and mischmetal alloys may further increase due to the use of wind turbines by state and federal governments to generate more electricity without using coal or foreign oil.

To achieve low cost and exceptional strength, the addition of mischmetal is required to control the shape of residual castings or moldings. Its expected benefits are due to the replacement of impurities in the material and the ability to retain spherical forms with increasing temperatures. The inclusion of mischmetal results in a greater resistance to buckling under stress and more uniform mechanical strength in the longitudinal and transverse directions. The inclusion of mischmetal is intended to reduce hydrogen-induced cracking in steel. Steel that is free from hydrogen cracking is best suited for deep-well drilling for oil and gas and drilling operations that occur below the surface of the earth.

Mischmetal is widely used in the automotive industry, construction of bridges, off-road vehicles, earth-digging tractors, rollover bars, offshore drilling towers where light and strong structures are required, large-diameter gas pipelines, and shipping containers that require both high strength and stress resistance for arctic and rough-road environments.

8.2.2.1 Cobalt Magnets for Automobile Components and Accessories

Cobalt magnets yield maximum savings in material costs, minimum use of material needed for thin curved sections, low labor costs, and satisfactory magnetic performance for automobile components and accessories. Automobile design engineers have given serious consideration to the use of these magnets in automobile equipment to reduce the size and weight of electric motors. Cobalt magnets are best suited for electric and hybrid electric vehicles. General Motors approved a new process for fabrication of thin curved sections using this magnetic material for electric motors and generators. The process offers the potential to maximize the use of expensive cobalt magnetic materials, particularly for direct-current (DC) electric motors and generators, by producing thin arc segments that do not require the diamond grinding process and do not involve unnecessary waste of this magnetic material or breakage during the fabrication process. This magnetic material was recommended for DC motors and generators because of its relatively low-cost and low-weight aspects, which are considered essential for automobile applications.

The processing of this magnetic material may be accomplished in part by substituting the mischmetal for most of the samarium in rare-earth cobalt magnet alloys without significant sacrifice of the motor performance. Samarium-cobalt (Sm_2O_7) periodic permanent magnets (PPMs) were introduced in the 1970s for deployment in traveling-wave tubes, which were the most critical elements of electronic warfare systems widely deployed in the Cold War period. Motors using polycrystalline ferrite-based permanent magnets, such as $MgFe_2O_4$ and $MnFe_2O_4$, are widely used in automobiles for air-conditioning blowers, windshield-wiping motors, rear-window defrosters, and heat blowers.

The Sm_2Co_7 magnet alloys have demonstrated a magnetic energy product (BH_{max}) that is five times greater than that of ferrite magnets, which has considerably improved resistance to demagnetization. Furthermore, these magnets can operate reliably at temperatures as high as 300°C while retaining high structural properties. They have also demonstrated significant reduction in weight and size in automobiles, leading to high fuel efficiency.

The greatest resistance to the use of samarium cobalt magnets is the high cost of cobalt. The cost of samarium cobalt magnets varies from 25 to 50 times that of ferrite magnets, with the current estimated price of cobalt varying between $60 and $80 per pound. Because of the high price of cobalt, manufacturing engineers are investigating cheaper metals, such as copper, manganese, iron, titanium, and chromium, to replace as much of the cobalt content as possible. Regardless of the cobalt cost, electric and hybrid electric automobile manufacturers are increasing the use of samarium cobalt magnets because they provide maximum magnetic energy product, a significant reduction in the weight and size of the automobile components, and improved fuel efficiency.

8.2.2.2 Cast Steels with Mischmetal Content for Continuous Casting

Some European steel-producing companies have already successfully used mischmetal in the form of steel-clad wire in continuous casting. The companies claim that residual sulfide has been effectively converted to a globular form, leading to a solidification change with a sharp reduction in internal cracking of the steel slab. The actual mechanism responsible for the sharp reduction in internal cracking of the slab is still being investigated.

By adding mischmetal wire to the mold, which is generally downstream of the tunic nozzle, clogging of the nozzle will not occur, as it often does when ladle additions are made. Furthermore, secondary reactions with refractory linings are avoided by using the protective shrouding, and reoxidization due to exposed turbulence of the tapping stream is effectively prevented. The wire-feeding rates can be calculated on the basis of residual sulfur content and shape-change requirements to yield the desired properties in the finished steel. Almost all high-strength low-alloy (HSLA) steel is made by ladle or the ingot mold. When a steel company is required to manufacture HSLA steel, the use of mischmetal in wire form should be advantageous.

8.2.2.3 Improvement in Ferrite Stainless Steels

Mischmetal has been used to develop certain austenitic steels for almost 35 years to make them forgeable. Research activities were undertaken to make ferrite stainless steels much less costly by argon-oxygen-decarburization. Materials scientists claim that this technique provides better corrosion resistance, particularly at elevated temperatures, and improved formability. Furthermore, the use of mischmetal may significantly improve oxidation resistance, even at higher temperatures.

8.2.2.4 Improvements in Modular and Nonmodular Graphite Iron Materials Using Mischmetal

Graphite cast irons generally have properties between gray iron and modular iron. These graphite irons may be produced by adding mischmetal decarburization and promoting the shape change of the cast irons. This method will provide high thermal conductivity and improved material strength, close to 80% of the strength of fully modular iron, achieved with additions such as 0.1% mischmetal along with 0.012% sulfur. Typical applications of such material include shrouding linings, automobile engine exhaust manifolds, eccentric molds, brake drums, and other auto components.

8.2.2.5 Use of Mischmetal to Eliminate Environmental Problems

Most of the magnesium should lead to the use of mischmetal for the treatment of both vermicular and graphite cast irons. Because the use of mischmetal has demonstrated the elimination of environmental problems arising from the techniques used and the elimination of oxide fumes, great demand is expected for mischmetal in these applications.

8.2.2.6 Production of Free-Machining Steel

Machining of steel casting and molds is not only time consuming but also very expensive. Research by Japanese scientists revealed that the use of less than 0.2% mischmetal promotes the formation of spheroidal graphite in steel; along with 90% carbon, it becomes possible to produce free-machining steels, thereby avoiding the poisonous atmospheres resulting from the use of screw machine steel technology. Screw machine steels containing sulfur have poor directional properties, and those containing lead, sulfur, tellurium, or selenium are not hot-rollable or hot-workable. Free-machining steels containing carbon that has been graphitized by mischmetal are reported to be both hot-workable and hot-rollable materials. These materials will have good directional properties.

Research efforts now directed towards this relatively new field have shown significantly improved V-notch toughness and ductility resulting from the maintenance of spherical residual sulfides. Mischmetal, which contains microcomponents of several rare earth elements, may lead to new applications for the electroslag remelting (ESR) steels.

8.2.2.7 Benefits of Electroslag Remelted Steels

The principal benefit of the ESR technique is the development of isotropic properties in the steel. The steels with isotropic properties will have relatively more cleanliness compared to vacuum-melted steels [1], excellent homogeneity, superior forgeability, improved toughness, and high impact resistance under harsh thermal and mechanical environments.

8.2.2.8 Energy Storage Devices

Lanthanum nickel ($LaNi_5$) is capable of storing as much hydrogen as an equal volume of hydrogen at room temperature and low pressure. This rare earth compound will release hydrogen safely whenever needed. Furthermore, the storage of gaseous hydrogen in conventional forged steel cylinders at a pressure of 2,300 psi takes three times as much space as storage of the same quantity of hydrogen in $LaNi_5$. Because hydrogen holds tremendous promise as a prime gaseous fuel of the future, there is high probability of demand for the use of intermetallic compound materials related to LaNi$_5$, in which mischmetal replaces lanthanum as a means of improving the cost-effectiveness.

8.2.3 Applications of Samarium Metal

Samarium is the fifth member of the lanthanide series. Several decades ago, it would have been very difficult to name any applications other than samarium cobalt; its demand was therefore a few pounds per year for other applications. Then, in the early 1970s, the rare earth cobalt magnet was first proposed as a replacement for the platinum cobalt magnet for high-power microwave traveling wave-tube amplifiers (TWTAs). Extensive laboratory and field tests demonstrated successful and reliable operations of these tubes under operating temperatures as high as 300°C. The demand for the samarium cobalt permanent magnets rose sharply, which increased cobalt prices fourfold over a short duration. Since then, the samarium cobalt magnet has attained a very high position in the permanent magnet market.

Sintered permanent magnets of samarium cobalt (Sm_2Co_5) demonstrate much better performance than both ferrite and alnico magnets. Comparative properties for the various types of magnets are summarized in Table 8.2. As shown, the maximum magnetic energy product for Sm_2Co_5 is significantly higher (by a factor of 4–6) than the other magnetic materials due to the high coercivity of the samarium cobalt alloy. This high coercivity is due to its hexagonal structure, which exhibits a high degree of uniaxial crystalline anisotropy.

Samarium cobalt magnets are used in a wide range of products, such as electronic watches, high-power DC motors, high-fidelity equipment, and magnetic bearings. These magnets could be useful in dental and surgical procedures. Unfortunately, their high price is sometimes a major hurdle. A less expensive type of magnet was designed based on a modified composition, which was first achieved by partial substitution of the samarium by mischmetal. Alloys of this type offer optimum magnetic energy and product-to-cost ratio. More importantly, the modified magnets provide the industry with a means of expanding production while conserving the supply of costly samarium oxide.

Table 8.2 Comparison of the Properties of Various Magnets

Magnet	Maximum Energy Product (kJ/m³)	Remanence (G)	Coercivity (kA°/m)
Sm_2Co_5 sintered	160	9,000	660
Anisotropic ferrite	26	3,700	240
Alnico-5	40	12,000	52

Samarium oxide—the precursor to the mischmetal-to-alloy product—is produced in the amount of approximately 150 to 200 tons per year. Even so, when expressed in terms of samarium metal, this quantity is relatively small. Therefore, it is desirable to optimize the available sources by whatever means possible. Table 8.3 illustrates one solution to the shortage problem—the production of alloys containing a much greater percentage of mischmetal, which is cheaper than samarium oxide.

The data presented in Table 8.3 provide multiple choices for the samarium and mischmetal contents in the selection of alloy composition for rare earth permanent magnets with minimum cost and reasonable magnet performance. The samarium content can be descreased while increasing the mischmetal content. The cost of mischmetal (a mixture of several rare earth elements) fluctuates between $6 to $10 per pound, which is very low compared to other rare earth elements.

The costs of typical rare earth elements are shown in Table 8.4. The high cost of cobalt could prevent its commercial applications. Metals such as manganese, copper, iron, titanium, and chromium are generally used to replace as much cobalt as possible. Note that the use of some conventional metals can maximize the magnetic energy available from cobalt.

Table 8.3 Worldwide Production of Samarium Oxide Expressed in Terms of Derived Alloys (Metric Tons per Year)

SmO	Metal	$SmCo_5$	$Sm_{0.67}$ $MM_{0.33}Co_5$	$Sm_{0.50}$ $MM_{0.50}Co_5$	$Sm_{0.25}$ $MM_{0.75}Co_5$
150	116	336	500	665	1,324
200	155	450	666	889	1,769
400	310	900	1,333	1,778	3,537

Note: MM, mischmetal.

Table 8.4 Typical Cost of Rare Earth Elements

Rare Earth Element	Cost per Pound ($)
Cobalt	40–55
Lanthanum	70
Neodymium	120
Praseodymium	229
Samarium	250

Note: Costs are estimated and are subject to change depending on the demand and supply of the rare earth elements, particularly cobalt and samarium.

Table 8.5 Thermal and Mechanical Properties of Rare Earth Elements

Property	Samarium	Neodymium	Cobalt
Density (g/cm³)	7.7	7.05	8.9
Melting point (°C)	1,072	1,024	1,495
Tensile strength (psi)	5,105	10,940	32,000
Curie temperature (K)	800	320	1,131
Cost per pound ($/lb.)	250	120	45

Some important thermal and mechanical properties of rare earth materials that are widely used in the development of permanent magnets are summarized in Table 8.5, along with the average cost. The demand for samarium permanent magnets will likely continue to rise because this rare earth magnetic material is widely used in the design and development of electric motors and generators for electric cars, hybrid electric vehicles, and TWTAs. The substitution of the samarium with mischmetal will reduce the cost of these permanent magnets, which should likewise reduce the cost of electric cars and hybrid electric vehicles.

8.2.4 Deployment of Rare Earth Oxides to Reduce the Cost of Samarium Magnets by Using Mischmetal Alloy

Both samarium oxide's content and price control the cost of samarium-based magnets. The use of mischmetal alloys offers an optimum ratio of magnetic energy product to alloy cost. In addition, these alloys provide a vital means of expanding the production of such alloys while conserving the supplies of expensive samarium oxide. This particular oxide, which is a precursor to the metal-to-alloy ratio, has a limited source for its availability (currently approximately 150–200 metric tons per year, with the potential for 300–400 metric tons per year). However, when expressed in terms of samarium metal, this quantity is relatively small. Therefore, it is essential to optimize the available sources of samarium oxide by whatever means possible so that there is no shortage of this oxide when the demand is high. The worldwide production of samarium oxide, expressed in terms of three derived alloys containing greater quantities of mischmetal, is shown in Table 8.6 [2].

The new magnet alloy Sm_2Co_{17} (also referred to as 2:17) does not offer enhanced magnetic properties compared with $SmCo_5$, but it has lower material costs. By decreasing the samarium content and increasing the mischmetal content in the alloy, a lower magnet cost is possible with acceptable magnetic properties. Such an alloy is best suited for commercial product applications, such as torque electric motors and generators for electric cars and hybrid electric vehicles. PPM-focused permanent magnets used in TWTAs, which are critical components of military

Table 8.6 Worldwide Production Rate of Samarium Oxide Expressed in Terms of Various Alloys Containing Higher Percentages of Mischmetal (Metric Tons per Year)

Oxide	Metal	$SmCo_5$	$Sm_{0.67}$ $MM_{0.33}Co_5$	$Sm_{0.50}$ $MM_{0.50}Co_5$	$Sm_{0.25}$ $MM_{0.75}Co_5$
150	116	336	500	665	1,324
200	155	450	666	889	1,769
400	310	900	1,332	1,778	35

Note: MM, mischmetal.

radars and electronic warfare equipment, must be made from the $SmCo_5$ alloy. $SmCo_5$ offers improved magnetic properties, structural integrity, and reliability at operating temperatures as high as 300°C.

8.2.5 Rare Earth Magnets for Commercial, Space, and Military Applications

Samarium cobalt ($SmCo_5$ and Sm_2Co_{17}) permanent magnets offer very high energy products (exceeding 400 kJ/m³). These rare earth magnets are best suited for a host of commercial and military applications. As stated in previous chapters, neodymium and samarium permanent magnets are widely used in the design and development of electric motors and generators for electric cars and hybrid vehicles. Because of their high energy product, these rare earth magnets allow for a significant reduction in the weight and size of motors and generators, leading to better gas mileage for electric and hybrid electric vehicles.

To avoid the high cost of samarium cobalt magnets, praseodymium cobalt ($PrCo_5$) permanent magnets, which have energy products greater than 200 kJ/m³, have been investigated. However, the inherent instability of such magnets limits their use to specific commercial applications.

The potential applications of neodymium cobalt and samarium cobalt magnets can be summarized as follows:

■ Samarium cobalt magnets can be used for high-power microwave sources and TWTAs where operating temperatures are as high as 300°C.
■ Samarium cobalt magnets are best suited for magnetic bearings used in high-speed turbines.
■ Samarium cobalt magnets offer improved reliability at elevated temperatures, high reliability under harsh thermal and mechanical environments, and high electrical efficiency for motors and generators.
■ Because of their lower material costs, neodymium cobalt magnets are widely used for wind turbine generators, hydroelectric turbine components, and a

host of commercial and industrial applications where operating temperatures seldom exceed 80°C.

■ Pure samarium cobalt ($SmCo_5$) is best suited for applications where improved reliability, high structural integrity, and consistent performance under harsh thermal environments are principal performance requirements. TWTAs use $SmCo_5$ permanent magnets because uniform electronic beam focusing and improved reliability are of critical importance under high temperatures and military threat. Currently, there is no substitute for rare earth magnets used in TWTAs.

■ Neodymium cobalt ($NdCo_5$) magnets are vital for applications involving high-power radar transmitters operating at unattended locations because of their high magnetic energy product, leading to improved reliability, high electrical efficiency, and structural integrity.

A host of commercial applications of samarium cobalt permanent magnets expressed in terms of derived alloys—namely, $Sm_xMM_{1-x}Co_5$, where the contents of samarium oxide and mischmetal (MM) vary, as shown in Table 8.6—are possible due to substantial savings in material costs. By defining the density, maximum safe operating temperature, tensile strength, modulus of elasticity, and maximum energy product parameters for the distinct Sm_2Co_{17} materials (shown in Table 8.5), the potential applications for commercial and industrial fields can be investigated.

Applications of Sm_2Co_{17} for space systems must be carefully investigated, because this material provides improved reliability, a significant reduction in weight and size, enhanced electrical efficiency, and high structural integrity, which are considered the most vital performance requirements for space detection, surveillance, and tracking sensors.

A newer generation of rare earth magnetic materials have been developed, based on the 2:17 alloys (Sm_2CO_{17}). They have not demonstrated enhanced magnetic characteristics compared with $SmCo_5$, but they offer substantial savings on the raw material costs. The magnets contain various percentages of samarium oxide and mischmetal. The magnetic materials have different magnetic properties, but the material costs are significantly less than the basic material $SmCo_5$. Specific applications and material costs for 2:17 alloy materials have not yet been identified.

8.2.6 Compositions of 2:17 Alloys

The 2:17 alloys that are most widely used are strictly based on compositions of the type $Sm(Co,Cu,Fe,M)_x$, where M indicates zirconium, titanium, or hafnium, and the variable x has a value anywhere between 7.0 and 8.5. Comparative material data on this and other types of RCo_5 magnets (where R represents a rare earth metal) are provided in Table 8.7. The data in the table indicate that the compositions of the material $Sm(Co,Cu,Fe,M)$ offer the least material cost and maximum energy product (BH_{max}).

Table 8.7 Comparative Data for RCo_5 and Sm_2Co_{17} Magnets

Composition	Content (%)		BH_{max} (kJ/m³)	B_R (G)	BH_C (kA°/m)	Relative Material Cost
	Sm	Co				
$SmCo_5$	33.5	66.5	160	9,000	660	1.00
$Sm_{0.67}MM_{0.33}Co_5$	22.8	66.5	110	7,500	520	0.85
$Sm_{0.50}MM_{0.55}Co_5$	17.1	66.5	200	10,000	800	1.15
$Sm(Co,Cu,Fe,M)_x$	25.5	50.0	240	11,200	450	0.80

Note: BH_{max}, maximum energy product; B_R, remanence; BH_C, coercivity.

Japanese industry is deeply involved in promoting the use of newer magnets in applications such as stepping motors for quartz-analog watches, high-performance audio pickups, and high-performance loudspeakers. Other industrial countries are using low-cost new magnetic materials in the design and development of miniaturized commercial devices, where low cost, compact size and minimum assembly weight are the principal design requirements.

8.3 Critical Applications of Rare Earth Elements in High-Technology Areas

This section discusses the unique applications of rare earth elements, such as the capture of high thermal neutrons in order to achieve thermal control in nuclear reactors, neutron radiography, sources for cancer therapy, and efficient lighting sources.

8.3.1 Application of Rare Earth Elements for Nuclear Shielding

The rare earth materials described here play critical roles in providing the reliable and safe operation of nuclear reactors. Gadolinium, samarium, europium, and dysprosium play critical roles in thermal neutron absorption, similar to neutron shielding [3]. Neutron shielding in a nuclear reactor involves three distinct functions:

- Slowing down the fast neutrons
- Capturing the slowed-down neutrons
- Absorpting all forms of gamma radiation, including inelastic scattering, and capturing gamma rays formed by the interaction between fast and slow neutrons with nuclei in the shield

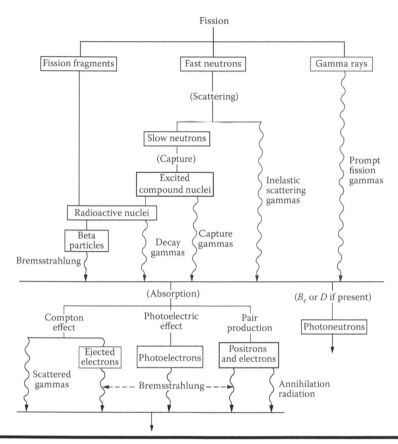

Figure 8.1   **Various radiation levels during the fission process in a nuclear reactor.**

The fission product in a nuclear reactor includes both the slow and fast neutrons. After considering the circulatory and heat exchanger, the nuclear fuel continues to emit delayed neutrons. More than 90% of neutrons liberated during the fission process are expelled promptly, but they still have radioactivity that must be reduced through the shielding technique. Delayed neutrons represent less than 1% of the total fission neutrons, but these neutrons may induce radioactivity as a result of neutron capture in the material used as a secondary coolant. Some form of shielding is essential for the attenuation of escaping nuclear radiation, which includes alpha and beta particles, gamma rays, beta rays, and protons and neutrons of various energy levels observed during the fission (Figure 8.1).

8.3.1.1 Thermal Neutron Shielding Materials

The shielding material must be selected from some specific rare earth metals, such as gadolinium, samarium, europium, or dysprosium, based on their ability to absorb

Table 8.8 Thermal Neutron Absorption Cross-section of Rare Earth Elements and Element Cost

Shielding Material	Thermal Neutron Absorption Cross-section (barns/atom)	Element Cost ($/lb.)
Gadolinium	40,000	250
Samarium	5,600	250
Europium	4,300	4,000
Dysprosium	1,100	19

the maximum number of thermal neutrons. The selected rare earth shielding material must be capable of absorbing the thermal neutron effectively. Essentially, the shielding material must provide a high thermal neutron capture cross-section, as shown in Table 8.8.

The barns per atom of the rare earth element play the most critical role in the absorption of the thermal neutron. Based on absorption capability and element cost, gadolinium is the most suitable material for thermal neutron shielding material. Gadolinium oxide is an essential component of certain fuel systems, which provide rapid control under emergency conditions.

8.3.2 Application of Neutron Radiography

Neutron radiography plays a critical role in nondestructive testing. Neutron radiography has become the most practical tool in the aerospace, nuclear, and engineering industries for nondestructive testing applications. Gadolinium and dysprosium in the form of thin foils are considered the most cost-effective materials for nondestructive testing.

Neutron radiography shares certain common features with x-ray radiography. In both systems, an imaging beam is passed through the test specimen and the attenuation beam is then detected in such a way to produce the image of the structural details of the specimen under test. In neutron radiography, the recording system is a standard x-ray film; however, because the x-ray film is relatively insensitive to neutrons, it is necessary to use the converter foils to produce the image.

In the most cost-effective method, a gadolinium thin foil (approximately 0.001 inch thick) is placed in direct contact with the x-ray film. Upon exposure to the attenuated thermal neutron beam, the gadolinium converter foil absorbs thermal neutrons and promptly emits the beta radiation, leading to activation of the film. When examination of the irradiated or unradiated specimen is required due to use of such a highly active fuel rod, it is necessary to use a somewhat more effective film technique other than gadolinium thin foil.

A dysprosium thin film (approximately 0.0025 inch thick) is exposed to an unattenuated neutron beam in the absence of the x-ray film. The activated dysprosium foil is then removed from the neutron beam and its beta decay is used to produce an autoradiograph in contact with x-ray film. The gamma cross-sections of the foil are small and will lead to prompt interactions.

8.3.3 Sealed-Tube Neutron Generators for Cancer Therapy

Some rare earth materials, such as holmium, erbium, thulium, lutetium, and ytterbium, are mostly used in laboratory research applications because of their low abundance and high material cost. Therefore, their use in applications outside the research laboratory is limited. Both thulium and erbium are widely used as target materials in sealed-tube neutron generators, designed particularly for use in cancer therapy.

8.3.4 Rare Earth Metals for the Lighting Industry

Despite its relatively rarity, scandium is currently used by the lighting industry. Certain halides have been demonstrated to change the spectral characteristics of mercury vapor lamps. It was found that mixed halides based on scandium iodine could provide spectral characteristics close to natural daylight. Because scandium iodine is extremely hygroscopic, it is not possible to use it under normal production conditions. Therefore, an alternate method was selected, which was to form the iodine in situ by reacting a small amount of scandium metal with elemental iodine. This method has been refined over the years to show excellent results. Today, much of the scandium used is supplied in the form of small discs of uniform size and weight that are matched exactly to a lamp manufacturer's specifications.

Metal halide lamps are considered to be one of the most efficient forms of lighting technology available. These lamps deliver at least 50% more light energy than conventional mercury lamps; compared with incandescent lamps of similar wattage ratings, their light output is at least three times as great. While initially used for outdoor and industrial lighting purposes, they were introduced for home use because of their substantial cost and energy savings.

Research and development activities are being undertaken to find new commercial and industrial applications involving rare earth metals and related alloys. Current research and development efforts are focusing on lower product costs and higher device performance. The author is optimistic that more applications using rare earth elements will be available in the near future.

8.4 Superconducting Ternary Rare Earth Compounds and Their Magnetic Properties

Superconducting ternary rare earth compounds offer a unique opportunity to investigate the coexistence of superconductivity and long-range ordering of rare earth magnetic moments [4]. There has been great interest in the interactions between superconductivity and long-range magnetic order in the rare earth molybdenum selenides ($RE_xMo_6Se_8$) and rhodium borides ($RERh_4B_4$).

Research studies to explore the interaction between superconductivity and long-range magnetic order were carried out on alloys containing low concentrations of rare earth impurity ions with partially filled 4f electron shells dissolved in superconducting rare earth elements or compounds. Because of these impurities, the random distribution throughout the lattice generally resulted in clustered magnetic structures with ill-defined magnetic ordering structures and other physical properties that are difficult to interpret.

Later studies indicated that two series of isostructural ternary rare earth compounds—the rhombohedral rare earth molybdenum chalcogenides, $RE_xMo_6S_8$ and $RE_xMo_6Se_8$ ($x = 1.0$ or 1.2) and the tetragonal rare earth rhodium borides, $RERh_4B_4$—exhibit the superconductivity phenomenon. The ternary compounds of rare earth ions with partially filled 4f electron shells that exhibit superconducting characteristics seem to be promising materials for investigations of the coexistence of superconductivity and the associated magnetic order. Rare earth ions in the ternary compounds are distributed periodically throughout the lattice; therefore, ordering of the magnetic moments via the interaction must be long range rather than the diffuse "spin-glass" type encountered in dilute substitutional rare earth alloys. Investigations have been conducted regarding the interaction between the superconducting phenomenon and long-range magnetic order in the ternary rare earth compounds, such as $RE_xMo_6Se_8$, $RE_xMo_6Se_8$ ($x = 1.0$ or 1.2), $RE_xMo_6S_8$, and $RE_xRh_4B_4$. Note these rare earth ternary compounds will be designated as A, B, C, and D ternary compounds, respectively, throughout this section.

8.4.1 Magnetic Moments of Rare Earth Ions in Ternary Compounds

Magnetic moments and superconducting transition temperatures are strictly dependent on the magnetic ions of the ternary compound materials. The two series of rhombohedral ternary rare earth molybdenum sulfide and selenide compounds—namely, compound A and compound B—are superconducting, even though these compounds contain 7% magnetic rare earth ions. A plot of the superconducting transition temperatures (T_c) of the rare earth compounds A and B (when $x = 1.0$ or 1.2) can be seen in Figure 8.2.

In Figure 8.2, the dashed lines indicate the linear interpolations between the transition temperatures of the corresponding nonmagnetic lanthanum (La) and lutetium (Lu) compounds. The partially filled 4f electron shells in the absence of pair-breaking interactions are a result of the magnetic moments of the rare earth ions. It can be seen in Figure 8.2 that the systematic variations of transition temperatures with rare earth elements can be described qualitatively by the de Gennes

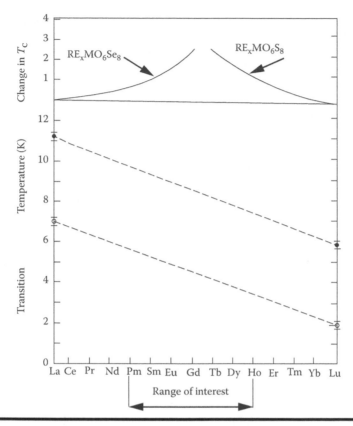

Figure 8.2 **Rare earth elements ranging from lanthanum (La) to lutetium (Lu). (Modified from McCarthy, G.I., and J.J. Rhyne. 1977.** *The Rare Earths in Modern Science and Technology.* **New York: Plenum Press.)**

factor $[(g - 1)^2 J(J + 1)]$, where g is the Landé factor and J is the total angular momentum of the Hund's rule for ground state [4].

The variation in transition temperature is shown by the solid lines at the top of Figure 8.2. The solid lines represent values of transition temperature variations, which can be expected from the Abrikosov-Gorkov theory. These values are relative to the measured values for the compound $Gd_xMo_6S_8$. Gadolinium lies in the middle of the rare earth metals, ranging from La to Lu, as shown in the figure. The transition temperatures of rare earth ternary compounds are between the two dashed lines, which cover the values over the entire rare earth element range from La to Lu.

The equation for the variation of transition temperature (ΔT) can be written as follows:

$$\Delta T_c = [C(g - 1)^2 J(J + 1)], \tag{8.1}$$

where *g* is the Lane's g-factor, *J* is the total angular momentum in the ground state, and *C* is a constant that is a function of *g* and *J* parameters. This equation has been used to compute the change in the transition temperatures (ΔT_c), which is indicated by the solid lines in Figure 8.2.

The effects of superconductivity on the magnetic properties of rare earth metals have been discussed great detail in various graduate-level textbooks, with particular emphasis on anomalous temperature-dependent exchange scattering due to the Kondo effect, as evidenced by Kondo-like anomalies in their normal-state physical properties.

8.4.2 Impact of Exchange Interactions in Superconducting Compounds

Any exchange interchange will cause pair breaking. In superconducting compounds A and C, this leads to magnetic ordering at sufficiently low cryogenic temperatures via the complex interactions between the compounds. Gadolinium molybdenum selenide ($Gd_{1.2}Mo_6Se_8$) was initially investigated because it appeared to offer the greatest potential for demonstrating the simultaneous occurrence of superconductivity and long-range magnetic order. Furthermore, a selenide compound was selected because of the significantly higher transition temperatures of rare earth molybdenum selenides relative to the sulfides. Gadolinium was expected to exhibit the highest magnetic ordering temperature in the compound A series because, like the depression of the transition temperatures, the magnetic ordering temperature should scale with the de Gennes factor of the rare earth ion, which happens to be largest for gadolinium metal. In addition, gadolinium is an S-state ion ($L = 0$, $S = 7/2$, where L is the orbital angular momentum and S is the spin angular momentum), so the crystal field splitting of the $J = 7/2$ Hund's rule multiplet should be negligible.

8.4.3 Specific Heat as a Function of Temperature for Gadolinium-Based Compounds

Plots of specific heat as a function of temperature are very important for the rare earth compounds $Gd_{1.2}Mo_6Se_8$, $La_{1.0}Mo_6Se_8$, and $Lu_{1.2}Mo_6Se_8$. Specific heat plots for these compounds are shown in Figure 8.3. The lanthanum and lutetium compounds are nonmagnetic compounds that exhibit specific heat spikes or jumps at their respective superconducting transition temperatures. The gadolinium compound shows a specific heat jump at 5.5 K, which is associated with its transition into the superconducting state temperature.

The most remarkable feature of the specific heat of $Gd_{1.2}Mo_6Se_8$ (shown in Figure 8.2) is the pronounced lambda-type anomaly that occurs in the superconducting state at 3.5 K [4], which is followed by an upturn at lower temperatures. Note the excess entropy between 1.2 and 10 K, relative to the nonmagnetic

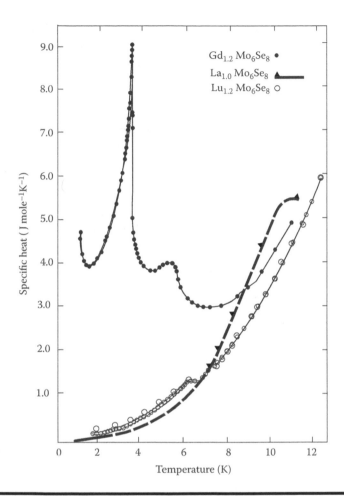

Figure 8.3 **Specific heat plots for gadolinium-, lanthanum-, and lutetium-based compounds as a function of transition temperature (K). (From McCarthy, G.I., and J.J. Rhyne. 1977.** *The Rare Earths in Modern Science and Technology.* **New York: Plenum Press.)**

isostructural $Lu_{1.2}Mo_6Se_8$ compound, which is estimated to be on the order of 40% of the entropy $S = 1.2R \ln 8$ per mole gadolinium (which would be associated with complete ordering of the Gd^{3+} magnetic moments). However, magnetic susceptibility measurements in the laboratory on $Gd_{1.2}Mo_6Se_8$ failed to reveal any feature at $T_\lambda = 3.5$ K, which is indicative of the magnetic order.

The presence of gadolinium 4f electrons appears to be responsible for the 3.5 K lambda-type specific heat anomaly in $Gd_{1.2}Mo_6Se_8$ (the anomaly does not occur in the nonmagnetic compounds, such as $La_{1.0}Mo_6Se_8$ and $Lu_{1.2}Mo_6Se_8$, in which the 4f electron shell is completely empty or completely filled). The origin of the

302 Rare Earth Materials

nature of the phase transition responsible for the lambda-type anomaly remains to be established. However, the inverse susceptibility, which appears to be at 0.8 K, does seem to be indicative of an antiferromagnetic order at this temperature that coexists with superconductivity. Specific heat measurements on compound A to temperatures as low as 0.1 K indicate a second specific heat capacity peak near superconductivity temperature 0.7 K, which apparently is associated with the magnetic transition inferred from magnetic susceptibility measurements collected in the superconducting state. This is possible in many of the $RE_xMo_6Se_8$ compounds because they are extreme type II superconductors, which permit nearly complete penetration of the magnetic fields in which the magnetization measurements were made.

The pronounced lambda-type specific heat anomalies in the superconducting state have been also observed in other $RE_xMo_6Se_8$ compounds involving Sm, Tb, Dy, and Er rare earth metals. Some of these lambda-type anomalies are accompanied by a corresponding feature in the magnetic susceptibility, which is indicative of long-range magnetic order. This type of behavior is best exemplified by the erbium-based compound $Er_{1.2}Mo_6Se_6$ (D compound), for which specific heat plots and inverse molar susceptibility curve as a function of temperatures are available. A very pronounced lambda-type specific heat anomaly occurs at temperature $T_\lambda = 1.07$ K, and there is a cup-like feature in the magnetic susceptibility near T_λ. Therefore, superconductivity and long-range ordering (probably antiferromagnetic) of the Er^{3+} magnetic moments evidently coexist in $Er_{1.2}Mo_6Se_8$.

Powder neutron diffraction data on $Er_xMo_6Se_8$ (where $x = 1.0$ or 1.2) show that magnetic Bragg peaks developed at 1.5 K cryogenic temperature, as illustrated in Figure 8.4, which is in agreement with the specific heat and susceptibility measurements. However, it is not yet possible to index the reflections and thereby determine the Er^{3+} magnetic structure.

8.4.4 Properties of Ternary Rare Earth Rhodium Borides

The superconducting and magnetic properties of the ternary rare earth rhodium borides ($RE_xRh_4B_4$, designated as compound E) are very unusual and therefore cannot be readily explained by considerations based on the exchange interaction between the spins of the conduction electrons and the magnetic moments of the rare earth ions. To illustrate the exchange interaction, a plot of the superconducting transition temperature (T_c) and the magnetic ordering temperature (T_m) of compound E is shown in Figure 8.4 [4]. As the 4f electron shell is progressively filled, samarium and neodymium rare earth compounds exhibit superconductivity; gadolinium, terbium, dysprosium, and holmium rare earth compounds are ferromagnetic; and erbium, thulium, and lutetium rare earth compounds display superconductivity characteristics. Linear extrapolation of the Curie temperatures for the rare earth dysprosium and holmium compounds of the erbium compound,

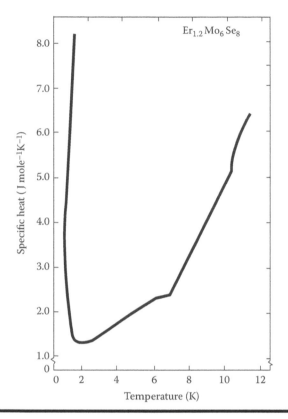

Figure 8.4 Specific heat for a rare earth compound shown as a function of temperature.

which is superconducting at 8.7 K [4], indicates that the long-range ordering of Er^{3+} magnetic moments might occur in the superconducting state close to 1 K.

The low-frequency alternating current (AC) magnetic susceptibility and the electrical resistivity measurements of the $ErRh_4B_4$ compounds reveal entirely different types of magnetic behavior. The superconductivity below the critical temperature $T_c1 = 8.7$ K is destroyed at a second lower critical temperature $T_c2 = 0.9$ K, as illustrated in Figure 8.5, which coincides with the onset of long-range magnetic order at a superconducting temperature of 0.9 K. Detailed information on the subject concerned can be obtained from the plots of the AC magnetic susceptibility and the electrical resistance of the $ErRh_4B_4$ rare earth compound as functions of temperature in zero applied magnetic field [5]. Both measurements indicate a normal superconducting state transition temperature T_c1 (8.7 K), followed by a loss of superconductivity at transition temperature T_c2 (0.9 K) [4]. In addition, the AC susceptibility (X_{AC}) exhibits a large positive value just below the transition temperature T_c2, which rapidly becomes less paramagnetic with decreasing temperature. This behavior indicates that a transition to a magnetically ordered state accompanies

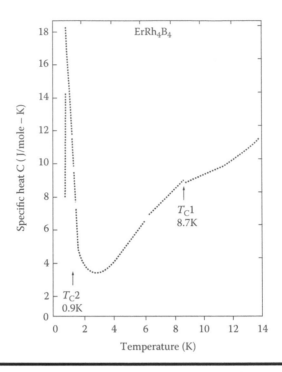

Figure 8.5 **Specific heat for the rare earth compound shown as a function of temperature at zero applied magnetic field, indicating the upper and lower critical temperature values.**

the superconducting-to-normal-state transition to a transition temperature T_c2. An interesting feature in the resistance data near the transition temperature T_c2 is a marked hysteresis, which occurs as the transition temperature and is transversed with increasing and decreasing transition temperatures. Plots of electrical resistance as a function of temperature under various fixed magnetic fields ranging from 0 to 15,000 G illustrate the evolution of the upper and lower critical temperatures or transition temperatures T_c1 and T_c2 with increasing magnetic field.

8.4.5 Specific Heat of ErRH₄B₄ as a Function of Temperature under Applied Magnetic Field

The specific heat of $ErRh_4B_4$ as a function of operating temperature under varying applied magnetic fields (ranging from 0 to 15,000 G) could play a critical role in the design and development of heat pumps. Experimental data on this particular rare earth compound as a function of temperature and applied magnetic field reveal a specific heat jump at the upper critical temperature T_c1 equal to 8.7 K, followed by a pronounced lambda-type anomaly at the lower critical temperature T_c2 equal to 0.9 K, as illustrated in Figure 8.5 [4].

An estimated entropy at the applied magnetic field between 0.7 and 15,000 G is expected for complete ordering of the magnetic moments of Er^{3+} ions, which possess the full 16-fold Zeeman degeneracy of the Er^{3+} Hund's rule for ground state. The low value of the entropy of ordering of a broad feature in the specific heat in the vicinity of a superconducting temperature of 8 K resembles a Schottky anomaly, suggesting that the 16-fold degeneracy of the Er^{3+} J equals 15/2 multiplet and is partially a crystalline electric field. The magnetic susceptibility measurements of $ErRh_4B_4$ indicate that the overall splitting of the crystal field levels is relatively small between the cryogenic temperature 7 K and room temperature (300 K). The magnetic susceptibility follows a Curie-Weiss law with an effective magnetic moment of 9.62 ± 0.15 µB, which is close to the free-ion Hund's rule value of 9.58 µB. Although the disappearance of superconduction at a second lower critical temperature equal to T_c2 has been observed, the phenomenon has always been restricted to matrix-impurity systems that simultaneously exhibit superconductivity and the Kole effect.

8.5 Crystallographic Qualities of Rare Earth Crystals and Their Unique Properties

Rare earth materials represented by the formula RFe_2 (where R stands for rare earth metals) are of significant interest due to their very large magnetostrictive strains, which have been observed in some forms of pure compounds and their associated alloys. Research indicates that single crystals, such as rare earth metals, must investigate the magnetic properties as a function of crystallographic orientation [5]. These magnetic properties are very useful in determining the optical device parameters that may lead to the design and development of magnetoacoustic- and surface acoustic-wave devices. Development of these devices will lead to significant improvement in electro-optics sensors for defense and space applications. This section summarizes the available experimental results, which may help in the preparation and characterization of a series of crystals. Results of laboratory measurements are reported elsewhere.

Close examination and assessment of the crystallographic quality of x-ray and neutron techniques have demonstrated that tri-arc grown crystals exhibit a mosaic spread that is one-tenth of that of levitation-growth crystals. The thermal and thermodynamic factors that are responsible for the growth of crystals and their implications will be discussed here, with a focus on commercially available rare earth metals such as holmium, terbium, and dysprosium metals of higher purities that do not require further purification. Some researchers believe that rare earth metals can be used in the growth of crystals without purification; however, the experimental results obtained seemed to be directly related to the crystal growth technology deployed, not correlated to the rare earth material per se. High-purity commercial iron was used in its initially received and reunified form. Repurification was accomplished using an electron-beam melting process in a vacuum (negative pressure) ranging from 0.2×10^{-6} to 5.0×10^{-6} psi.

8.5.1 Crystal Growth Using the Czochralski Method

The Czochralski method starts from ions of rare earth metal and iron ($R\mathrm{Fe}_{1.98}$) of various compositions. Compositions include HoFe_2, $\mathrm{Ho}_{1-x}\mathrm{Tb}_x\mathrm{Fe}_2$ ($0.02 < x < 0.25$) and $\mathrm{H}_{1-y-z}\mathrm{Tb}_y\mathrm{Dy}_z\mathrm{Fe}_2$ ($0.13 < y < 0.20$) and ($0.02 < z < 0.22$), which were grown in the same way as in the starting compositions. Parameter x indicates the composition for terbium, $1 - x$ is holmium, y is terbium in the complex composition, z is dysprosium, and $1 - y - z$ is holmium in the complex composition. Both the tri-arc technique [3, 4] and the frequency levitation method with a Hukin cold crucible in an Arthur D. Little Model MP crystal-growing furnace have been employed.

Electronic beam analysis demonstrated that the relative rare earth containing ions in grown crystals are the same as in the starting compositions. Crystal growth takes place in each instant under the atmosphere of gettered argon, which is maintained at an absolute pressure of 16 psi. To maintain uniform crystal growth, a seed pull rate of about 5 nm/s must be employed throughout the growth process. Such seed pull rates are commonly used in the growth process. Pulling can be terminated when approximately half the original charge has been crystallized to ensure good crystal structural characteristics.

8.5.2 Quality Control Techniques

Quality control specification requirements must be satisfied to ensure high-quality crystal growth. Visual inspection of the rare earth metals must be performed to confirm the authenticity of materials used. Routine examination of the grown crystals must be carried out using the most reliable reflection techniques and optical methods, such as use of Polaroid Type 57 film as the recording medium. Accurate crystal examination can be accomplished by measuring diffraction coefficients leading to preparation of the diffraction rocking curves on a selected basis. Independent reflections can be measured with high accuracy using a two-axis diffractometer equipped with counterstabilization and a crystal rotating steps provision. Some researchers have successfully used inelastic neutron scattering involving several tri-arc grown crystals, such $\mathrm{Ho}_{0.88}\mathrm{Tb}_{0.12}\mathrm{Fe}_2$, HoFe_2, and ErFe_2.

8.5.3 Structural Properties and Applications of Tri-Arc Crystals

Optical crystals grown by the tri-arc technique are typically not greater than 1 cm in diameter and 2 cm in length. With moderate care, one can prepare single crystals with a spread of approximately 0.1° or 1.74 mrad as measured by neutron-diffraction rocking curves. A typical rocking curve (shown in Figure 8.6) is slightly skewed

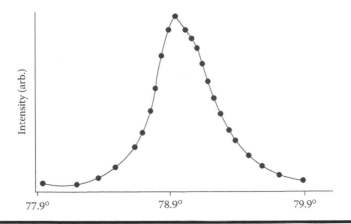

Figure 8.6 **Neutron diffraction rocking curve for a tri-arc grown HoFe$_2$ crystal peak. (From Milstein, J.B. 1975.** *Crystallographic Quality of RFe$_2$ Crystals Containing Holmium.* **Washington, DC: Naval Research Laboratory.)**

but otherwise featureless, implying the presence of low-angle grain boundaries. Rocking curves have been measured for the compositions HoFe$_2$ and Ho$_{1-x}$Tb$_x$Fe$_2$, which showed clean Laue patterns. Their small size generally limits their studies to fundamental magnetic properties, although crystals grown by the levitation technique have reached commercial-quality levels [5].

Fluctuations in the diffracted intensity as the crystal is rocked indicate the presence of a number of low-angle grain boundaries, leading to a total mosaic spread ranging from 1 to 3°. Laue photos of such specimens generally show diffraction spots with splitting comparable to the rocking curve of half-width. Specimens of this type are acceptable for research studies when the direction of the applied magnetic field may not be fully known to better than a few degrees or when the researcher wishes to observe bulk effects in the crystal structure.

8.5.4 Applications of Levitation-Grown Crystals

Levitation-grown crystals are well suited for scientific research studies. They can provide results that are typical of oriented single crystals, which can be readily differentiated from the results obtained using polycrystalline specimens of the same nominal composition. The magnetomechanical coupling coefficient for such crystals is larger than that for polycrystalline crystals with the same compositions.

The systematic variations observed in the diffraction rocking curves are not caused by the invariant features of the crystal growth experiments conducted in the laboratory. The systematic variations can be affected by various factors, such as starting materials, purity level of the materials used, seed quality, operator ability to handle the research investigation, and certain growth parameters, such as seed pull and rotation rates. The thermal environments present in each system and the consequent differences in kinetics and thermodynamic factors might explain the accuracy and the authenticity of results obtained during research studies.

8.5.5 Molten Regions during the Crystal Growth Procedure

Several steps are involved in crystal growth. Typically, thermal energy is supplied to the melt. The heat is withdrawn from the melt following the crystal through the cooled copper hearth where the melt rests. Note that this particular method results in a molten state involving two cooler solidified masses—a suitable thermal geometry that can be obtained in a float zone or pedestal growth. The tri-arc developed during the crystal growth process provides an appreciable amount of electronic charge in the molten region. It is estimated that roughly half the charge is used in the molten region during the growth procedure. The tri-arc also provides assistance to remove the dross that appears to be on the melt.

8.5.6 Levitation Process

The levitation process requires vigorous radiofrequency excitation. The charge is developed on the surface. The crystal-shaped geometry does not provide field intensification for the melt. The upper surface of the globular melt is very complex and the coolest portion of the molten material can be easily observed. There is a technique that can remove the dross from the surface of the crystal. The dross tends to appear near the growth region, typically located at the top of the crystal.

8.5.7 Comparison between the Tri-Arc and Levitation Methods

The tri-arc method for crystal growth can be characterized as a vertical zoning process, whereas the levitation process is considered to be a normal freezing process that requires appreciable stirring of the melt. Under such conditions, thermodynamic factors alone lead one to predict that crystals grown by the levitation method will be purer than crystals grown by the tri-arc method. Impurities are more soluble in molten RFe_2 material than in solid RFe_2 material (i.e., those for which the segregation coefficient K_{seg} is less than the unity) and less soluble with respect to impurities that have a segregation coefficient greater than unity.

During the termination of crystal growth in either method, the unpulled portion of the charge contains an appreciably larger concentration of the contaminant than either the grown crystal or the starting melt. Because oxygen is the primary impurity present, one must conclude that the segregation factor is less than unity. Therefore, it seems logical to conclude that the levitation method will yield the most pure material. Nevertheless, lower crystallography quality has been observed in this particular case, primarily in the form of a higher density of low-angle grain boundaries. This higher density of low-angle grain boundaries affects the quality of the crystal grown. It is most likely that, under some unidentified impurity conditions where the segregation coefficient is greater than unity, it is the main cause of the observed differences [5].

Some crystal experts believe that there are possible kinetic bases for the apparent difference in crystallographic quality produced by these two distinct methods. Four distinct possibilities might yield the observed results:

- Decoration of grain boundaries due to impurities in the material
- Growth of spurious grains as a result of bulk nucleation
- Constitutional supercooling phenomenon
- Degradation of crystallographic quality after crystal growth

Dross is effectively removed from the growth region in the tri-arc method, but no dross removal occurs in the levitation method. The particulate material that floats on the surface of a melt could produce some crystallographic defects upon incorporation of such particles into the crystal at the growth interface.

The vertical thermal profiles of the melts in the two methods described here are radically different. In the tri-arc method, the heat energy is introduced at the upper surface of the melt and is extracted at the bottom, yielding a circumstance that is stable with regard to convective forces. However, in the levitation method, the top of the melt appears to be the coldest region, which yields a thermal regime that is highly susceptible to the convective flow. In such a regime, it is highly possible that bulk nucleation occurs in a cool region. In a cool environment, the nuclei are then transported by convection close to the growth interface, where some may be incorporated into the growing crystal, thereby downgrading the crystallographic quality of the crystal. If the thermal gradient in the ratio of salt-to-crystal growth rate is less than a constant determined by the material impurities, such as concentration, segregation coefficient, or diffusion constant, a constitutional supercooling effect and thus cellular growth may result.

Researchers are constantly working to reduce the impact of material impurities. The crystal growth experience is of critical importance. In the tri-arc method, only radiation from the highly directional arcs impinges on the growing crystal. However, in the levitation method, the melt level drops as the dial is removed; thus, a longer length of grown crystal is subjected to the radiofrequency field to heat the melt. Heating the melt with a radiofrequency field tends to maintain the crystal at elevated temperatures and hence softer conditions. Under these conditions, it is possible that mechanical or thermal stresses tend to deform the crystal growth, thereby reducing the low-angle grain boundaries. The low-angle grain boundaries are responsible for the quality control of the crystal during its growth and the overall performance of the crystal.

Growing a crystal is a very complex and expensive job. Normally, crystal growth requires an extended duration ranging from several days to several months, depending on materials used and the desired performance specifications. It is recommended to use both the tri-arc and levitation methods, using starting materials of high purity to meet quality control and intended performance parameters. If one can preclude formation, the contribution of the first kinetic mechanism can be eliminated, which can save time as well as the cost of crystal growth. By using commercially available materials and starting materials of high purity, higher quality crystals might be yielded by the levitation method.

The last kinetic mechanism might be eliminated by introducing a suitable lift, such that the height of the growth interface remains constant relative to the radiofrequency field [5]. Much of the processing in the growth crystal with the radiofrequency field can be eliminated. There appears to be no convenient mechanism by which the convective flow in the levitated melt can be eliminated. This is particularly true for the levitation method under zero-gravity environments. Furthermore, the levitation method does not appear to provide an opportunity to stably increase the thermal gradient in the melt. The melt becomes more unstable at increased gradient levels.

Studies performed by the author on the rare earth iron ($R\text{Fe}_2$) crystal reveal that such crystals are of critical importance for research applications where fundamental magnetic properties are dependent on the crystallographic orientation. These crystals are best suited for determining optical device performance parameters. Potential applications of these crystals include magnetoacoustic transducers, surface acoustic-wave devices, and other electro-optic sensors.

8.6 Role of Rare Earth Metals and Oxides in the Design of Nanotechnology Sensors

Some rare earth metals and rare earth oxides play critical roles in the design and development of microelectromechanical systems (MEMS) and nanotechnology devices and sensors [6]. Some rare earth metals, such as lithium, cadmium, telluride, and their oxides, are widely used in the design and development of nanotechnology devices such as implantable medical devices, microbatteries, multilayer high-efficiency solar cells, and very long-wave infrared focal planar array detectors using the ternary alloy compound mercury-cadmium-telluride ($\text{Hg}_{1-x}\text{Cd}_x\text{Te}$). The mole fractional parameter x has an optimum value of 0.22 when using this particular complex rare earth compound consisting of cadmium and telluride oxides.

8.6.1 Nanotechnology Devices, Sensors, and Their Potential Applications

This section discusses the use of thin films of rare earth elements and their oxides in the design and development of various nanotechnology-based devices and their applications. The performance capabilities of various devices using thin film and nanotechnology are described.

8.6.1.1 Solar Cells Using Nanotechnology and High-Quality Thin Films

8.6.1.1.1 Cadmium Sulfide Solar Cells

Solar cells made from thin films of cadmium sulfide (CdS) offer many advantages, but they have poor conversion efficiencies (ranging from 5 to 7%) [7]. The efficiency of CdS solar cells is expected to exceed 8% in the near future using high-quality thin films of CdS. The maximum solar energy conversion efficiency data for solar cells using various semiconductor and rare earth materials can be seen in Figure 8.7. CdS solar cells offer 50% lower material costs, high stowage efficiency, higher solar-array design flexibility, optimum spacecraft configuration, and 100 times more space radiation resistance than silicon devices. The major disadvantages of this material include low conversion efficiency, moderate toxicity, and higher material density.

8.6.1.1.2 Cadmium Telluride Solar Cells

Research studies indicate that cadmium telluride (CdTe) solar cells are best suited for commercial solar installations. The conversion efficiency of these devices has improved significantly, from 6% in 2005, to 8% in 2006, and to 10% in 2008 (possibly close to 11% under field operating conditions). The price of a CdTe cell is better

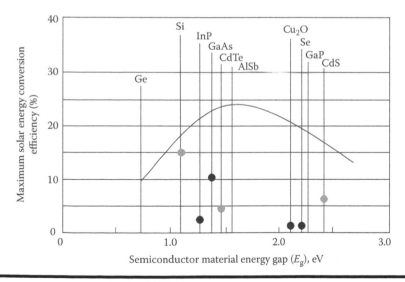

Figure 8.7 **Maximum solar energy efficiency for solar cells made from various semiconductor materials and rare earth oxides. As the operating wavelength is increased, the quantum efficiency decreases, regardless of cryogenic temperature and Hg:Cd:Te detector size.**

than $1.18 per peak watt, which is approximately 50% cheaper than standard crystalline silicon solar cells. Conversion efficiencies of solar cells using semiconductor material and rare earth oxide films are shown in Figure 8.7. The critical performance parameters and advantages of CdTe solar cells are summarized as follows:

■ CdTe solar cells are cost effective and are best suited for domestic applications.
■ The production of CdTe solar modules can be achieved with minimum cost and complexity.
■ The deployment of CdTe thin-film technology produces a high energy yield across a wide range of climatic conditions with excellent low-light solar response and temperature response coefficients.
■ The estimated operating life of CdTe solar cells is approximately 25 years; solar panels using these cells will be cost-effective for residential and small business installations.

8.6.1.2 Focal Planar Array Detectors Using Thin Films of Rare Earth Materials and Nanotechnology

Detectors are classified into various categories, such as infrared detectors, photodiode detectors, quantum detectors, optical detectors, infrared detectors, and ternary compound detectors fabricated from thin films of the rare earth ternary compounds Hg:Cd:Te and Pb:Sn:Te. Detectors using thin films of ternary compound alloys require cryogenic cooling to enhance detector sensitivity, responsivity, response time, and noise-equivalent power capability and to reduce dark current level. Performance capabilities and unique characteristics of optical detectors and focal planar array detectors using thin films of rare earth materials can be summarized as follows:

■ Optical detectors using thin films of superconducting yttrium barium copper oxide are capable of measuring infrared radiation at 13 μm or higher under cryogenic cooling of 77 K.
■ The low-frequency ($1/f$) noise current is as low as 5×10^{-15} A/Hz$^{0.5}$ at a wavelength of 5 μm and a cryogenic temperature of 120 K. This low-frequency current can be reduced at lower cryogenic temperatures.
■ They provide noise-equivalent power as low as 7×10^{-12} W/Hz$^{0.5}$.
■ Such optical detectors provide a response time that is better than 50 ps.
■ A superconducting hot-electron bolometer using thin films of niobium nitride offer measurements at gap frequencies as high as 1,400 GHz.

The cryogenically cooled focal planar array detector Hg$_{1-x}$:Cd$_x$:Te, when designed and developed for the detection of very-long wavelength infrared (VLWIR) signals, is best suited for detection of intercontinental ballistic missile (ICBM) signatures during the missiles boast phase, cruise phase, and terminal phase with a high probability of detection. The infrared signals in these three distinct phases could vary from approximately 14 to 17 μm. VLWIR array detectors are widely used in military and space applications for surveillance of long-range tactical and strategic missiles, detection of space targets, infrared search and tracking sensors, airborne missile seekers, tracking of ICBM missiles (during launch phase, cruise phase, and terminal phase), and discrimination between decoy and hostile missiles. Discrimination of real targets from decoys and debris is greatly improved by using cryogenically cooled multicolor Hg:Cd:Te focal planar array detectors [8].

Rare earth materials used in the fabrication of Hg:Cd:Te focal planar arrays must be virtually free from impurities to significantly reduce the tunneling dark currents. The optimum performance of these focal planar array detectors requires a reverse bias of about 50 mV, dark current less than 100 μA, quantum efficiency close to 58%, and a cryogenic temperature of 78 K. A cutoff wavelength of 17 μm for the focal planar array detector Hg:Cd:Te is possible when the cryogenic temperature is 40 K. Quantum efficiency is greater than 60%, as shown in Figure 8.8.

The $Hg_{1-x}:Cd_x:Te$ focal planar array detector can be designed to sense long-wavelength infrared (LWIR) signals, which would make it suited for national missile defense systems, theatre missile defense systems, and strategic missile defense systems because the infrared signals range from 8 to 14 μm. Texas Instruments has developed focal planar arrays capable of operating over short wavelength regions (1.3–2.5 μm) operating close to 300 K, mid-wavelength regions (3.0–5.5 μm) operating at 200 K, and extended long-wave regions (8–14 μm) operating at 80 K. The performance levels of these focal planar arrays vary with doping level (parameter x), absorber thickness, interface shunting, and operating temperature.

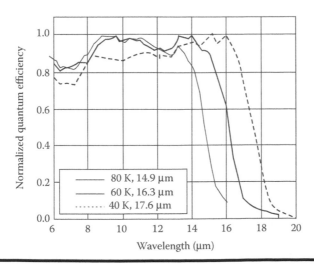

Figure 8.8 Normalized quantum efficiency of an 80-μm diameter detector as a function of wavelength at three distinct cryogenic temperatures.

The performance parameters of these focal arrays, such as dark current, thermal noise or Johnson noise level, induced noise level due to background flux density variation, flicker noise due to corner noise, figure of merit, surface leakage current, and noise-equivalent power are strictly dependent on the parameter x and the cryogenic cooling temperature. Potential applications of these planar arrays operating in the short-wavelength and mid-wavelength regions include infrared line scanners, forward-looking infrared sensors, and imaging sensors, which are vital for forward battlefield and border-security monitoring applications.

Figure 8.8 indicates that a smaller detector size and a cryogenic temperature close to 40 K offer maximum quantum efficiency and optimum detector performance over a wide range of wavelengths (6–17 μm). This figure exhibits the quantum efficiencies of an 80-μm diameter Hg:Cd:Te detector as a function of infrared wavelength at three distinct cryogenic temperatures [8].

8.6.2 Miscellaneous Devices Using Nanotechnology and Rare Earth Materials for Defense and Commercial Applications

The following sections briefly describe devices using nanotechnology and rare earth materials that are widely used in defense and commercial applications.

8.6.2.1 Nanotechnology- and MEMS-Based Materials for Armor

American scientists have designed, developed, and evaluated armor using MEMS and nanotechnology to provide needed protection for soldiers on the battlefield. These lightweight body garments have demonstrated adequate protection for the chest, legs, and arms in hostile environments. Bulletproof vests were fabricated with nanotechnology-based carbon nanotubes (CNTs). Battlefield statistics show that approximately 75 to 80% of soldiers are seriously injured from shrapnel and excessive bleeding. Field tests indicate that nanotechnology-based CNT fibers have significantly reduced these injuries and collateral damage. Ballistic tests show that the bound multiple layers of fabric made from CNTs and plastic thin sheets are better at stopping bullets than conventional bulletproof vests.

8.6.2.2 Nanotechnology and Rare Earth Materials for Analog-to-Digital Converters

Niobium is widely used in the design and development of analog-to-digital (A/D) converters, which are critical elements of various military, commercial, and space systems. Low-temperature superconductors, such as pure niobium (T_c = 9.4 K), niobium nitride (T_c = 16.0 K), and niobium tin (T_c = 18.1 K), have been widely used in the design and development of A/D converters. Particular applications include superconducting quantum interference (SQUID) magnetometers for the detection of underwater targets or a sensitive SQUID for mapping neural activity of magnetic parameters in the brain and to obtain magnetocardiograms and magnetoencephalograms. A SQUID device consists of one or more Josephson junctions and their associated electronics.

Major applications and advantages of SQUID sensors can be summarized as follows:

- SQUID sensors are compact devices that are best suited for military and medical diagnostic applications.
- A SQUID magnetometer measures the functional activities of human organs, rather than providing structural information that is often provided by magnetic resonance imaging (MRI) and computed tomography (CT).
- SQUID sensors using low-temperature superconducting technology offer optimum sensitivity at low cryogenic temperatures. SQUID-based gradiometers using superconducting detection coils permit biomagnetic measurements in unshielded environments with reasonably good accuracy and reliability.
- By mapping neural activity, it is possible to locate electrical signals of interest in the brain within a few millimeters, which can provide vital information on neurological disorders including epileptic seizures and other brain-related disorders.

The most critical performance parameters of a SQUID are as follows:

- Sensibility, which is defined as the ratio of flux density in the pickup coil of the sensor to the amount of flux linkage
- Flux density noise level
- Flux noise level
- Bandwidth
- Slew rate
- Resolution

These performance parameters improve significantly when the niobium is 100% free from impurities and the operating temperature is maintained well below the transition temperature of pure niobium metal (T_c = 4.4).

8.6.2.3 Microheat Pipes Using a MEMS Device and Nanotechnology-Based Heat Pipes

Microheat pipes with hydraulic diameters ranging from 80 to 100 μm are used in designing a microheat pipe system composed of a microminiaturized heater and several miniaturized heat pipes, depending upon the heat removal requirements. Temperature microsensor arrays are installed along the heat pipes to monitor the temperature levels at strategic points. The entire system package is very small. The

MEMS-device is best suited for military airborne systems and space sensors, where minimum weight and size are the principal design requirements.

8.6.2.4 Electrochemical Actuation Mechanism Using Nanotechnology

Research studies performed at the University of Cincinnati and the University of North Carolina revealed that CNT arrays can play critical roles in the design and development of electrochemical actuators. CNT arrays are considered smart materials, which have demonstrated high mechanical strength and electrical conductivity in addition to unique piezoresistive and electrochemical actuation capabilities. Potential applications of nanotechnology-based CNTs have spurred intensive research activities in the areas of processing, device fabrication, rare earth oxide material characterization, and development of miniaturized electrochemical actuators.

8.6.2.5 Miniaturized Nanotechnology-Based Laser Scanning Systems

Nanotechnology-based components and MEMS devices have been used in the design and development of a laser scanning image system, which are best suited for many commercial and military applications. Critical design issues include the cross-scan technique, optical errors, polygonal facet optical errors, image distortion, and stationary ghost images in the image format. The laser scanning mechanism technology is considered to be very reliable. It offers high image quality regardless of the operating environment.

8.6.2.6 Mischmetal and Nanotechnology-Based CNTs for Helicopter Components

This classic example uses a rare earth metal or mischmetal (a complex alloy compound consisting of several rare earth elements) for its bearings and nanotechnology-based CNTs in the fabrication of rotor blades. The use of mischmetal will minimize wear and tear of the rotor and will provide reliable performance over extended durations. Scientists at the U.S. Army Research Laboratory are exploring the use of CNTs in the fabrication of rotor blades. Nanotubes have been shown to dissipate energy when embedded in small samples of composite materials, which improves the damping qualities of the materials. Researchers are looking for the same effect when embedding CNTs in large quantities in helicopter blades. The maintenance records of helicopters deployed in the Iraq and Afghanistan wars indicate that significant maintenance hours were consumed, leaving relatively few hours for military missions.

A compromise between stability and vibrations is required when designing the helicopter blades. Blades with excellent stability tend to transmit more vibrations to the helicopter body, thereby requiring the need for more maintenance hours and therefore additional maintenance costs. Blades that limit vibrations have stability issues that limit the helicopter's performance. This trade-off essentially prevents the development of a next generation of helicopters with improved payloads, speeds, and cost. Nanotechnology may solve these outstanding problems while maintaining reliable operation and the structural integrity of the rotor blades.

Engineers are planning to place nanotubes throughout the blade structure by concentrating them close to the hub to minimize the damping effects. It is theorized that the friction at the nanotube–resin interface will dissipate a reasonable amount of energy, thereby improving the damping effects. The improvement in damping may make the blades more stable, while the CNT fibers in the composite resin will provide improved strength and stiffness, which are of critical importance for rotor reliability and helicopter survival.

8.7 High-Power, High-Temperature Electronic Components Using Rare Earth Oxides

Off-the-shelf or conventional electronic components will not meet the performance requirements of electronic components such as large power capacitors, inductors, and resistors. High-power, high-temperature inductors and resistors are likely available in the market, but at higher costs; also, they may not be readily available to meet specific performance requirements.

To accelerate the availability of such capacitors with unusual performance requirements, critical technologies are needed, such as high-temperature multilayer ceramic capacitors; high-voltage, high-temperature, lead-lanthanum-zirconate-titanate ceramic; and nickel-cofired (sodium and potassium) niobium oxide (NbO_3) ceramics. The last two ceramic compounds contain rare earth materials—lanthanum, nickel, and niobium oxide. This section summarizes the important properties of these ceramic materials and their performance capabilities, with emphasis on reliability, cost, and complexity.

A summary of high-temperature, high-power capacitors using current is provided in Figure 8.9. This figure identifies the capacitors currently using ceramics, thin-film technology, and extra-high voltage materials. The rating voltages (ranging from 10 to 20,000 V) and capacities (ranging from 0.0001 to 100 μF) are specified for each capacitor.

8.7.1 High-Power, High-Temperature Capacitors for Various Electronic Modules

High-energy, high-temperature capacitors are best suited for silicon carbide modules, inverters for solar power systems, electrical generators for wind power

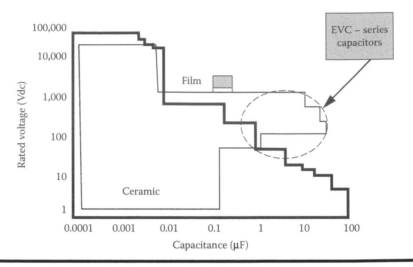

Figure 8.9 **High-voltage, high-temperature power capacitors using various material technologies, such as ceramic, film, and multilayer ceramic capacitors. Black indicates ceramic capacitors, gray indicates film capacitors, and dashed lines indicates elevated voltage capacitor (EVC) technology, respectively. Typical voltage range: 500 to 5,000 V; temperature range: 125 to 285°C.**

installation, electronic circuit boards operating at temperatures ranging from 240 to 280°C, and other electronic circuits operating under high-power, high-temperature environments, such as electrostatic energy storage devices.

8.7.2 Rare Earth Materials for High-Temperature, High-Energy Capacitors

Research conducted by Japanese scientists in 2003 on high-temperature, high-energy capacitor materials revealed that composite ceramic materials containing rare earth oxides of neodymium and titanium meet the performance requirements of discrete electronic components. These ceramic composite materials are also known as laminated ceramics. Their electrical and mechanical properties can be tailored by adjusting the thickness and composition of different layers—namely, the top layer, interface layer, and bottom layer. The most suitable composite ceramic materials involving oxides of rare earth materials include the following:

■ Bismuth titanium oxide ($Bi_4Ti_3O_{12}$)
■ Barium neodymium titanium oxide ($BaNd_2Ti_4O_{12}$)

These composite ceramic materials are also known as the BNT materials, where B stands for barium or bismuth, N stands for neodymium, and T stands for titanium.

8.7.2.1 Optimum Sintering Temperatures for Various Layers

The preparation of composite or laminated ceramics is a complex job. Critical fabrication procedures such as tape casting, sequential slip casting, electrophoretic deposition, and colloidal techniques are sintered at an appropriate temperature. It is very difficult to sinter such compact layers sufficiently because the compositions of each layer generally require different optimum sintering temperatures.

The ferroelectric interface layer $Bi_4Ti_3O_{12}$ (known as the BIT layer) has a positive temperature coefficient of the dielectric constant (D). It has been used to modify the negative temperature coefficient of the top dielectric ceramic $BaNd_2Ti_4O_{12}$ (the BNT layer). The optimum sintering temperature of this BNT layer is approximately 1,350°C, which is much higher than the optimum sintering temperature of the BIT layer (~1,100°C) [9]. When a mixture of $Bi_4Ti_3O_{12}$ and $BaNd_2Ti_4O_{12}$ is sintered at the optimum temperature of 1,300°C, they chemically react, resulting in the formation of a solid solution of $Ba(Nd,Bi)_2Ti_4O_{12}$ [9]. Therefore, $BaNd_2Ti_4O_{12}/Bi_4Ti_3O_{12}$, which is a laminated composite ceramic, could not be prepared by a conventional or general method.

The spark plasma sintering (SPS) technique could be densified to approximately 96% of the theoretical density at a lower sintering temperature of 900°C over a period of approximately 10 minutes. This rapid sintering technology can be successfully used to prepare intermetallic compounds, such as nanostructured materials, nonequilibrium ceramics, transparent ceramics, and other compound materials that are difficult to sinter using conventional methods. The SPS technique combined with heat treatment is best suited to prepare sandwich-type composite ceramic compounds, such as BNT/BIT/BNT ceramic layers. This method is divided into three distinct steps:

1. The preparation of the $BaNd_2Ti_4O_{12}$ ceramic using a conventional sintering method.
2. Calcination and prepressed functions for the $Bi_4Ti_3O_{12}$ layer powders that are sandwiched between the top and bottom BNT ceramic pellets: The SPS technique is used at the sintering temperature of 900°C for a period of 10 minutes to synthesize the BNT/BIT/BNT composite ceramic specimen. The calcined BNT and BIT powders must be individually ground and pressed into pellets under a pressure of approximately 100 MPa or 145 psi.
3. A heat treatment to reoxidize the partially reduced BNT/BIT/BNT composite ceramics.

8.7.2.2 Techniques and Equipment Required to Measure the Important Characteristics of the Specimen

The following methods and equipment can be deployed to measure the important characteristics of composite ceramic specimens:

- The densities of the specimen can be calculated from the dimensions and weight of the samples. The sandwiched BNT/BIT/BNT composite ceramics generally have high relative density. Furthermore, the heat treatment may slightly increase the densities of the SPS sample.
- Field-emission scanning electron microscopy (SEM), electron probe micro-analysis, and x-ray diffraction equipment can be used to characterize the microstructures and the product phases.
- The permittivity or dielectric measurements can be made at a frequency of 1 MHz on the silver-plated pellets using an impedance analyzer.
- An SEM micrograph of the cross-section can reveal whether the BNT layer and the BIT layer are well bonded. A well-bonded interface can be attributed to the spark plasma sintering.
- The relative radial diffraction patterns of the BNT and BIT layers clearly identify a well-bonded interface between the upper and lower BNT layers.
- The X-ray diffraction patterns can verify that the interface between the layers is absolutely clear and there is no significant diffusion.

8.7.2.3 Dielectric Properties of BIT and BNT Ceramics and BNT/BIT/BNT Composite Ceramic Layers

The dielectric constant and temperature coefficient of the composite dielectric layers BNT, BIT, and BNT/BIT/BNT can be computed using the following expressions [9]:

$$[1/(T_c)_{BNT/BIT/BNT}] = [V_{BNT}/D_{BNT}] + [V_{BIT}/D_{BIT}] \tag{8.2}$$

$$(T_c)_{BNT/BIT/BNT} = [D_{BNT/BIT/BNT}] [(V_{BNT}/D_{BNT}) (T_c)_{BNT} + (V_{BIT}/D_{BIT}) (T_c)_{BIT}] \tag{8.3}$$

D indicates the dielectric constant of the material, T_c is the temperature coefficient of the dielectric constant, V is the specimen volume, BNT represents the top and bottom composite dielectric layers, and BIT represents the composite dielectric interface layer. Typical values of dielectric constants for BIT and BNT ceramics and BNT/BIT/BNT composite ceramics as a function of temperature are summarized in Table 8.9.

The BNT/BIT/BNT composite ceramic has 10.7 vol% of BIT and BNT ceramic materials. The calculations in Table 8.9 assume a temperature coefficient of dielectric constant of +420 ppm/°C and –137 ppm/°C for the BIT ceramic and the BNT ceramic, respectively. Both the dielectric constant and the temperature coefficient do not obey the mixing rules. For example, the dielectric constant for the composite ceramic BNT/BIT/BNT was found to be slightly higher than the calculated values using Equation 8.2. The values of the temperature coefficient of

Table 8.9 Dielectric Constants for BIT and BNT Ceramics and BNT/BIT/BNT Composite Ceramics as a Function of Temperature

Temperature (°C)	BIT Ceramic	BNT Ceramic	BNT/BIT/BNT Composite Ceramic
20	122.8	89.3	93.8
40	123.5	88.6	92.8
60	124.6	88.3	95.3
80	126.2	88.1	97.5
100	127.7	88.0	99.2

Note: Values of dielectric constants are estimated and may vary by ±10%. B, barium or bismuth; N, neodymium; T, titanium.

the dielectric constant for the same composite ceramic were found slightly smaller than that calculated using Equation 8.2. The author feels that the differences are due to not obeying the mixing rules.

Studies performed by the author indicate that the composite ceramic is best suited for the design and development of capacitors for possible applications in high-power electronics, such as inverters or high-power solid-state amplifiers, where reliability and structural integrity are of critical importance under high-current environments. These laminated ceramics have demonstrated excellent mechanical properties, such as resistance to permanent deformation, hardness, and fracture toughness under high-temperature environments. These material properties are considered to be essential for high-temperature, high-energy capacitors for power electronics operating under harsh thermal and mechanical environments.

8.7.2.4 Capacitance and Voltage Ratings for High-Energy Capacitors Using Multilayer Technologies

Power capacitors come with various capacities and voltage ratings depending on the material technology deployed, as illustrated in Figure 8.9. Single-layer ceramic capacitors with capacitance ranging from 0.0001 to 0.01 μF can be rated up to approximately 35,000 V. The performance of power capacitors using multilayer technology is significantly improved in terms of power handling capability, voltage rating, reliability, and structural integrity of the device under harsh thermal and mechanical environments.

8.7.2.5 Unique Dielectric Materials for High-Temperature, High-Energy-Density Capacitors

Dielectric materials with ultrahigh permittivity have been developed for high-energy-density, high-temperature capacitors and surface acoustic-wave devices. Specific details on these materials and their applications are summarized in Table 8.10. These dielectric materials offer the following outstanding characteristics:

■ High resistivity at elevated temperatures
■ Linear or near-linear permittivity-voltage dependence
■ High-temperature chemical and structural resistance to degradation

High-performance multilayer ceramic capacitors have demonstrated rated voltages as high as 5,000 V, capacitance ranging from 0.5 to 100 μF, and operating temperature ranging from 120 to 250°C. These capacitors are best suited for the engine room, engine control units, and exhaust auxiliary systems, where operating temperatures as high as 200°C are common. The latest high temperature ceramic material developed by the industry is classified as $Na(Ta_xNb_{1-x})O_3yLiF$, which has a continuous operating temperature ranging from 150 to 300°C. This complex ceramic material contains sodium (Na), oxides of tantalum (Ta) and niobium (Nb), and lithium fluorine (LiF). The parameter x varies from 0.2 to 0.4, whereas the parameter y varies from 0.2 to 0.5. This particular ceramic material containing two rare earth materials possesses a wide range of dielectric constants as a function of operating temperature, as shown in Table 8.11. This particular capacitor material offers high operating temperatures, with a dielectric constant of 550 at the maximum temperature. Note the capacitance value could change if the operating temperature changed even by 2% or so.

Table 8.10 High-Energy-Density, High-Temperature Ceramic Materials and Their Applications

Dielectric Material	Dielectric Type	Permittivity or Dielectric Constant	Applications
Alkaline earth zirconates	Linear	45–190	High-power electronic elements
Bi(Me) O_3-PMN-PT	Relaxer	7,000–10,000	Surface acoustic-wave devices
$BiMeO_3$-$BaTiO_3$ alkali-niobates	AFE-inadequate	550–2,200	Electronic components

Note: Bi, bismuth; $BaTiO_3$, barium titanate; Me, metal; PMN, lead manganese niobium; PT, lead titanate.

**Table 8.11   Dielectric Constant of Na(Ta$_x$Nb$_{1-x}$)
O$_3$-yLiF-Laminated Ceramic Material as a
Function of Temperature**

Temperature (°C)	Dielectric Constant
0	2,000
100	1,500
200	927
300	695
400	550

8.7.2.6 Performance Characteristics of Various Types of High-Voltage Capacitors

This section summarizes the performance characteristics and applications of high-voltage capacitors using mica paper and stacked mica, polypropylene, and film materials. Such capacitors are manufactured by Cornell Dubilier Corporation for power electronics. The typical performance characteristics and applications of each category are briefly summarized in Table 8.12.

8.7.2.7 Next-Generation Ultracapacitor

Computer simulation studies at the Massachusetts Institute of Technology (MIT) in 2013 demonstrated that a nanotube-enhanced ultracapacitor should be able to store more ions than conventional activated-carbon capacitors, thereby achieving higher energy storage capability. MIT scientists tested several materials and found one that provided a satisfactory performance—a compound consisting of a tungsten layer, then a thin layer of aluminum acting as a conducting element, and finally a top layer of iron oxide that acted as a catalyst for the fabrication process. The research team started a company (FastCap Systems) to commercialize the nanotube-enhanced capacitor product. They designed and developed their first CNT-coated chip for ultracapacitors, which can store twice as much energy and deliver 7 to 15 times as much electrical power as a conventional device. The new ultracapacitor uses low-cost, domestically abundant materials and a manufacturing process similar to those used by the solar industry companies. This ultracapacitor may be ideal for electric and hybrid electric vehicles, combining fuel efficiency with high performance and significantly lower costs.

8.7.2.8 Operating Principle of Nanotube-Enhanced Ultracapacitors

When voltage is applied across the two plates, it induces an excess of negative charge or electrons on the top plate and an excess of positive charge on the

Table 8.12 Performance Characteristics of High-Voltage Capacitors Using Specific Capacitor Materials

Material Used	Applications	Capacitance Range	DC Rating (V)
Stacked mica	Microwave transmitters	47 pF to 1 µF	Up to 30,000
Mica	High-frequency RF sources	0.5–2,200 pF	Up to 30,000
Double metalized	Medium-power electronics	0.01–2 µF	Up to 3,000
Foil/metalized hybrid	Medium-power electronics	0.01–1.4 µF	Up to 2,000
Film/foil	Low-power electronic components	0.01–0.47 µF	Up to 1,600
Foil/metalized hybrid	Low-power combining electronics	0.001–0.022 µF	1,000–2,000
Oil-filled	Filter and blocking applications	0.1–15 µF	600–5,000
Oil-filled	High RMS current capacitors	0.25–20 µF	600–2,000
PLZT ceramic	Frequency converters	1–100 µF	400 (rated)

Note: DC, direct current; PLZT, lead lanthanum zirconate titanate; RF, radiofrequency; RMS, root mean square.

bottom plate. As a result, the nanotubes are coated by ions with a positive charge. When the two plates are connected by an external loop of wire, electrons will flow through that external circuit from the negative to the positive electrode, powering an electricity-consuming device along the way. Over time, both plates will lose their charge, and the positive and negative ions will break away and mix back into the electrolyte. The liquid electrolyte fills the space between the top and bottom plate, while a porous separator located in the middle keeps the plates from electrically shorting.

8.7.2.9 Advantages and Disadvantages of Nanotube-Enhanced Ultracapacitors

This section summarizes the advantages and disadvantages of a nanotube-enhanced ultracapacitor. The advantages of this product can be summarized as follows:

- Electrical double-layer capacitors or ultracapacitors are capable of providing high electrical power by delivering electrical energy quickly and recharging in seconds.
- These capacitors can withstand cold temperatures, shocks, and vibrations.
- The capacitors can be charged or discharged hundreds of thousands of times before they wear out.
- These devices contain certain earth-abundant and nontoxic materials; therefore, they pose no environmental problems compared to today's batteries, such as lithium batteries. Furthermore, these devices do not release toxic gases in contrast to some other batteries.
- Because they can be charged in seconds, they may be ideal for electric automobiles. Other batteries used by electric and hybrid electric vehicles require a long charging time, ranging from 2 to 12 hours depending on the battery type and its storage capacity.

The disadvantages of this product can be summarized as follows:

- This product has one serious drawback—it has low energy capacity. At an equivalent physical size, this ultracapacitor can store only roughly 5% as much electrical energy as a lithium-ion battery.
- Due to its extremely limited energy storage capability, its potential for other applications could be limited.

8.8 Summary

This chapter summarized the potential commercial, industrial, and space applications of the mischmetal alloy, with particular emphasis on cost, structural integrity, and product quality. The manufacturing capacities of various countries were reported. Potential applications of mischmetal include refractory linings, high-speed turbine blades, protective shrouding, jet engine nozzles, high-speed bearings, critical components of hydroelectric turbines, deep-well drilling tools for gas and oil exploration, and a host of industrial and defense products requiring high mechanical strength, tensile strength, flexural strength, structural integrity, and reliability under harsh thermal and mechanical operating environments. The critical properties of mischmetal alloy—namely, toughness under severe shock and vibrations, resistance to long-term shelf life, improved ductibility, and high reliability—make this rare earth alloy most attractive for industrial, defense, and space systems applications.

The applications of rare earth cobalt magnets were discussed in great detail. Major beneficiaries of these magnets include various commercial, industrial, defense, medical, and space systems. Currently, rare earth–based magnets are widely used by electric motors and generators, which are the critical elements of electric and hybrid electric vehicles, airborne radar transmitters, TWTAs in electronic warfare equipment, and medical diagnostic systems such as MRI and CT. The rare earth magnets used in the design of MRI and CT equipment offer crisp and high-quality images, which play critical roles in medical assessment and patient medical diagnosis.

Major industrial applications of mischmetal include cast steel containers, steel-clad wires to avoid internal cracking, and steel slabs. Mischmetal is used to provide excellent mechanical strength, improved structural integrity, and high shelf-life under severe thermal conditions, shocks, and vibrations. These requirements are essential for heavy-duty industrial systems, space sensors, and battlefield defense products. Isotropic properties will provide homogeneity, forgeability, toughness, and optimum impact resistance. The use of mischmetal to replace lanthanum from the intermetallic compound $LaNi_5$ provides maximum cost-effectiveness.

Rare earth–based samarium cobalt permanent magnets are widely used in TWTAs and high-power radar transmitters because they offer optimum magnetic performance under harsh operating environments at elevated temperatures close to 300°C. The cost of such magnets can be reduced if more mischmetal content and less cobalt metal are used in the design and development of such magnets. The cost of mischmetal is approximately \$5 to \$10 per pound, whereas the cost of cobalt metal is close to \$40 per pound. Therefore, if cost reduction is the principal objective, one must seriously consider increasing the mischmetal content. The thermal and mechanical properties of samarium, cobalt, and mischmetal were summarized in this chapter.

Another important application of mischmetal alloy involves the manufacturing of ferrite stainless steel, which has demonstrated better corrosion resistance, improved formability, and enhanced oxidation resistance at elevated temperatures. Isotropic properties in the steels are improved due to mischmetal deployment. Deployment of mischmetal in modular and nonmodular graphite iron materials has confirmed significant structural improvements, such as high thermal conductivity and improved mechanical integrity under shocks and vibrations. Significant improvements have been observed in automobile product applications using mischmetal alloy, including the following:

■ Shrouding linings
■ Automobile exhaust manifolds
■ Brake drums
■ Exhaust pipelines
■ Significant reduction in environmental problems

Rare earth metals such as gadolinium, neodymium, and other metals can provide nuclear shielding. Nuclear shielding plays a critical role in thermal neutron absorption. The absorption cross-sections were provided for rare earth metals that are suited for nuclear shielding. The application of sealed-tube neutron generators was discussed, with particular emphasis on cancer therapy.

The important magnetic characteristics of superconducting ternary rare earth compounds, such as $RE_xMo_6S_8$ and $RE_xMo_6Se_6$, were summarized as a function of cryogenic temperatures. Specific heat values for the gadolinium-based compound and other similar rare earth metals were provided as a function of temperature. Critical magnetic properties of ternary rare earth–rhodium bromides were summarized as a function of superconducting temperatures and applied magnetic fields.

Crystallographic quality control requirements for rare earth crystals and applications of these crystals were summarized, with emphasis on cost and complexity. Crystals with better crystallographic characteristics are best suited for magnetostrictive devices. The benefits and applications of Czochralski crystal using tri-arc and levitation growth methods were briefly summarized, with emphasis on quality control, structural properties, and crystal dimensional accuracy. Performance comparison data for the crystals using tri-arc and levitation growth techniques were provided, with emphasis on vertical zoning process, stirring of melt, thermodynamic factors, and segregation coefficient.

The roles of rare earth metals and their oxides in the design and development of nanotechnology-based sensors and devices were discussed, with emphasis on cost and performance. Design aspects and performance capabilities of solar cells using rare earth thin films, such as CdTe and CdS were summarized, with emphasis on efficiency, cost, longevity, and reliability. Advantages and performance parameters of focal planar array (FPA) detectors using thin films of mercury-cadmium-telluride (Hg:Cd:Te) were summarized as a function of wavelength and cryogenic temperature. The cooling of these detectors is of critical importance because the lowest values of dark current and noise-equivalent power, optimum response time, and detectivity are only possible at lower cryogenic temperatures. These FPA detectors offer optimum performance at VLWIR signals. These VLWIR detectors are best suited for the detection of IR signals of ICBMs during the boost phase, cruise phase, and terminal phase. These detectors are equally good for the detection and tracking of long-range tactical and strategic missiles. These VLWIR FPA detectors play critical roles in the discrimination between the target and its decoy and between the target and debris. The discrimination is significantly improved when cryogenically cooled multicolor Hg:Cd:Te FPA detectors are used.

Cutting-edge, nanotechnology-based carbon nanotubes (CNTs) are critical for armor to protect soldiers deployed in the battlefield or hostile regions. Deployment of CNT technology is very suitable for the design of electrochemical actuating mechanisms, where compact size, high reliability, and minimum weight are the principal design requirements. CNTs are considered to be smart devices because they have demonstrated high mechanical strength, significantly improved electrical conductivity, and excellent piezoresistivity. These microdevices are especially suited for missile, airborne radar, and space sensor applications where minimum weight and size and high reliability are of critical importance. Nanotechnology-based devices and components are used in the design of laser-scanning image systems, which are most ideal for some commercial and industrial system applications. Lately, CNTs with minor mischmetal content have been used in helicopter blades and bearings where ultrahigh reliability, high torsional strength, and minimum vibrations are of major concern.

Nanotechnology-based thin films of niobium are widely used in the design and development of A/D devices, which are critical elements for defensive and offensive weapon systems, such as high-resolution forwarding-looking radars, underwater detection systems, and airborne electronic warfare equipment. This rare earth metal is also used in the design and development of SQUID devices. Superconducting SQUID-based magnetometers and gradiometers are best suited for accurate measurement of brain-related magnetic and electrical parameters and for precision mapping of neural activities. SQUID sensors are considered most ideal in the design and development of MRI and CT equipment, which are widely used by medical experts to obtain high-resolution images of human organs. Optimum SQUID performance parameters—namely, flux density noise level, instantaneous bandwidth, slew rate, and resolution—are strictly dependent on cryogenic cooling. Niobium-based SQUID sensors require cryogenic cooling at 4.2 K if high electrical performance and reliability of data measurement are critical requirements.

Certain rare earth oxides are ideal for the design of high-temperature, high-voltage capacitors, which are best suited for high-power electronic system applications. Rare earth–based composite ceramics—namely, barium-neodymium-titanium oxide and bismuth-neodymium-titanium oxide—are most ideal for high-temperature, high-voltage power capacitors. Multilayer ceramic capacitors (MLCCs) are selected for deployment in high power electronic systems, where operating temperatures ranging from 240 to 300°C have been observed. MLCCs offer reliable performance at elevated temperatures, in addition to high mechanical integrity under severe shock and vibration environments.

Major benefits and potential applications of MLCCs consisting of a barium interface layer of titanium oxide (BIT) and top and bottom layers of BNT materials were discussed in great detail. Potential applications of composite (laminated) ceramics such as BNT/BIT/BNT were briefly mentioned, with emphasis on equations for the computation of dielectric constants and temperature coefficients as a function of operating temperature.

References

1. Hirschhorn, I.S. 1980. Trends in the industrial uses for mischmetal. In *The Rare Earths in Modern Science and Technology*, eds. G.J. McCarthy, J.J. Rhyne, and H.B. Silber. New York: Springer, 527–532.
2. Gschneidner, K. 1981. Industrial applications of rare earth elements. Presented at the American Chemical Society Symposium, Washington, DC.

3. Glasstone, S. 1955. *Principles of Reactor Engineering*. Princeton, NJ: Van Nostrand Company.
4. McCarthy, G.I. and J.J. Rhyne. 1977. *The Rare Earths in Modern Science and Technology*. New York: Plenum Press.
5. Milstein, J.B. 1975. *Crystallographic Quality of RFe₂ Crystals Containing Holmium*. Washington, DC: Naval Research Laboratory.
6. Jha, A.R. 2012. *MEMs and Nanotechnology-Based Sensors and Devices of Communications, Medical, and Aerospace Applications*. Boca Raton, FL: CRC Press.
7. Jha, A.R. 2010. *Solar Cell Technology and Applications*. Boca Raton, FL: CRC Press.
8. Jha, A.R. 1985. *Technical Report on Application of Hg:Cd:Te Focal Planar Array Detectors for Detection of LWIR and VLWIR Signals*. Cerritos, CA: Jha Technical Consulting Services.
9. Wu, Y.J. N.A. Fumi, et al. 2003. *Sandwiched $BaNd_2Ti_4O_{12}/Bi_4Ti_3O_{12}/BaNd_2Ti_4O_{12}$ Ceramics Prepared by Spark Plasma Sintering*. Philadelphia, PA: Elsevier Science, 4088–4092.

Index

Milton Keynes UK
Ingram Content Group UK Ltd.
UKHW031126141024
449569UK00006B/403

9 781466 564022